# Deep Braced Excavations and Earth Retaining Systems

## 深基坑开挖与挡土支护系统

■ 主 编　仇文岗

■ 副主编　王 尉　章润红　侯中杰

重庆大学出版社

## 内容提要

*Deep Braced Excavations and Earth Retaining Systems* collects the selected publications on deep braced excavation from the Editor. Apart from that, it also contains the basic design theories and principles in analysis of basal heave stability, toe stability, strut forces, retaining wall and ground deformations, auxiliary measures and instrumentation, observation methods and back analysis, etc. Aimed at both theoretical explication and practical application, this book covers a large scope. From basic to advanced, it tries to attain theoretical rigorousness and consistency. Each chapter is followed by problems and solutions so that the book can be readily taught at senior undergraduate and graduate including PhD students. Professors, research students, design engineers as well as staff work for consultation in the field of civil engineering, especially geotechnical engineering can benefit from the book.

**图书在版编目(CIP)数据**

深基坑开挖与挡土支护系统 = Deep Braced
Excavations and Earth Retaining Systems：英文／仉
文岗主编. -- 重庆：重庆大学出版社，2020.4
ISBN 978-7-5689-2018-6

Ⅰ.①深… Ⅱ.①仉… Ⅲ.①深基坑—工程施工—研
究生—教材—英文②深基坑支护—研究生—教材—英文
Ⅳ.①TU473.2②TU46

中国版本图书馆 CIP 数据核字(2020)第 017867 号

深基坑开挖与挡土支护系统
SHENJIKENG KAIWA YU DANGTU ZHIHU XITONG
主　编　仉文岗
策划编辑：林青山

责任编辑：王　婷　贾兴文　　　版式设计：颜丽娟
责任校对：谢　芳　　　　　　　责任印制：赵　晟
特约编辑：贾兴文
*
重庆大学出版社出版发行
出版人：饶帮华
社址：重庆市沙坪坝区大学城西路 21 号
邮编：400031
电话：(023) 88617190　88617185(中小学)
传真：(023) 88617186　88617166
网址：http://www.cqup.com.cn
邮箱：fxk@ cqup.com.cn（营销中心）
全国新华书店经销
POD：重庆新生代彩印技术有限公司
*
开本：787mm×1092mm　1/16　印张：22.25　字数：724 千
2020 年 4 月第 1 版　　2020 年 4 月第 1 次印刷
ISBN 978-7-5689-2018-6　定价：59.00 元

# Preface

Excavation is an essential segment of foundation engineering, for example, in the construction of the foundations or basements of high rise buildings, underground oil tanks, subways or mass rapid transit systems, etc. Though books on general foundation engineering introduce the basic analysis and design of excavations, they are usually too general to cope with the engineering practice. With economic development and urbanization, excavation goes deeper and larger in scale, sometimes it is carried out in difficult soils or complicated built environment. These conditions require advanced analysis, design methods and construction technologies.

The main author has focused on the studies of soil behavior and deep braced excavation problems since working as research fellow in Nanyang Technological University, and has published many reputable journal and conference papers concerning this subject. After joining Chongqing University in 2016, Prof Zhang lectured the postgraduate course "Excavation and Retaining Systems" in English to master students from overseas (the photo below). This textbook is mainly based on the lecture notes as well as the scientific journal papers published recently.

This textbook is aimed at both theoretical explication and practical application from fundamentals to advanced technologies, trying to attain theoretical rigorousness and consistency. Besides, this textbook, including the best methods currently used in engineering practice, also tries to cope with the problems in designing.Therefore, it can be readily used for analysis and design in excavation practice.

School of Civil Engineering, Chongqing University, China
Wengang Zhang

# Contents

1 **Overview** ································································································· 1

2 **Overall Stability** ·················································································· 4

   2.1  Introduction ···················································································· 4

   2.2  Types of Factors of Safety ······························································· 5

   2.3  Basal Heave Stability ········································································ 6

   2.4  Push-in Stability ··············································································· 17

   2.5  Overall Shear Failure of Cantilever Walls ··········································· 25

3 **Earth Pressure and Strut Force** ···························································· 28

   3.1  Introduction ···················································································· 28

   3.2  Lateral Earth Pressure in Braced Excavations ······································· 29

   3.3  Parametric Study ·············································································· 38

   3.4  One-Strut Failure Analysis ································································· 42

4 **Retaining Wall and Bending Moment** ··················································· 51

   4.1  Introduction ···················································································· 51

   4.2  Wall Types ····················································································· 51

   4.3  Stress Analysis Method ······································································ 61

   4.4  Design of Retaining Walls ································································· 67

5  **Ground Movements** ································································· 71

    5.1  Introduction ································································· 71

    5.2  Sources of Ground Movements ································· 72

    5.3  Ground Movement Predictions Adjacent to Excavations ················ 85

    5.4  Damage to Buildings ······················································· 92

6  **Finite Element Method** ··············································· 99

    6.1  Introduction ································································· 99

    6.2  Basic Principles ··························································· 100

    6.3  Determination of Initial Stresses ································· 107

    6.4  Modeling of an Excavation Process ······························ 109

    6.5  Mesh Generation ··························································· 110

    6.6  Excavation Analysis Method ·········································· 113

    6.7  Example: Excavation in Sand ······································· 117

7  **Soil Constitutive Models** ··········································· 130

    7.1  Introduction ································································· 130

    7.2  Linear Elastic Perfectly Plastic Model (Mohr-Coulomb Model) ··········· 134

    7.3  Hardening Soil Model (Isotropic Hardening) ························· 141

    7.4  Hardening Soil Model with Small-strain Stiffness (HS-small) ·········· 147

    7.5  The Soft Soil Model ······················································· 155

    7.6  Modified Cam-Clay Model ············································· 163

8  **Dewatering of Excavations** ········································· 166

    8.1  Introduction ································································· 166

    8.2  Dewatering Methods ····················································· 168

    8.3  Well Theory ································································· 171

    8.4  Pumping Test ································································· 183

    8.5  Dewatering Plan for an Excavation ································· 187

    8.6  Dewatering and Ground Settlement ································· 191

9  **Soil Improvement by Grouting** ··································· 200

    9.1  Introduction ································································· 200

    9.2  Grouting Equipment ····················································· 200

    9.3  Grouting Methods ························································· 201

    9.4  Ground Improvement Design ········································· 207

10 **Adjacent Building Protection** ·············································· 211

    10.1 Introduction ·············································· 211
    10.2 Building Protection by Utilizing the Characteristics of Excavation-induced Deformation
    ·············································· 211
    10.3 Building Protection by Utilizing Auxiliary Methods ·············· 214
    10.4 Building Rectification Methods ·············· 222

11 **Instrumentation and Monitoring** ·············································· 226

    11.1 Introduction ·············································· 226
    11.2 Element of a Monitoring System ·············· 227
    11.3 Measurement of Movement ·············· 228
    11.4 Measurement of Stress and Force ·············· 234
    11.5 Measurement of Water Pressure and Groundwater Level ·············· 240
    11.6 Plan of Monitoring Systems ·············· 244
    11.7 Application of Monitoring Systems ·············· 245

12 **Back Analysis for Excavation** ·············································· 248

    12.1 Introduction ·············································· 248
    12.2 General Procedure of Back Analysis in Excavation Issues ·············· 250
    12.3 Deterministic Method ·············· 253
    12.4 Probabilistic Methods ·············· 264

13 **Excavation Failure Case Analysis** ·············································· 276

    13.1 Introduction ·············································· 276
    13.2 Nicoll Highway Collapse, Singapore, 2004 ·············· 276
    13.3 Xianghu Mctro Station Collapse, Hangzhou, China, 2008 ·············· 279
    13.4 Guangzhou Haizhu City Square Foundation Pit Collapse ·············· 283
    13.5 Shanghai Metro Line 4 Seepage ·············· 286
    13.6 Other Cases ·············· 289

**Appendix** ·············································· 290

    Appendix I  Symbols and Abbreviations ·············· 290
    Appendix II  Database of Propped and Anchored Deep Excavation ·············· 297

**Reference** ·············································· 327

| No. | Date | Project | Method description |
|---|---|---|---|
| 13 | 9/2013 | Wuhan Metro Line No. 3 | Water leakage |
| 14 | 5/2019 | Overhead Chamber | The foundation pit collapse, induced by excessive wall deformation and water pipe leakage |

# 1　Overview

With the development of urbanization, the scale of urban construction is expanding, and the contradiction between urban development and shortage of land resources is becoming increasingly inevitable. The increase of urban population, development of industry, and popularity of automobile also bring a series of problems of space demands. In order to solve problems such as traffic congestion and population congestion, many big cities have developed urban underground space. The 21st century is "an era of development of underground space". The utilization of underground space includes subways, underground pipe corridors, underground shopping malls, underground garages, and civil defense basements. They are characterized by diversification, integration, and large-scale. At present, underground space has been developed in many cities all around the world.

Within a built-up environment, the construction safety of a deep excavation becomes more crucial with the ever-increasing building density. There are many excavation accidents reported in China recent years, as shown in Table 1.1.

**Table 1.1　Excavation accidents reported in China in recent years**

| No. | Date | Project | Accident description |
|---|---|---|---|
| 1 | 7/1/2003 | Shanghai Metro Line 4 | TBM tunnel water seepage |
| 2 | 4/20/2004 | Singapore Metro Line | Open tunnel collapse |
| 3 | 8/3/2004 | Guangzhou Metro Line 5 | Gas pipeline drilled, 10000 people evacuated |
| 4 | 7/21/2005 | Guangzhou Zhuhai City Plaza | The foundation pit collapse |
| 5 | 2/5/2007 | Nanjing Metro | Gas explosion |
| 6 | 3/10/2007 | Shenzhen Metro Line 1 | Induced ground surface settlement |
| 7 | 3/28/2007 | Beijing Metro Line 10 | Collapse |
| 8 | 12/16/2007 | Nanjing Metro Line 2 | Induced road collapse |
| 9 | 11/15/2008 | Hangzhou Metro Line 1 | Foundation pit collapse |
| 10 | 3/4/2009 | Shenzhen Metro Line 3 | TBM induced ground collapse |
| 11 | 8/2/2009 | Xi'an Metro Line 1 | Waller trench induced collapse |
| 12 | 12/30/2012 | Wuhan Metro Line No. 4 | Strut failure induced collapse |

continued

| No. | Date | Project | Accident description |
|-----|------|---------|---------------------|
| 13 | 9/2013 | Wuhan Metro Line No. 3 | Water leakage |
| 14 | 6/8/2019 | Greenland Central Square, Nanning, Guangxi | Foundation pit collapse induced by excessive wall deformation and water pipe leakage |

This book reviews state-of-the-art design of excavation support systems in soils. Chapter 2 is on the overall stability of excavation including basal heave stability and toe stability. It illustrates the types of factors of safety and analysis methods for basal heave including the bearing capacity method, the negative bearing capacity method, and the slip circle method, the finite-element method with strength reduction technique, and the Goh's method (Goh, 1994). Push-in failure and overall shear failure of cantilever walls are also mentioned.

Chapters 3 to 5 analyze the response of excavation including strut force, types of retaining wall and deflection of wall, and ground movements around excavations. It is of vital importance to predict and control the ground movement of a deep excavation during construction to ensure the minimal structural damage to nearby buildings and utilities. Chapter 3 illustrates how to calculate the strut and anchor forces, and helps to understand the factors that influence the strut forces as well as the load transfer mechanism when the retaining system is subjected to one strut failure. Lateral Earth Pressure including apparent pressure diagrams (APD) is also discussed in Chapter 3. Chapter 4 introduces several types of retaining wall, such as soldier piles, sheet piles, column piles and diaphragm walls. Both of their merits and disadvantages are listed and the designs of the retaining walls can also be referred to. Stress analysis method including simplified methods, beam on elastic foundation method and finite element method are also mentioned. Chapter 5 analyzes the sources of ground movements including the wall installation, excavation in front of the wall, groundwater flow resulting in loss of ground and consolidation caused by changes in water pressures due to seepage through and/or around the wall, and other site-specific sources of movements. Ground movement predictions methods are presented, such as Peck's method (Peck 1969a), Clough and O'Rourke's method (Clough and Rourke, 1990), Bowles's method (Bowles, 1998), Ou and Hsieh method (Ou and Hsieh, 1993) and some new methods. In addition, the damage to buildings caused by excavation induced ground movement is also assessed and some protective measures are proposed.

Chapters 6 and 7 present the prevailing numerical analysis. Chapter 6 presents the basic principles of finite element method, and explains the determination of initial stresses, mesh generation and boundary conditions. Excavation analysis methods are also illustrated with a detailed example of excavation in sand. Chapter 7 illustrates several types of soil constitute models on their applications and limitations; the mechanism of each models are also presented, including Linear Elastic model (LE), Mohr-Coulomb model (MC), Hardening Soil model (HS),

Hardening Soil model with small-strain stiffness (HSsmall), Soft Soil model (SS), Soft Soil Creep model (SSC), and Modified Cam-Clay model (MCC).

Chapters 8 to 13 illustrate some practice in excavation construction, such as dewatering of excavation, jet grouting and other improvement methods, as well as the protection of adjacent buildings and instrumentation. Chapter 8 is on the dewatering methods of excavation such as open sumps or ditches, the well point method, and the deep well method. Several widely used well formulas are discussed for reference and application. Then two types of pumping tests (the step drawdown and constant rate tests) are demonstrated, and dewatering plan for an excavation is also displayed. At last, the influence of dewatering on the ground settlement is discussed. Chapter 9 presents several main methods of soil improvement in excavations such as the chemical grouting method, the deep mixing method, the jet grouting method and the compaction grouting method. Some grouting equipment are also introduced. Chapter 10 introduces methods to protect adjacent buildings near excavation during construction. Excavation-induced allowable settlement is discussed, and several methods of utilizing the characteristics of excavation-induced deformation by optimizing the design parameters are discussed, including decrease of the unsupported length of the retaining wall, decrease of the creep influence, taking full advantage of corner effect, increase of stiffness of the retaining-strut system, and understanding of the characteristics of ground settlement. Building protection by utilizing auxiliary methods as well as the method of underpinning and the building rectification methods is also presented. Chapter 11 illustrates a monitoring system used to monitor the excavation conditions and ensures the safety of excavation. The element of a monitoring system is showed, and monitoring objectives are presented with the detail application of monitoring system. Chapter 12 presents the use of method of back-analysis on excavation. It shows the common procedure of back-analysis in excavation issues, and the deterministic method is explained in detail. Probabilistic methods including maximum likelihood method, bayesian method, and first-order reliability method are illustrated.

At last, a brief review and analysis of the design practice and geotechnical control of excavation failure case is given in Chapters 13.

Readers are expected to be familiar with the basic concepts of Soil Mechanics and Foundation Engineering. The target audience is mainly the undergraduates and graduate students, faculty and practicing professionals in the fields of Civil and Geotechnical Engineering.

# 2 Overall Stability

## 2.1 Introduction

Failures or collapses of excavations are disastrous at excavation sites. At worst, they endanger the safety of workers and the adjacent properties. Their influential zones are usually so large that much ground settlement may be introduced and adjacent properties within the influential zone of settlement may be damaged significantly. Due to its significant damage, to avoid failures or collapses is of vital importance and stability analyses are therefore firstly required.

Failure of an excavation may arise from the stress on the support system exceeding the strength of its materials when, for example, the strut load exceeds the buckling load of struts or the bending moment of the retaining wall exceeds the limiting bending moment. Failure can also arise from the shear stress in soil exceeding its shear strength.

When the shear stress at a point in soil exceeds or equalizes the shear strength of soil at the point, the point is in the failure or the impending state. When many failure points connect and form a slip line, the failure surface is thus produced. Once the failure surface is produced, the excavation failure or collapse will occur. This is called the overall shear failure.

Basal heave and the push-in are two main overall failure modes of excavation. As shown in Figure 2.1(a), the push-in is caused by the excessive earth pressures, reaching the limiting state on both sides of the retaining wall, which is thereby moved a large distance toward the excavation zone (especially the part embedded in soil) until reaching the full-zone failure. The analysis views the retaining wall as a free body and the external forces on the wall and internal forces of the wall are in equilibrium. The factor of safety against push-in or penetration depth of the wall can thus be obtained. When push-in is caused, with different extents of movement of the embedded part of the retaining wall, the earth pressure on the retaining wall varies. Thus there are fixed earth support method and free earth support method for analysis.

The basal heave arises from the weight of soil outside the excavation zone exceeding the bearing capacity of soil below the excavation bottom, causing the soil to move and the excavation bottom to heave so much that the whole excavation collapses. Figure 2.1(b) is a possible form of basal heave. When analyzing the basal heave, we should assume several possible basal heave failure surfaces and find their corresponding factors of safety according to mechanics. The surface having the smallest factor of safety is the most likely potential failure surface or critical failure surface. With the variable forms of critical failure surfaces, there exist many analyzing methods.

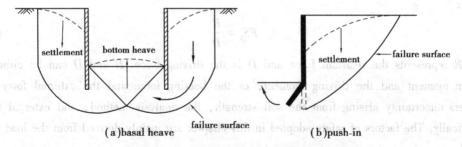

settlement | bottom heave

(a)basal heave

settlement — failure surface

failure surface

(b)push-in

**Figure 2.1 Overall shear failure modes**

As discussed above, the mechanisms of push-in and basal heave are different. Basically, push-in refers to the stability of the retaining wall. Push-in also causes soil near the wall to heave. Basal heave refers to the stability of the soil below the excavation bottom, and its failure surface may pass through the bottom of the retaining wall or through the soil below the bottom of the retaining wall. When basal heave occurs, the soil around the excavation bottom will mostly heave. Nevertheless, when it occurs to a soft clay ground, the earth pressure on both sides of the wall may also reach the limiting state, from which a push-in failure is also possible.

# 2.2 Types of Factors of Safety

There are basically three methods to determine the factor of safety for stability analysis: the strength factor method, the load factor method, and the dimension factor method, which are given as follows:

①*Strength factor method*: The method considers the soil strength involving much uncertainty and has the strength reduced by a factor of safety. If the factor of safety for the strength factor method is represented as $FS_s$, the parameters for the effective stress analysis are as follows:

$$\tan \varphi'_m = \frac{\tan \varphi'}{FS_s} \tag{2.1a}$$

$$c'_m = \frac{c'}{FS_s} \tag{2.1b}$$

the parameter for the undrained analysis is

$$s_{u,m} = \frac{s_u}{FS_s} \tag{2.2}$$

After conducting a force equilibrium or a moment equilibrium analysis with the after-reduction parameters $c'_m, \varphi'_m$ or $s_{u,m}$ derived as above, we can design the penetration depth. The method locates the factor of safety at the source where the largest uncertainty arises and is therefore quite a reasonable method. Since the after-reduction parameters will lead to a smaller $K_p$ and a larger $K_a$, the distribution of earth pressures on the retaining wall will be skewed. As a result, the method is applicable only to stability analysis and cannot be applied to deformation analysis or stress analysis.

②*Load factor method*: The factor of safety for the load method, $FS_1$, can be defined as follows:

$$FS_1 = \frac{R}{D} \tag{2.3}$$

where $R$ represents the resistant force and $D$ is the driving force. $R$ and $D$ can be either the resistant moment and the driving moment, or the bearing force and the external force. $FS_1$ considers uncertainty arising from the soil strength, the analysis method, and external forces synthetically. The factors of safety adopted in this chapter are mainly derived from the load factor method.

③*Dimension factor method*: Suppose that retaining walls are in the limiting state and the soil strengths are fully mobilized. With the force equilibrium (the horizontal force equilibrium, the moment equilibrium or other type of force equilibrium), the penetration depth of retaining walls in the limiting state can be found. The penetration depth for design is

$$H_{p,d} = FS_d H_{p,cal} \tag{2.4}$$

where $FS_d$ = factor of safety for the dimension factor method; $H_{p,cal}$ = penetration depth computed from the limit equilibrium.

The factor of safety is usually defined as a ratio of the resistant force to driving force or as a factor to reduce the strength. Eq. 2.4 is too much empirically oriented and cannot properly express the meaning of the factor of safety, leading to unreasonable results and is not recommended. If applied, cross checking by other methods is necessary.

# 2.3 Basal Heave Stability

The analyses of the basal heave failure are only applicable to clayey soils (On the other hand, the bearing capacity of sandy soils is quite large to avoid bearing failure; thus it is not necessary to examine the possibility of occurrence of basal heave failure for excavations in sandy soils.). Since $\varphi = 0$ for clay, the failure surfaces of bearing capacity failures in clay (e.g. the slope stability problems, the ultimate bearing capacity problems of foundations, etc.) are circular arc surfaces. The basal heave failure due to excavation is also a kind of bearing capacity failure and might also have a main circular arc failure surface. The analysis method for basal heave varies with the assumed shapes of failure surfaces near the ground or excavation surface, though the main failure surface is still a circular arc. The analysis method for basal heave assumes many possible failure surfaces and finds their corresponding factors of safety according to mechanics. The one with the smallest factor of safety is the most likely potential failure surface. Many analysis methods have been proposed for basal heave, of which the most commonly applied are (1) **Terzaghi's method**, (2) **Bjerrum and Eide's method**, and (3) **the slip circle method**. This section will categorize these methods into the bearing capacity method, the negative bearing capacity method, and the slip circle method according to their characteristics.

## 2.3.1 Bearing capacity method

As shown in Figure 2.2, the soil weight above the level of the excavation surface (plane **abc**) can be seen as the load to cause excavation failure. Supposing a trial failure surface caused by the soil weight within the width of $B_1$ acts on plane **abc** as is shown in Figure 2.2(a), we can find the ultimate load for the width of $B_1$ following Terzaghi's bearing capacity method with the shear strength along **bd** considered. The ratio of the ultimate load to the weight of soil within the width of $B_1$ is the factor of safety for the trial failure surface. Then increase the value of $B_1$(which denotes increasing of the size of trial failure surfaces) and find the corresponding factor of safety accordingly until the trial failure surface covers the whole excavation (i.e. $B_1 = B/\sqrt{2}$), as shown in Figures 2.2(b) and 2.2(c). Since the weight of $B_1$-wide soil on each side of the excavation zone may produce failures, the schematic diagram to calculate the factor of safety is illustrated in Figure 2.2(d). Following the principle of virtual work, the factor of safety induced from Figure 2.2(c) and that from Figure 2.2(d) would be identical. The factor of safety against basal heave ($F_b$) for the excavation is the smallest one among the safety factors corresponding to the trial failure surfaces.

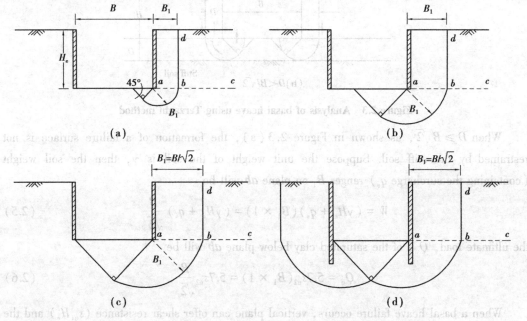

(a)                                    (b)

(c)                                    (d)

**Figure 2.2　Analysis of basal heave by the bearing capacity method**

Terzaghi (1943) did not adopt the above method, where the smallest factor of safety is taken to be the factor of safety against basal heave. Instead, he directly assumed the trial failure surface where $B_1 = B/\sqrt{2}$ (i.e. $X = B/\sqrt{2}$) is the critical failure surface and its corresponding factor of safety is the factor of safety against basal heave, as shown in Figure 2.3. According to Terzaghi's bearing capacity theory, the bearing capacity of saturated clay under plane **ab** can be denoted as

$P_{max} = 5.7s_u$. When the soil weight above plane $\textbf{\textit{ab}}$ is greater than the soil bearing capacity, the excavation will fail. Besides, the failure surface will be restrained by stiff soils. Let $D$ represent the distance between the excavation surface and the stiff soil, and we can discuss Terzaghi's method in two parts: $D \geqslant B\sqrt{2}$ and $D < B/\sqrt{2}$.

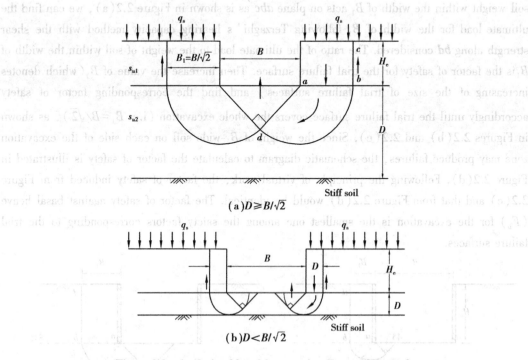

(a)$D \geqslant B/\sqrt{2}$

(b)$D < B/\sqrt{2}$

**Figure 2.3    Analysis of basal heave using Terzaghi method**

When $D \geqslant B\sqrt{2}$, as shown in Figure 2.3(a), the formation of a failure surface is not restrained by the stiff soil. Suppose the unit weight of the soil is $\gamma$, then the soil weight (containing the surcharge $q_s$) ranges $B_1$ on plane $\textbf{\textit{ab}}$ will be

$$W = (\gamma H_e + q_s)(B_1 \times 1) = (\gamma H_e + q_s)\frac{B}{\sqrt{2}} \qquad (2.5)$$

the ultimate load, $Q_u$, of the saturated clay below plane $\textbf{\textit{ab}}$ will be

$$Q_u = 5.7s_{u2}(B_1 \times 1) = 5.7s_{u2}\frac{B}{\sqrt{2}} \qquad (2.6)$$

When a basal heave failure occurs, vertical plane can offer shear resistance ($s_{u1}H_e$) and the factor of safety against basal heave ($F_b$) will be

$$F_b = \frac{Q_u}{W - s_{u1}H_e} = \frac{5.7s_{u2}B/\sqrt{2}}{(\gamma H_e + q_s)B/\sqrt{2} - s_{u1}H_e} = \frac{1}{H_e} \times \frac{5.7s_{u2}}{\gamma + \left(\dfrac{q_s}{H_e}\right) - \dfrac{s_{u1}}{0.7B}} \qquad (2.7)$$

where $s_{u1}$ and $s_{u2}$ represent respectively the undrained shear strengths of the soils above and below the excavation surface; $q_s$ denotes surcharge on the ground surface.

When $D<B/\sqrt{2}$, under such a condition, the failure surface will be restrained by the stiff soil, as shown in Figure 2.3(b), and its factor of safety ($F_b$) will be

$$F_b = \frac{Q_u}{W - s_{u1}H_e} = \frac{5.7s_{u2}D}{(\gamma H_e + q_s)D - s_{u1}H_e} = \frac{1}{H_e} \times \frac{5.7s_{u2}}{\gamma + \left(\dfrac{q_s}{H_e}\right) - \dfrac{s_{u1}}{D}} \qquad (2.8)$$

For most excavation cases, Terzaghi's factor of safety ($F_b$) should be greater than or equal to 1.5 (Mana and Clough, 1981; JSA, 1988).

Assuming that the penetration depth of the retaining wall is deep enough, the failure surface may be formed as illustrated in Figure 2.4(a), which is one of the possible failure modes. According to the analysis on the basis of the principle of virtual work, the factor of safety for a failure surface as illustrated in Figure 2.4(a) is close to that of Eqs 2.7 and 2.8. The only difference is that the failure surface in Figure 2.4(a) ranges wider (with the extra failure surface be) and the average soil strength on the failure surface is higher than that in Figure 2.4(b), assuming the undrained shear strength of clay increases with the increase of depth.

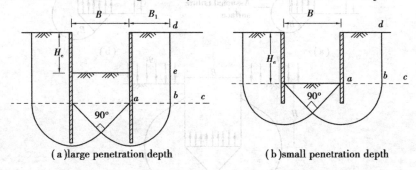

(a)large penetration depth          (b)small penetration depth

**Figure 2.4   Relation between the embedded part of retaining wall and failure surface**

As shown in Figure 2.4(b), assuming the penetration depth of the retaining wall is not deep enough, the calculation of the factor of safety will still follow Eqs 2.7 or 2.8. That is to say, the value of the factor of safety against basal heave has nothing to do with the existence of the retaining wall according to the equations. However, theoretically speaking, the retaining wall with high stiffness may be capable of restraining basal heave failures. Thus, the actual factor of safety should be greater than the result from Eqs. 2.7 or 2.8 though there does not exist a suitable way to estimate it.

The bearing capacity method or Terzaghi's method is suitable for shallow excavations, where the excavation width ($B$) is larger than the excavation depth ($H_e$). For deep excavations, $B<H_e$, the bearing capacity method or Terzaghi's method may not yield reasonable results because the method assumes that the failure surface extends up to the ground surface and that the shear strength of clay is fully mobilized all the way to the ground surface, neither of which is necessarily true for deep excavations.

### 2.3.2 Negative bearing capacity method

The negative bearing capacity method assumes that the unloading behavior caused by excavation is analogous to the building foundation subject to an upward loading and that the shape of the failure surface is similar to the failure mode of the deep foundation. Then, using the bearing capacity equation for the deep foundation, we can obtain the ultimate unloading pressure. The factor of safety is the ratio of the ultimate unloading pressure to the unloading pressure. As shown in Figure 2.5, assuming there are various failure surfaces for analysis (representing different $B_1$-values) with their separate corresponding factors of safety, the smallest factor of safety among them is the factor of safety against basal heave for the excavation.

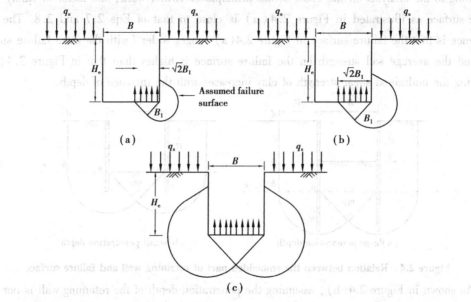

Figure 2.5   Analysis of basal heave failure by the negative bearing capacity method

Like Terzaghi's method, Bjerrum and Eide (1956) did not yield the factor of safety against basal heave by finding the smallest, as just mentioned. Instead, they assumed the failure surface where the radius of the circular arc is equal to $B/\sqrt{2}$ is the critical failure surface and the corresponding factor of safety is the one against basal heave [see Figure 2.5(c)]. The factor of safety can be expressed as follows:

$$F_b = \frac{s_u N_c}{\gamma H_e + q_s} \tag{2.9}$$

where $q_s$ is the surcharge on the ground surface and $N_c$ is Skempton's bearing capacity factor as shown in Figure 2.6.

Since $N_c$ has taken into account the effects of the embedment depth of foundations and excavation size, Eq. 2.9 is equally valid for shallow and deep excavations, as well as rectangular excavations.

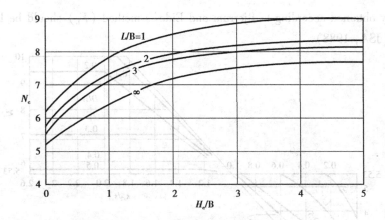

**Figure 2.6　Skempton's bearing capacity factor (Skempton, 1951)**

Like Terzaghi's method, when there exists stiff soil below the excavation surface, the failure surfaces assumed by Bjerrum and Eide's method would also be restrained by the stiff soil. According to Reddy and Srinivasan's study (1967), NAVFAC DM 7.2 (1982) modified Bjerrum and Eide's method to apply the method to the excavations where there are stiff soils below the excavation surfaces or there are two layers of soils. As shown in Figure 2.7, the Extended Bjerrum and Eide's method can be expressed as follows:

$$F_b = \frac{s_{u1} N_{c,s} f_d f_s}{\gamma H_e} \qquad (2.10)$$

where

$\gamma$ = unit weight of the soil;

$H_e$ = excavation depth;

$s_{u1}$ = undrained shear strength of the upper clay;

$s_{u2}$ = undrained shear strength of the lower clay;

$N_{c,s}$ = bearing capacity factor that does not consider the excavation depth. This can be determined according to Figure 2.7(a) or 2.7(b);

$f_d$ = depth correction factor, which can be found in Figure 2.7;

$f_s$ = shape correction factor, which can be estimated by the following equation: $f_s = 1+0.2B/L$.

If the penetration depth of the retaining wall is deep enough, Bjerrum and Eide's method computes the factor of safety in a way similar to Terzaghi's method. That is to say, the failure surface will be formed in a deeper level, similar to what is illustrated in Figure 2.4(a). Under such conditions, Eq. 2.9 is still workable to estimate the factor of safety with the slight differences of average soil strengths on failure surfaces. When $H_p$ is not large enough, the calculation of the factor of safety will still follow Eq. 2.9. That is to say, the value of the factor of safety against basal heave has nothing to do with the existence of the retaining wall according to the equations.

The negative bearing capacity method or Bjerrum and Eide's method take into account the effects of excavation shape, width, and depth. Therefore, the methods are applicable to various shapes of excavations, shallow excavations as well as deep excavations. For most excavations, the factor of safety obtained according to Bjerrum and Eide's method ($F_b$) should be larger than or equal to 1.2 (JSA, 1988).

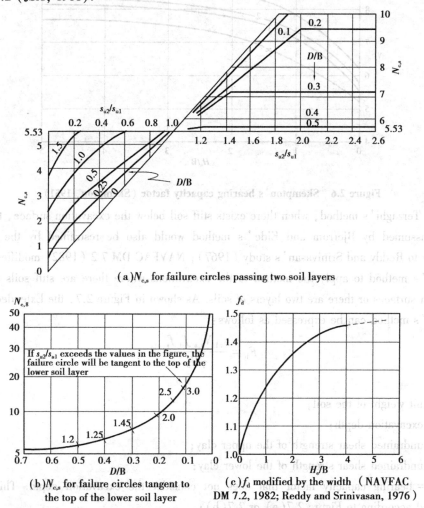

(a)$N_{c,s}$ for failure circles passing two soil layers

(b)$N_{c,s}$ for failure circles tangent to the top of the lower soil layer

(c)$f_d$ modified by the width (NAVFAC DM 7.2, 1982; Reddy and Srinivasan, 1976)

**Figure 2.7   Extended bjerrum and eide's method**

## 2.3.3   Slip circle method

Let the trial failure surfaces of the basal heave failure be assumed to be basically circular arcs, and separately compute the ratios of the resistant moments to the driving moments for the trial circular arc failure surfaces. The smallest factor of safety among them is then the factor of safety against basal heave for the excavation. The method is designated as the slip circle method. The center of the circle with the slip circle method can be set at the lowest level of strut, at the excavation surface, or not be set at any specific position, for example, points $A$, $B$, $O$ as shown in Figure 2.8. The slip circle method, without setting the center at a specific position, is to try out the circles for various positions and sizes and to find the corresponding factors of safety. The circle with the smallest factor of safety is the critical circle.

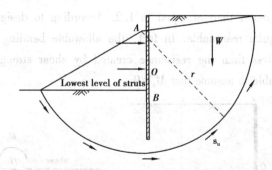

**Figure 2.8  Location of the center of a failure circle for slip circle method**

Theoretically, the critical circle has the smallest factor of safety though few adopt it for analysis, considering it causes a lot of complication in computation. According to the analysis results, the factor of safety corresponding to the failure circle whose center is set at the lowest level of strut ( point $O$ ) is smaller than that at the excavation surface ( point $B$ ) and is close to the factor of safety of the critical circle ( Liu et al., 1997 ). Thus, the circular arc failure surface whose center is set at the lowest level of strut is often adopted for analysis.

Suppose the failure surface of the basal heave failure is a combination of a circular arc which centers at the lowest level of strut and a vertical plane above the lowest level of struts, as shown in Figure 2.9( a ). Let the shear strength on the vertical failure plane [ line $bc$ in Figure 2.9( a ) ] be ignored, and take the retaining wall and soil below the lowest level of struts as well as above the circular arc as a free body, as shown in Figure 2.9( b ). If we regard the soil weight above the excavation surface in back of the retaining wall as the driving force and the shear strength along the failure surface as the resistant force, the factor of safety against basal heave, the ratio of the resistant moment to the driving moment with regard to the point at the lowest level of strut will be

$$F_b = \frac{M_r}{M_d} = \frac{X \int_0^{(\pi/2)+\alpha} s_u (X d\theta) + M_s}{W(X/2)}$$
(2.11)

where

$M_r$ = resistant moment;

$M_d$ = driving moment;

$M_s$ = allowable bending moment of the retaining wall;

$s_u$ = undrained shear strength of clay;

$X$ = radius of the failure circle;

$W$ = total weight of the soil in front of the vertical failure plane and above the excavation surface, including the surcharge on the ground surface.

Equation 2.11 is a commonly used slip circle method. The original source of the slip circle method is untraceable. Nevertheless, TGS ( 2001 ) and JSA ( 1988 ) adopted the method in their building codes. Both assume that $M_s = 0$ and recommend that the factor of safety against basal heave ($F_b$) should be greater than or equal to 1.2. According to design experience in some countries, the value is quite reasonable. In fact, the allowable bending moment value of the retaining wall $M_s$ is far less than the resistance created by shear strength. Thus, to simplify computation, it is reasonable to assume that $M_s = 0$.

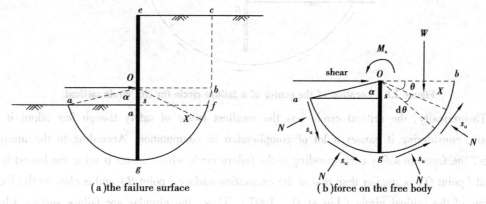

(a)the failure surface          (b)force on the free body

**Figure 2.9   Analysis of basal heave by the slip circle method**

In analysis, we should try out different radii of circles and find the one with the smallest factor of safety as the critical circle, which represents the factor of safety against basal heave. In fact, a failure circle cannot pass through the embedded part of a retaining wall. Thus, for soils with constant strength or strength increasing with depth, the circle passing through the bottom of a retaining wall is the critical circle with the smallest factor of safety. Therefore, it is rational to let $X = s + H_p$, where $s$ is the distance between the lowest level of strut and the excavation surface and $H_p$, the penetration depth of a retaining wall. Eq. 2.11, therefore, can be used to compute the penetration depth of a retaining wall. Nevertheless, if there exist soft soils below the bottom of a retaining wall, the circle passing through the bottom of a retaining wall is not necessarily the critical failure circle, with the smallest factor of safety. Thus, we have to try out different values of $X$ to find the failure circle with the smallest factor of safety, as shown in Figure 2.10.

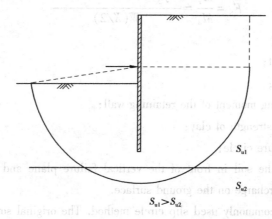

$$S_{u1} > S_{u2}$$

**Figure 2.10   Analysis of basal heave in layered soft soils**

### 2.3.4 Finite-element method with strength reduction technique

Besides the limit equilibrium methods discussed above, finite element method (FEM) combined with strength reduction (SR) technique is also commonly used to calculate the factor of safety against basal heave. More information about theory and application of FEM can be found in Chapter 6. The factor of safety is defined here as the number by which the original shear strength parameters must be divided in order to bring the excavation to the point of failure, as given in Eqs. 2.12a and 2.12b.

$$\tan \varphi'_m = \frac{\tan \varphi'}{F_b} \tag{2.12a}$$

$$c'_m = \frac{c'}{F_b} \tag{2.12b}$$

This definition of the factor of safety is exactly the same as that used in limit equilibrium methods, namely the ratio of restoring to driving moments.

SR technique allows to apply different factors of safety to the $c'$ and $\phi'$ terms. In common practice, however, the same factor is applied to both terms. To find the true $F_b$, it is necessary to initiate a systematic search for the value of $F_b$ that will just cause the excavation to fail. This is achieved by the program solving the problem repeatedly using a predefined sequence of $F_b$ values. Moreover, the criteria to define the failure should be chosen properly. There are three commonly utilized criteria as follows:

①A plastic zone going through the slope from the toe to the top, as shown in Figure 2.11. At failure, a band is formed within the slope, in which all elements are in the plastic state and the band would extend through the slope from the toe to the top.

②A significant increase in the nodal displacement within the mesh.

③Non-convergence of the solution.

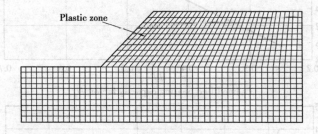

**Figure 2.11 Plastic zone inner a slope under limit state**

Pei (2010) suggested these three criteria are consistent and would result in similar $F_b$. Most of commercial FE codes, such as Plaxis(2017), adopt the third as the criteria to define the failure.

Compared to traditional limit equilibrium methods, the advantages of FEM with SR technique can be summarized as follows:

①No assumption needs to be made in advance about the shape or location of the failure surface. Failure occurs naturally through the zones within the soil mass in which the soil shear strength is unable to sustain the applied shear stresses.

②Since there is no concept of slices in the FEM, there is no need for assumptions about slice side forces. The FEM preserves global equilibrium until failure is reached.

③As soil constitutive models are employed in FEM, the stress-strain relation is considered.

④It is capable to simulate the soil-structure interaction.

## 2.3.5 Goh's method

Considering the influence of clay thickness, wall embedment depth and wall stiffness, Goh (1994) carried out a large number of parametric FE analyses and developed a simple design procedure to estimate $F_b$, the equation is given:

$$F_b = \frac{s_u N_h}{\gamma H} \mu_t \mu_d \mu_w \qquad (2.13)$$

where

$H$ = excavation depth;

$\gamma$ = unit weight of the soil;

$s_u$ = undrained shear strength of the clay;

$N_h$ = modified bearing capacity factor, which can be found in Figure 2.12(a);

$\mu_t$ = clay thickness modification factor, which can be found in Figure 2.12(b);

$\mu_d$ = wall embedment modification factor, which can be found in Figure 2.12(c);

$\mu_w$ = wall stiffness modification factor, which can be found in Figure 2.12(d).

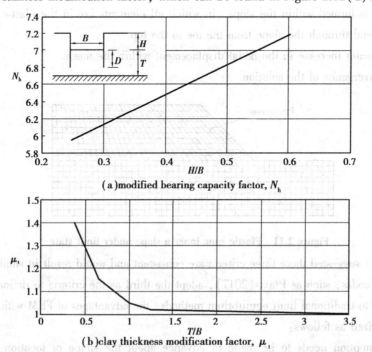

(a)modified bearing capacity factor, $N_h$

(b)clay thickness modification factor, $\mu_t$

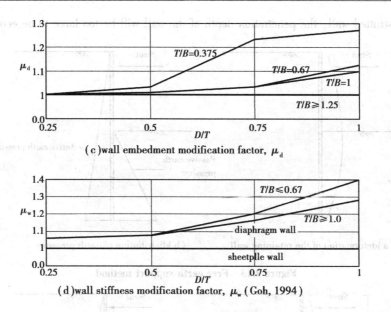

(c)wall embedment modification factor, $\mu_d$

(d)wall stiffness modification factor, $\mu_w$ (Goh, 1994)

**Figure 2.12　Goh's design charts**

However, the parametric analyses only investigate the wide excavations under plane strain condition, so the proposed method is only reliable when the excavation is long ($L \gg B$) and wide ($B/H > 1$).

# 2.4　Push-in Stability

There are two analysis methods for the push-in failure: free earth support method and fixed earth support method. As shown in Figure 2.13(a), the free earth support method assumes that the embedment of the retaining wall is allowed to move to a certain distance under the action of lateral earth pressure. Therefore, the earth pressure on the retaining wall in the limiting state can be assumed as shown in Figure 2.13(b).

The fixed earth support method is to assume that the embedment of the retaining wall seems to be fixed at a point below the excavation surface. The embedded part may rotate about the fixed point, as shown in Figure 2.14(a). Thus, when the retaining wall is in the limiting state, the lateral earth pressure around the fixed point on the two sides of the retaining wall does not necessarily reach the active or passive pressures, as shown in Figure 2.14(b).

If a cantilever wall is designed based on the free earth support method, no fixed point is supposed to exist in the embedded part of the wall, as discussed above. The external forces, only passive and active forces, on the retaining wall are not to come to equilibrium. Therefore, the free earth support method is not applicable to cantilever walls. On the other hand, if the free earth support method is applied to a strutted wall, the forces acting on the wall will include both the passive and active forces and the strut load. With external forces on the wall coming to equilibrium, the method is applicable to a strutted wall. On the other hand, if we apply the fixed earth support method to a strutted wall, the penetration depth of the wall will be too large to be economical.

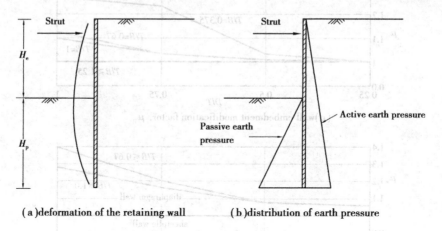

(a)deformation of the retaining wall      (b)distribution of earth pressure

**Figure 2.13　Free earth support method**

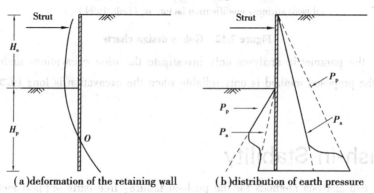

(a)deformation of the retaining wall      (b)distribution of earth pressure

**Figure 2.14　Fixed earth support method**

For a strutted wall, the free earth support method is the commonly used analysis method. As shown in Figure 2.15(a), the earth pressures on the outer and the inner sides of the retaining wall in the braced excavation will reach the active and the passive earth pressures respectively in the limiting state. Take the retaining wall below the lowest level of strut as a free body and conduct a force equilibrium analysis [Figure 2.15(b)], and we can then find the factor of safety against push-in as follows:

$$F_p = \frac{M_r}{M_d} = \frac{P_p L_p + M_s}{P_a L_a} \tag{2.14}$$

where

$F_p$ = factor of safety against push-in;

$M_r$ = resisting moment;

$M_d$ = driving moment;

$P_a$ = resultant of the active earth pressure on the outer side of the wall below the lowest level of strut;

$L_a$ = length from the lowest level of strut to the point of action $P_a$;

$M_s$ = allowable bending moment of the retaining wall;

$P_p$ = resultant of the passive earth pressure on the inner side of the retaining wall below the excavation surface;

$L_p$ = length from the lowest level of strut to the point of action $P_p$.

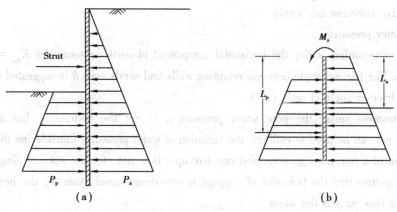

**Figure 2.15  Analysis of push-in by gross pressure method**

Equation 2.14 is generally called the gross pressure method. JSA (1988) and TGS (2001) suggested $F_p \geqslant 1.5$. Nevertheless, when assuming $M_s = 0$, $F_p \geqslant 1.2$. Eq. 2.14 can be used either to obtain the factor of safety against push-in for a certain depth of wall or the required penetration depth of a retaining wall with a certain value of safety factor.

As to short-term behavior of cohesive soils, their earth pressures should be obtained following the total stress method while parameter of soil strength should be derived from the UU test, the FV test, or the CPT test. If the cohesive soil is 100% saturated, it should be analyzed assuming $\varphi = 0$. Rankine's earth pressure can be employed:

$$\sigma_a = \sigma_v K_a - 2cK_a \qquad (2.15)$$

$$\sigma_p = \sigma_v K_p + 2cK_p \qquad (2.16)$$

where

$\sigma_a$ = total active earth pressure (horizontal) acting on the retaining wall;

$\sigma_p$ = total passive earth pressure (horizontal) acting on the retaining wall;

$c$ = cohesion;

$K_a$ = coefficient of active earth pressure;

$K_p$ = coefficient of passive earth pressure.

When cohesive soils are completely saturated, $\varphi = 0$ and $c$ equals the undrained shear strength; that is, $K_a = K_p = 1$.

For cohesionless soils, the excess pore water pressure dissipates quickly as soon as shearing occurs. As a result, the analysis should follow the effective stress method. Supposing there exists friction between the retaining wall and the surrounding soil, the earth pressure for design can be represented as follows:

$$\sigma'_a = (\sigma_v - u)K_a - 2c'K_a \qquad (2.17)$$

$$\sigma'_p = (\sigma_v - u)K_p + 2c'K_p \qquad (2.18)$$

where

$\sigma'_a$ = effective active earth pressure acting on the retaining wall;

$\sigma'_p$ = effective passive earth pressure acting on the retaining wall;

$c'$ = effective cohesion intercept;

$u$ = porewater pressure.

For most cohesionless soils, the horizontal component of earth pressure are $K_{a,h} = K_a \cos \delta$, $K_{p,h} = K_p \cos \delta$. The friction angle between retaining walls and sandy soils $\delta$ is suggested to be 0.5 ~ 0.67 effective friction angle of soils ($\varphi'$).

For cohesionless soils, the pore water pressure $u$ is not the hydrostatic but affected by seepage. Flow net can be used to estimate the variation of water pressure. Considering the difficulty of the depiction of a non-homogeneous and non-isotropic flow net, for the sake of simplification, we sometimes assume that the behavior of seepage is one dimensional, that is, the head loss per unit length of a flow path is the same.

As shown in Figure 2.16(a), the difference of the total heads between the upstream water level (outside the excavation zone) and the downstream water level is $(H_e + d_i - d_j)$. The length of the path of the water flowing from the upstream water level along the retaining wall down to the downstream water level is $(2H_p + H_e + d_i - d_j)$. Assuming that the datum of the elevation head is set to be the same with the upstream water level, the total head ($h$) at a distance of $x$ from the upstream water level would be

$$h = 0 - \frac{x(H_e + d_i - d_j)}{2H_p + H_e - d_i - d_j} = \frac{x(H_e + d_i - d_j)}{2H_p + H_e - d_i - d_j} \qquad (2.19)$$

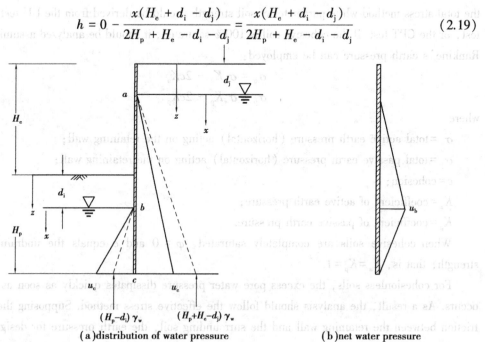

(a)distribution of water pressure      (b)net water pressure

**Figure 2.16  Simplified analysis method for seepage**

Let $h_e$ be the elevation head and $h_p$ pressure head. The pressure head at a distance of $x$ from the upstream water level would be

$$h_p = h - h_e = \frac{x(H_e + d_i - d_j)}{2H_p + H_e - d_i - d_j} - (-x) = \frac{2x(H_p - d_i)}{2H_p + H_e - d_i - d_j} \quad (2.20)$$

Thus, the water pressure, $u_x$ at a distance of $x$ from upstream water level would be

$$u_x = \frac{2x(H_p - d_i)\gamma_w}{2H_p + H_e - d_i - d_j} \quad (2.21)$$

The largest net water pressure is to be found at $b$, whose value would be

$$u_b = \frac{2(H_e + d_i - d_j)(H_p - d_i)\gamma_w}{2H_p + H_e - d_i - d_j} \quad (2.22)$$

Seepage will decrease the porewater pressure on the active side (lower than the hydrostatic water pressure) and increase it on the passive side (higher than the hydrostatic water pressure). As discussed earlier, we can derive the rates of increase of porewater pressure per unit length on the active and passive sides as follows:

$$u_a = \frac{u_c}{H_p + H_e - d_j} \quad (2.23)$$

where $u_a$ and $u_p$ represent the rate of increase of the porewater pressure per unit length on the active and passive sides, respectively.

Thus, the water pressure $x$ below the water level at the active and passive sides are separately $u_x = \mu_a x$ (the active side) and $u_x = \mu_p x$ (the passive side).

Therefore, $\mu_a$ and $\mu_p$ can be seen as the modified unit weights of water on the active side and passive side, respectively. As a result, the total lateral earth pressure at a depth of $z$ below the ground surface on the back of the wall would be

$$\sigma_h = \sigma'_v K_a + u = \left[\sigma_v - \frac{(z - d_j)}{H_p + H_e - d_j}u_c\right]K_a + \frac{(z - d_j)}{H_p + H_e - d_j}u_c \quad (2.24)$$

Similarly, the total lateral earth pressure at a distance of $z$ below the ground surface on the front of the wall would be

$$\sigma_h = \sigma'_v K_p + u = \left[\sigma_v - \frac{(z - d_j)}{H_p - d_j}u_c\right]K_a + \frac{(z - d_j)}{H_p - d_j}u_c \quad (2.25)$$

Alternatively, the active and passive earth pressures on the two sides of the retaining wall are expressed in terms of net values. By conducting a force equilibrium of the net forces on the retaining wall below the lowest level of strut, Equation 2.12 is also available to compute the factor of safety against push-in. This method is called net pressure method. Nevertheless, according to Burland and Potts (1981), and Ou and Hu (1998), any slight difference of the penetration depth will lead to a large change in the factor of safety computed from the net pressure method. That is to say, the factor of safety is too sensitive to the penetration depth and this method is not good for stability analysis of the push-in failure.

## Example 2.1

An excavation in clay goes 9.0 m into the ground ($H_e = 9.0$ m). The groundwater outside the excavation zone is at the ground surface level while that within the excavation zone is at the level of the excavation surface; $\gamma_{sat} = 17.0$ kN/m³. The undrained shear strength $s_u = 45$ kN/m². Suppose the excavation width $B = 10$ m and the excavation length $L = 30$ m. Compute the factor of safety against basal heave according to Terzaghi's method and Bjerrum and Eide's method, respectively.

Solution:

According to Terzaghi's method,

$$F_b = \frac{1}{H_e} \times \frac{5.7 s_u}{\gamma - (s_u/0.7B)} = \frac{1}{9} \times \frac{5.7 \times 45}{17 - 45/(0.7 \times 10)} = 2.69$$

According to Bjerrum and Eide's method,

$$\frac{L}{B} = \frac{30}{10} = 3.00$$

$$\frac{H_e}{B} = \frac{9}{10} = 0.90$$

According to Fig 2.6, $N_c = 7.1$,

$$F_b = \frac{s_u N_c}{\gamma H_e} = \frac{45 \times 7.1}{17 \times 9} = 2.09$$

## Example 2.2

The same as Example 2.1 except that there exists a stiff clayey layer 4.0 m below the excavation surface and the undrained shear strength of the clay is 80 kN/m². Compute the factor of safety against basal heave according to Terzaghi's method and Bjerrum and Eide's method, respectively.

Solution:

According to Terzaghi's method,

$$F_b = \frac{1}{H_e} \times \frac{5.7 s_u}{\gamma - (s_u/D)} = \frac{1}{9} \times \frac{5.7 \times 45}{17 - (45/4.0)} = 4.96$$

According to Bjerrum and Eide's method,

$$\frac{s_{u2}}{s_{u1}} = \frac{80}{45} = 1.778$$

$$\frac{D}{B} = \frac{4}{10} = 0.4$$

$$\frac{D}{H_e} = \frac{4}{9} = 0.444$$

According to Fig 2.7 (a) and (b), $N_{c,s} = 6$,

$$\frac{H_e}{B} = \frac{9}{10} = 0.9$$

According to Fig 2.7 (c), $f_d = 1.22$

$$\frac{B}{L} = \frac{10}{30} = 0.3$$

$$f_s = 1 + 0.2 \times B/L = 1 + 0.2 \times 0.3 = 1.06$$

$$F_b = \frac{s_{ul} N_{c,s} f_d f_s}{\gamma H_e} = \frac{45 \times 6 \times 1.22 \times 1.06}{17 \times 9} = 2.28$$

## Example 2.3

An excavation in completely saturated clay goes 9.0 m into the ground ($H_e = 9.0$ m) and the lowest level of struts is 2 m above the excavation surface. The groundwater outside the excavation zone is at the ground surface level while that within the excavation zone is at the level of the excavation surface. $\gamma_{sat} = 17.0$ kN/m³. The undrained shear strength $s_u = 45$ kN/m². Assume the penetration depth $H_p = 2$ m and ignore the allowable bending moment of the retaining wall $M_s$, compute the factor of safety against push-in.

Solution:

When cohesive soils are completely saturated, $K_a = K_p = 1$

Active earth pressure:

At the lowest level of strut

$$\sigma_{a1} = \sigma_v K_a - 2cK_a = 17 \times 7 - 2 \times 45 = 29 \text{ kN/m}^2$$

At the toe of the retaining wall

$$\sigma_{a2} = \sigma_{a1} + 4 \times 17 = 97 \text{ kN/m}^2$$

Resultant of the active earth pressure below the lowest level of strut

$$P_a = \frac{(\sigma_{a1} + \sigma_{a2})}{2} \times 4 = \frac{28 + 97}{2} \times 4 = 252 \text{ kN/m}$$

Length from the lowest level of strut to the point of action $P_a$

$$L_a = \frac{2\sigma_{a1} + \sigma_{a2}}{3(\sigma_{a1} + \sigma_{a2})} \times 4 = \frac{2 \times 29 + 97}{3(29 + 97)} \times 4 = 1.64 \text{ m}$$

Passive earth pressure:

At the excavation bottom

$$\sigma_{p1} = \sigma_v K_p + 2cK_p = 17 \times 9 + 2 \times 45 = 243 \text{ kN/m}^2$$

At the toe of the retaining wall

$$\sigma_{p2} = \sigma_{p1} + 2 \times 17 = 277 \text{ kN/m}^2$$

Resultant of the passive earth pressure below the lowest level of strut

$$P_p = \frac{(\sigma_{p1} + \sigma_{p2})}{2} \times 2 = \frac{243 + 277}{2} \times 2 = 520 \text{ kN/m}$$

Length from the lowest level of strut to the point of action $P_p$

$$L_p = \frac{2\sigma_{p1} + \sigma_{p2}}{3(\sigma_{p1} + \sigma_{p2})} \times 2 = \frac{2 \times 243 + 277}{3 \times (243 + 277)} \times 2 = 0.98 \text{ m}$$

The factor of safety against push-in

$$F_p = \frac{M_r}{M_d} = \frac{P_p L_p + M_s}{P_a L_a} = \frac{520 \times 0.98 + 0}{252 \times 1.64} = 1.23$$

## Example 2.4

A 9.0 m deep excavation in a sandy ground and the lowest level of struts is 2.5 m above the excavation surface. The level of groundwater outside the excavation zone is at the ground surface level while that within the excavation zone is as high as the excavation surface. The unit weight of saturated sandy soils $\gamma_{sat} = 20 \text{ kN/m}^2$, the effective cohesion $c' = 0$, and the effective angle of friction $\varphi' = 30°$. Assume that the friction angles ($\delta$) between the retaining wall and soil on both the active and passive sides are $0.5\varphi'$ and the factor of safety against push-in, $F_p = 1.5$. Compute the required penetration depth ($H_e$).

Solution:

Let $z$ represent the depth from the ground surface and $x$ depth from the groundwater level

①Determine the coefficient of the earth pressure:

$$K_{a,h} = K_a \cos \delta = \tan^2\left(45° - \frac{\varphi'}{2}\right) \cos\left(\frac{\varphi'}{2}\right) = \tan^2\left(45° - \frac{30°}{2}\right) \cos\left(\frac{30°}{2}\right) = 0.32$$

$$K_{p,h} = K_p \cos \delta = \tan^2\left(45° + \frac{\varphi'}{2}\right) \cos\left(\frac{\varphi'}{2}\right) = \tan^2\left(45° + \frac{30°}{2}\right) \cos\left(\frac{30°}{2}\right) = 2.90$$

②Compute the effective active earth pressure on the wall:

According to Eq. 2.19, the porewater pressure at $x$ away from upstream water level would be

$$u = \frac{2x(H_p - d_i)\gamma_w}{2H_p + H_e - d_i - d_j} = \frac{2 \times 6.5 \times H_p \times 9.81}{2H_p + 9} = \frac{63.77H_p}{H_p + 4.5}$$

$$\sigma'_{a,h} = (\sigma - u)K_{a,h} = \left(20 \times 6.5 - \frac{63.77H_p}{H_p + 4.5}\right) \times 0.32 = 41.6 - \frac{20.41H_p}{H_p + 4.5}$$

At the bottom of the retaining wall ($z = 9 + H_p$, $x = 9 + H_p$)

$$u = \frac{2x(H_p - d_i)\gamma_w}{2H_p + H_e - d_i - d_j} = \frac{2 \times (9 + H_p) \times H_p \times 9.81}{2H_p + 9} = \frac{9.81H_p^2 + 88.29H_p}{H_p + 4.5}$$

$$\sigma'_{a,h} = (\sigma - u)K_{a,h} = \left(180 + 20H_p - \frac{9.81H_p^2 + 88.29H_p}{H_p + 4.5}\right) \times 0.32$$

$$= 57.6 + 6.4H_p - \frac{3.14H_p^2 + 28.25H_p}{H_p + 4.5}$$

③Compute the lateral effective passive earth pressure on the wall:

At the bottom of the wall

$$\sigma'_{p,h} = \left(20H_p - \frac{9.81H_p^2 + 88.29H_p}{H_p + 4.5}\right) \times 2.9 = 58H_p - \frac{27.47H_p^2 + 256.04H_p}{H_p + 4.5}$$

④Compute the maximum net water pressure at the excavation surface:

$$u_b = \frac{2(H_e + d_i - d_j)(H_p - d_i)\gamma_w}{2H_p + H_e - d_i - d_j} = \frac{2 \times 9 \times H_p \times 9.81}{2H_p + 9} = \frac{88.29H_p}{H_p + 4.5}$$

⑤Compute the driving moment and resistant moment for the free body below the lowest level of struts:

$$M_d = P_{a,h} L_a = \left(41.6 - \frac{20.41H_p}{H_p + 4.5}\right) \times \frac{(H_p + 2.5)^2}{2} + \left(57.6 + 6.4H_p - \frac{3.14H_p^2 + 28.25H_p}{H_p + 4.5}\right) \times$$

$$\frac{2(H_p + 2.5)^2}{2 \times 3} + \frac{u_b H_p}{2} \times \left(2.5 + \frac{H_p}{3}\right) + \frac{6.5u_b}{9} \times \frac{2.5^2}{2} + \frac{2.5u_b}{9} \times \frac{2 \times 2.5^2}{2 \times 3}$$

$$M_r = P_{p,h} L_p = \left(58H_p - \frac{27.47H_p^2 + 256.04H_p}{H_p + 4.5}\right) \times \frac{H_p}{2} \times \left(2.5 + \frac{2H_p}{3}\right)$$

⑥Determine the penetration depth ($H_p$):

$$F_b = \frac{M_r}{M_d} = 1.5$$

Then we have $H_p = 17.4$ m.

# 2.5　Overall Shear Failure of Cantilever Walls

Theoretically speaking, the overall shear failure analysis of a cantilever wall should include analyses of push-in failure and basal heave failure. However, the stability of the cantilever wall is rather weak and its application is usually confined to sand, gravelly soils or stiff clays.

On soft clay, the cantilever wall is not reliable enough. Since no basal heave failure is found in sandy gravel soils or stiff clays, as far as the cantilever wall is concerned, only analysis of push-in failure is required.

The stability of a cantilever wall counts on the soil reaction at a specific fixed point. The design is therefore confined to the fixed earth support method and the free earth support method is inapplicable. Figure 2.17(a) illustrates a cantilever wall rotating about point $O$ in a limiting state. Figure 2.17(b) shows the earth pressure on the retaining wall. For the simplification of analysis, assume the active and passive earth pressures above and below point $O$ are fully mobilized and therefore the earth pressure distribution is discontinuous around point $O$ as shown in Figure 2.17(c).

See the earth pressure distribution as shown in Figure 2.17(c), and $H_p$ and $L$ are unknown. With the horizontal force equilibrium and the moment equilibrium, we would obtain the required penetration depth. Since both the horizontal force equilibrium and the moment equilibrium will generate quadratic and cubic equations, it is not easy to solve the equations directly. The trial-and-error method is not less complicated, so it is necessary to simplify the analysis method for practical use.

Figure 2.17(d) illustrates the simplified earth pressure distribution where the concentration force, $R$ represents the difference between the passive earth pressure from outside the excavation zone and the active earth pressure from inside the excavation zone below the turning point $O$. It is necessary that $R$ exists to keep the horizontal force equilibrium. Based on the moment equilibrium against point $O$, we can find the value $d_0$. Because of the simplification of the analysis, $d_0$ should be slightly smaller than the actually required penetration depth and has to be increased properly (generally, 20%). The increment has to be examined to make $R$ satisfy the horizontal force equilibrium as shown in Figure 2.17(d). The detailed computing process can be found in Example 5.4. The simplified analysis method as shown in Figure 2.17(d) has been commonly adopted in engineering design (Padfield and Mair, 1984).

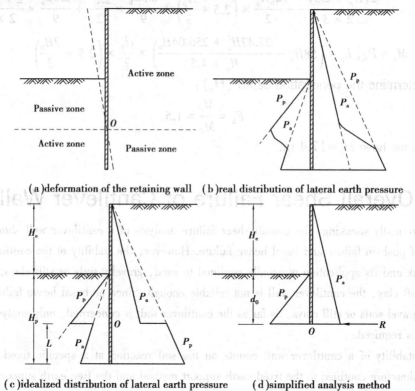

(a)deformation of the retaining wall　(b)real distribution of lateral earth pressure

(c)idealized distribution of lateral earth pressure　(d)simplified analysis method

Figure 2.17　Analysis of a cantilever wall by gross pressure method

Excavation in a sandy ground with groundwater on both sides of the retaining wall should be analyzed in terms of the distribution of net water pressure as discussed in Section 2.3, viewed as the driving force, following the method shown in Figure 2.17(d). In clayey soils, even though there exists groundwater, water pressure is not to be considered because the undrained analyses have considered the water pressure automatically.

Figure 2.18 shows the net earth pressure distribution. According to the characteristics of the deformation of the cantilever wall, we can see the earth pressures at point $b$ (excavation surface) on the front and back of the retaining wall should achieve the passive earth and the active earth pressures, respectively. The net value of the earth pressure at the point is $CD$ [Figure 2.18(b)]. Point $c$ is close to point $b$ and the earth pressures on the front and the back of the wall should also achieve the passive and active earth pressures separately. Thus, the net earth pressure at point $c$ should be

$$\sigma_{h,c} = [\gamma H_e + \gamma(z - H_e)]K_a - \gamma(z - H_e)K_p = \gamma H_e K_a - \gamma(z - H_e)\gamma(K_p - K_a)$$

where

$\gamma$ = unite weight of the soil;

$K_a$ = coefficient of the active earth pressure;

$K_p$ = coefficient of the active earth pressure.

We can see from the above equation that the slope of line CF is $1 : \gamma(K_p - K_a)$ (the vertical : the horizontal). The net earth pressure with the slope $1 : \gamma(K_p - K_a)$ maintains till point $d$, where the passive and active earth pressures "begin" not to be fully mobilized. Assuming the lateral movement of the bottom of the wall is large enough, the soil in front and back of the wall will reach the active and passive states, respectively, and the net value of the earth pressure here is $BG$. The soil strength between points $f$ and $d$ in front and in back of the wall may not be fully mobilized. TO simplify the analysis, we assume the net earth pressures between points $d$ and $f$ are of a linear relation, as line $FG$. Figure 2.18(b) shows an assumed net earth pressure distribution. According to the relation of the horizontal force equilibrium in Figure 2.18(b), we know

$$\text{Area } ACE - \text{Area } EFHB + \text{Area } FHBG = 0$$

We can find $L$ according to the above equation. Substitute $L$ in the moment equilibrium equation with regard to point $f'$ and we have the quadratic equation with $H_p$ as a single unknown variable. To solve $H_p$, the trial-and-error method is recommended. When performing trial and error, we usually begin from $H_p = 0.75 H_e$.

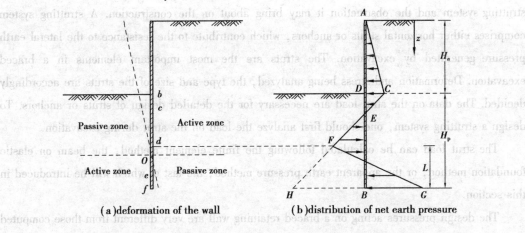

(a)deformation of the wall　　　　(b)distribution of net earth pressure

**Figure 2.18　Analysis of a cantilever wall by net pressure method**

# 3 Earth Pressure and Strut Force

## 3.1 Introduction

Problems of deep excavation, whether stability analysis, stress analysis or deformation analysis, entail the distribution of earth pressures. Though introductory books on soil mechanics or foundation engineering have discussed quite a few earth pressure theories along with many examples, a systematic organization is lacking, and some important points may not be sufficiently emphasized. In actual analyses, a wrong choice of earth pressure theory may lead to an uneconomical or even unsafe design. This chapter is going to do a systematic organization and to simplify the complicated calculations for excavation analyses and design. Most of the methods introduced here have been frequently used in engineering practice though they have not been introduced in general textbooks.

Other than the gravity retaining walls (the retaining wall alone can rarely resist the lateral pressure), supplementary strutting systems are also required. The selection of the strutting system depends on not only the magnitude of lateral pressure, but also the period it will take to install the strutting system and the obstruction it may bring about on the construction. A strutting system comprises either horizontal struts or anchors, which contribute to the resistance to the lateral earth pressure generated by excavation. The struts are the most important elements in a braced excavation. Deformation and stress being analyzed, the type and size of the struts are accordingly decided. The data on the strut load are necessary for the detailed design of struts or anchors. To design a strutting system, one should first analyze the load on the strut during excavation.

The strut load can be calculated following the finite element method, the beam on elastic foundation method, or the apparent earth pressure method, the last of which will be introduced in this section.

The design pressures acting on a braced retaining wall are very different from those computed from conventional walls where the pressure distribution is usually triangular. Because of the redistribution due to arching and the incremental nature of excavation and strut installation, the pressure distribution does not linearly increase with depth.

When excavation proceeds, the pressure distributions during different excavation stages are shown in Figure 3.1, which states as follows:

Stage 1: The wall is subjected to an active earth pressure and the wall deforms.

Stage 2: A strut is installed and preloaded. Generally, the wall and soil will not be pushed back to its original position, but since the strut force is larger than the active pressure, this causes an increase in the wall pressure.

Stage 3: The excavation in stage 3 causes a new lateral displacement. The soil moves out of the zone behind the first strut into the displacement between b and c.

Stage 4: The installation of the second strut in stage 4 will result in similar changes to the earth pressures.

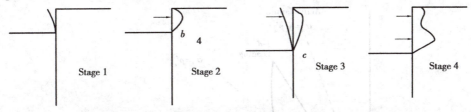

**Figure 3.1  Pressure distributions during different excavation stages**

# 3.2 Lateral Earth Pressure in Braced Excavations

## 3.2.1 In situ lateral stress

Horizontal (lateral) stress in soils is usually described and quantified by means of a lateral earth pressure coefficient, $K$:

$$K = \sigma_h'/p_v' \qquad (3.1)$$

where

$p_v'$ is the effective overburden pressure;

$\sigma_h'$ is the horizontal (lateral) effective stress at the same point within the soil mass.

The effective overburden pressure $p_v'$ is given by

$$p_v' = \int_0^z \gamma \mathrm{d}z + q - u \qquad (3.2)$$

where

$\gamma$ is the bulk density;

$z$ is the depth below;

$u$ is the pore water pressure;

$q$ is any uniform surcharge at the ground surface.

Unlike some other retaining walls, embedded walls usually retain predominantly natural ground. So the pre-existing or in situ horizontal (lateral) earth pressure, as modified by wall installation, is potentially important. The symbol $K_0$ is used to denote the earth pressure coefficient describing the initial in situ stress state in the ground before the wall is installed.

The in situ earth pressure coefficient may be of particular concern in a clay deposit, in which $K_0$, like the specific volume, depends on the stress history. Deposition (or burial under a glacier) corresponds approximately to one-dimensional (1D) compression, during which the horizontal effective stress $\sigma'_h$ increases in proportion to the effective overburden pressure $p'_v$. Clays may also become consolidated by desiccation (drying) on exposure to air, by vegetation or by freezing, where the effective stress is increased through a reduction in pore water pressure while the total stress remains constant, as shown in Figure 3.2.

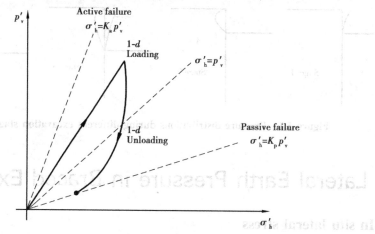

**Figure 3.2    Schematic stress history of an overconsolidated clay deposit**

On unloading (e.g. due to the erosion of overlying soil, re-saturation after desiccation, the melting of an overlying glacier or a rise in groundwater level), the horizontal effective stress, $\sigma'_h$, tends to remain "locked-in", decreasing proportionately less quickly than the effective overburden pressure, $p'_v$. So the in situ earth pressure coefficient $K_0$ in an overconsolidated clay stratum is usually greater than unity. In heavily overconsolidated clays, a zone of soil extending to a depth of several meters from the surface maybe be close to its limiting passive pressure because of geological unloading.

The earth pressure coefficient in granular material or weak rock is not uniquely related to the stress history of the deposit as it is in clay.

In situ lateral earth pressure may be estimated using an equation of the form:

$$K_0 = (1 - \sin \varphi').OCR^{\lambda}.(1 + \sin \beta) \qquad (3.3)$$

where

$\varphi'$ is the drained angle of shearing resistance;

$OCR$ is the over consolidation ratio;

$\beta$ is the slope angle (to the horizontal) of the soil surface.

For normally consolidated deposit ($OCR=1$) and level ground ($\beta=0$), this reduces to Jaky's empirical formula $K_0=(1-\sin\varphi')$.

## 3.2.2  Conventional earth pressure theory

When soil is removed from the front of an embedded retaining wall, the wall tends to move into the excavation. This will result in a reduction in the lateral stress in the ground behind the wall, eventually bringing it to the active condition in which the ground is at failure with the horizontal effective stress as small as it can be for the effective overburden pressure. In the ground that remains in front of the wall below formation level, the horizontal effective stress at failure is as large as it can be for the effective overburden pressure.

Approximations to these limiting pressures may be calculated by considering either the stresses in a zone of ground at failure, or the equilibrium of an assumed sliding wedge. The first approach, following Rankine (1857), is based on a stress state that can be in equilibrium without exceeding the limiting strength of the ground. In a uniform deposit, the limiting ratio of horizontal to vertical effective stresses is constant. That is, if the vertical effective stress increases linearly with depth, so will the limiting horizontal effective stress. The limits calculated in this way ensure stability, but may be unnecessarily severe, and are inherently safe. (assuming that the correct boundary conditions, ground strength parameters and pore pressures have been identified)

In the second approach, following Coulomb (1776), the force that must be exerted by a retaining wall to prevent a wedge of soil from sliding down an assumed slip surface is determined, and is then usually assumed to arise from a stress that increases linearly with depth. The limits obtained will prevent sliding along the slip surface assumed in the calculation, but may not be sufficient to prevent failure from occurring in some other unidentified mechanisms, so they could be unsafe.

The results obtained from Rankine's and Coulomb's earth pressure theories are identical under the same conditions (smooth wall surfaces, level grounds, and homogeneous cohesionless soils) though the two theories are quite differently based (Figures 3.3 and 3.4). The actual conditions, however, may not conform to the hypothetical terms: the wall surface may be rough and the ground surface in back of the wall may be of irregular shape with certain load. Rankine's theory can hardly apply under these conditions. Coulomb's theory can cope with these complicated conditions, though it is difficult to obtain a theoretical solution. However, experience overwhelmingly demonstrates that these more recent theories give accurate values of the limiting lateral stresses for use as a basis for design, assuming that variation in pore water pressures and strength properties, etc. in the ground around the wall have been identified and taken into account.

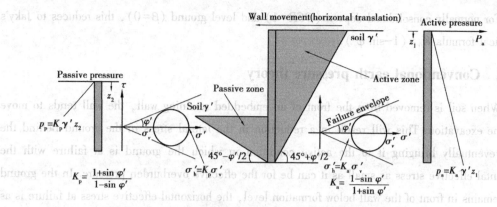

**Figure 3.3    Rankine plastic equilibrium for a frictionless wall or soil interface translating horizontally**

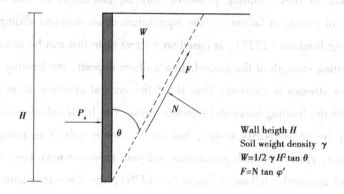

Wall heigth $H$
Soil weight density $\gamma$
$W=1/2\ \gamma\,H^2\tan\theta$
$F=N\tan\varphi'$

Res. Hor. $P_a=N\cos\theta-F\sin\theta$
            $=N(\cos\theta-\sin\theta\tan\varphi')$
Res. vert. $W=1/2\ \gamma\,H^2\tan\theta=N(\sin\theta+\cos\theta\tan\varphi')$
Eliminate $N$ and minimize $P_a$
$\theta=45°-\varphi'\ P_a=1/2\ \gamma\,H^2K_a$ where $K_a=(1-\sin\varphi')/(1+\sin\varphi')$

**Figure 3.4    Coulomb's method to calculate the limiting active force for a frictionless wall or soil interface translating horizontally**

In summary, the conventional approach to the design of an embedded retaining wall is something of a hybrid, with earth pressure coefficients derived using a mechanism-based approach to take account of factors such as wall interface friction, the profile of the retained ground surface and the presence of line loads used in an equilibrium analysis of the wall.

## 3.2.3    Apparent pressure diagrams (APD)

The empirical methods of Terzaghi and Peck (1967) and Peck (1969a) were obtained from field measurements of strut loads and are envelopes of the maximum measured pressures. They are considered to give conservative estimates of earth pressures and provide estimates of the maximum strut loads for design purposes. They are not the actual pressure distributions but correspond to the maximum values expected. These pressure envelopes are commonly referred to as apparent pressure diagrams (APD). Figure 3.5 illustrates the procedure of calculating the apparent earth pressure (by method of Tributary Areas). That is, the load on the struts can be approximately determined by the method of tributary areas (Terzaghi, Peck & Mesri, 1996), which consists in dividing in sections the pressure acting on the diaphragm wall. The division of these sections is performed in the middle of the distance between the struts, assigning the pressure value of each section to the strut that is in that section. As shown in Figure 3.5. The horizontal distance between struts here is considered to be a constant as b. From this way, strut load $Q$ can also be calculated easily.

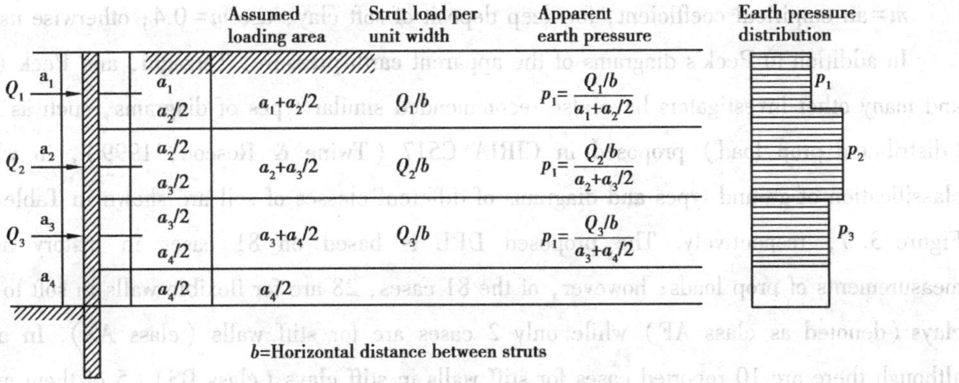

| Assumed loading area | Strut load per unit width | Apparent earth pressure | Earth pressure distribution |
|---|---|---|---|
| $a_1$, $a_2/2$ | $Q_1/b$ | $p_1=\dfrac{Q_1/b}{a_1+a_2/2}$ | $p_1$ |
| $a_2/2$, $a_3/2$ | $Q_2/b$ | $p_1=\dfrac{Q_2/b}{a_2+a_3/2}$ | $p_2$ |
| $a_3/2$, $a_4/2$ | $Q_3/b$ | $p_1=\dfrac{Q_3/b}{a_3+a_4/2}$ | $p_3$ |
| $a_4/2$ | | | |

$b$=Horizontal distance between struts

**Figure 3.5　Calculation of strut load**

Figure 3.6 shows diagrams of the apparent earth pressure established by Peck (1969b). As shown in the figure, when the soil in back of the wall mainly consists of sandy soils, the apparent earth pressure $P_a$ will be

$$P_a = 0.65 K_a \gamma H_e \qquad (3.4)$$

where

　　$\gamma$ = unit weight of sandy soils;

　　$H_e$ = excavation depth;

　　$K_a$ = Rankine's coefficient of earth pressure = $\tan(45° - \varphi/2)$.

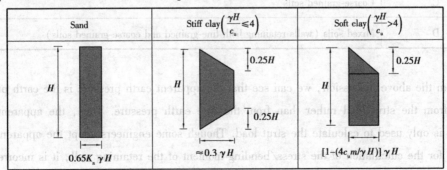

**Figure 3.6　Diagrams of the apparent earth pressure (Peck, 1969a)**

If the soil in back of the wall is stiff clay ($\gamma_H/c_u \leqslant 4$), the apparent earth pressure $P_a$ would be

$$P_a = 0.2\gamma H \sim 0.4\gamma H \text{ (the average is } 0.3\gamma H) \tag{3.5}$$

If the soil in back of the wall is soft to medium soft clay (i.e. $\gamma H/c_u > 4$), the apparent earth pressure, $P_a$, would be the larger of

$$p_a = [1 - 4c_u m/\gamma H]\gamma H \text{ or } p_a = 0.3\gamma H \tag{3.6}$$

where

$c_u$ = undrained shear strength of soil within the range of the excavation depth;

$m$ = an empirical coefficient, for deep deposit of soft clay, use $m = 0.4$; otherwise use $m = 1$.

In addition to Peck's diagrams of the apparent earth pressure, Terzaghi, and Peck (1967), and many other investigators have also recommended similar types of diagrams, such as the DPL (distributed prop load) proposed in CIRIA C517 (Twine & Roscoe, 1999), in which the classification of ground types and diagrams of different classes of soil are shown in Table 3.1 and Figure 3.7, respectively. The proposed DPL is based on 81 cases in history and field measurements of prop loads; however, of the 81 cases, 28 are for flexible walls in soft to medium clays (denoted as class AF) while only 2 cases are for stiff walls (class AS). In addition, although there are 10 reported cases for stiff walls in stiff clays (class BS), 5 of them are singly propped while 2 cases have two strut levels; only the remaining 3 cases have three levels of struts. At present, Peck's diagrams of apparent earth pressures are the most commonly applied in engineering design, though.

**Table 3.1  Classification of ground types**

| Soil Class | Description |
|---|---|
| A | Normally and slightly overconsolidated clay soils (soft to firm clays) |
| B | Heavily overconsolidated clay soils (stiff and very stiff clays) |
| C | Coarse-grained soils |
| D | Mixed soils (walls retaining both fine-grained and coarse-grained soils) |

From the above discussion, we can see that the apparent earth pressure is the earth pressure derived from the strut load rather than from the true earth pressure. Thus, the apparent earth pressure is only used to calculate the strut load. Though some engineers adopt the apparent earth pressure for the calculation of the stress/bending moment of the retaining wall, it is incorrect.

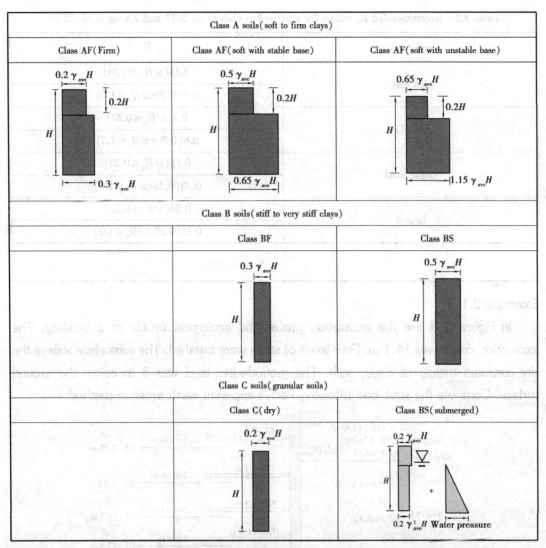

**Figure 3.7   DPL of different classes of soil** (Twine & Roscoe, 1999)

According to many documents and experience, nevertheless, the apparent earth pressure method is still useful for excavations not deeper than 10 m. As for deep excavations (over 20 m), the applicability of the method needs more examination. Recently, based on numerical studies as well as field measurements from a number of reported case histories, Goh et al. (2017) and Zhang et al. (2019) proposed updated empirical charts for determining strut loads for excavations in stiff wall systems, as shown in Table 3.2, which could be more practical in designing process of retaining structures.

**Table 3.2    Recommended $K_a$ values for stiff walls (Goh et al. 2017 and Zhang et al. 2019)**

| Soil types | $K_a$ |
|---|---|
| Soft clay | $0.45(z/H_e \leqslant 0.20)$ |
|  | $0.9(0.20 < z/H_e \leqslant 1.0)$ |
| Stiff clay | $0.4(z/H_e \leqslant 0.20)$ |
|  | $0.6(0.20 < z/H_e \leqslant 1.0)$ |
| Dense sand | $0.15(z/H_e \leqslant 0.20)$ |
|  | $0.30(0.20 < z/H_e \leqslant 1.0)$ |
| Gravel | $0.20(z/H_e \leqslant 0.20)$ |
|  | $0.35(0.20 < z/H_e \leqslant 1.0)$ |

## Example 3.1

In Figure 3.8 are the excavation profile and geological profile of a building. The excavation depth was 14.1 m. Four levels of struts were installed. The subsurface soils at the site consisted mainly of clayey soils. The groundwater level was 3 m below the ground surface. Compute the strut load following Peck's apparent earth pressure method.

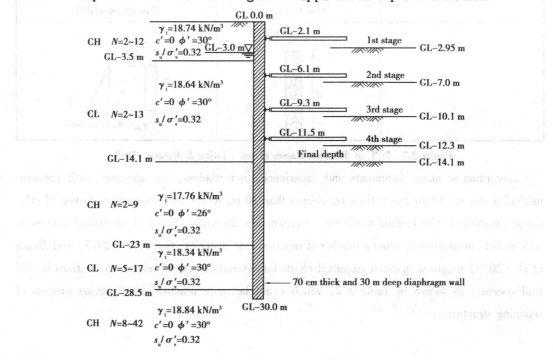

**Figure 3.8    Soil and excavation profile of Building $Q$ in Taipei**

Solution

Suppose the soil above the excavation surface was clay whose normalized undrained shear strength is $s_u/\sigma'_v = 0.32$. The average unit weight of soil is $\gamma = 18.64$ kN/m. We can compute the undrained shear strength at the depth of 7.05 m (14.1 m/2) below the ground surface as follows:

$$\sigma_v = 18.74 \times 3.5 + 18.64 \times (7.05 - 3.5) = 131.8 \text{ kN/m}^2$$

$$u = 9.81 \times (7.05 - 3.0) = 39.7 \text{ kN/m}^2$$

$$\sigma'_v = \sigma_v - u = 131.8 - 39.7 = 92.1 \text{ kN/m}^2$$

$$s_u = 0.32\sigma'_v = 29.5 \text{ kN/m}^2$$

$$\gamma H_e/s_u = 18.64 \times 14.1/29.5 = 8.9 > 4.0$$

The soil at the site can then be categorized as soft to medium soft clay according to Peck's apparent earth pressure diagrams. The influence depth of excavation can be defined to be as deep as the excavation width. Since the excavation width $B = 28.8$ m, the influence depth is about from the excavation surface down to $14.1 + 28.8 = 42.9$ m below it. Then the stress in the soil at a depth of 28.5 m (14.1 + 28.8/2) is

$$\sigma_v = 131.8 + 18.64 \times 7.05 + 17.76 \times (23 - 14.1) + 18.34 \times (28.5 - 23) = 522.2 \text{ kN/m}^2$$

$$u = 9.81 \times (28.5 - 3.0) = 250.2 \text{ kN/m}^2$$

$$\sigma'_v = \sigma_v - u = 522.2 - 250.2 = 272 \text{ kN/m}^2$$

$$s_u = 0.32\sigma'_v = 87 \text{ kN/m}^2$$

$$N_b = \gamma H_e/s_{u,b} = 18.64 \times 14.1/87 = 2.97 \text{ kN/m}^2$$

Assume that m = 1.0

$$P_a = \gamma H_e - 4s_u = 18.64 \times 14.1 - 4 \times 29.5 = 144.6 \text{ kN/m}^2$$

or

$$P_a = 0.3\gamma H_e = 78.8 \text{ kN/m}^2$$

The apparent earth pressure is $P_a = 144.6$ kN/m. The distribution of earth pressure is as shown in Figure 3.9. According to the half method, the load on each level of struts would be

The load on the 1st level of struts = 144.6×3.5/2+144.6×(2+2.1−3.5) = 339.8 kN/m

The load on the 2nd level of struts = 144.6×(4.0/2+3.2/2) = 520.6 kN/m

The load on the 3rd level of struts = 144.6×(3.2/2+2.2/2) = 390.4 kN/m

The load on the 4th level of struts = 144.6×(2.2/2+2.6/2) = 347 kN/m

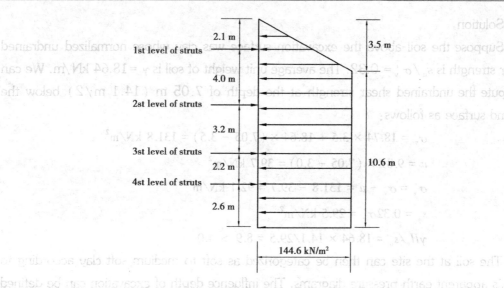

**Figure 3.9    Distribution of apparent earth pressure and locations of struts**

# 3.3    Parametric Study

A series of parametric study was carried out to look into a refined apparent pressure diagram for braced excavations in soft clay involving diaphragm wall by Chang and Wong (1996). Figure 3.10 shows the soil profile and geometry adopted. The effects on the strut force from factors such as soil modulus, undrained shear strength ($c_u$) profile, wall stiffness (EI), preloading level, thickness to hard stratum ($T$), excavation width ($B$), and number of strut levels are studied. More details can be found in their literature "Apparent pressure diagram for braced excavations in soft clay with diaphragm wall", and the following section presents the main findings.

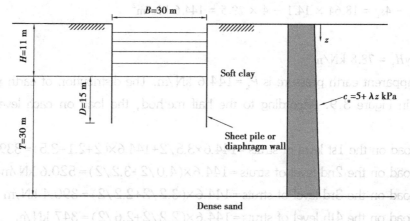

**Figure 3.10    Soil profile and geometry adopted for parametric study (Chang and Wong, 1996)**

### 3.3.1   Effect of soil modulus

The strut force exceedance ratio between the strut force is derived from the finite element analysis; the corresponding value computed from the reference APD, and the initial tangent modulus, $E_i$ quantified in terms of the $c_u$ values. The secant modulus, $E_u$, is approximately 60% of $E_i$.

It is observed that, when the $E_i/c_u$ ratio falls between 200 and 500, except the lowest strut, the reference APD underestimates the strut forces by as much as 100%. As the $E_i/c_u$ ratio increases to 1000, the reference APD becomes more applicable.

### 3.3.2   Effect of the $c_u$ profile and value

Variations of $c_u$ with depth have been shown to affect the performance of braced excavations (Wong, 1989). It found the strut force of the lowest strut falls within the applicable range of the reference APD even in a very soft clay with which the $c_u^*/\gamma H$ value is still very small. The reference APD actually increasingly overestimates the lowest strut force when the $c_u^*/\gamma H$ value enlarges. This situation further enhances when it is compounded by a high $E_i/c_u$ ratio.

Conversely, the reference APD appears to be consistently underestimating the strut force at higher strut levels regardless of $c_u^*/\gamma H$ value, and the situation of underestimation worsens when both of the ratios of $c_u^*/\gamma H$ and $E_i/c_u$ are low. Nevertheless, the trend of underestimation appears to level off around a $c_u^*/\gamma H$ value of 1.5. It is postulated that in the $E_i/c_u - c_u^*/\gamma H$ space, the area within the borders marked by $c_u^*/\gamma H$ of 1.5 and $E_i/c_u$ of 500, defines a non-applicable zone for the reference APD. Outside the zone, the reference APD is able to provide a reasonable estimation of strut forces together with a factor of safety of 1.5.

### 3.3.3   Effect of wall stiffness

Diaphragm walls ranging from 0.6 m to 1.5 m thick, which represent an almost 15 times increase in wall stiffness (EI), were studied. Strut forces derived indicate a wall stiffness dependency. The greater the wall stiffness is, the larger the strut forces are. The phenomenon is believed to be linked to the arching effect induced by wall displacement. Stronger arching effect associating the larger displacement of thinner walls reduces the corresponding strut forces. It is noted that the highest strut force exceedance ratio is approximately 2.5 at the S3 level, with the rest falling generally below 2.0, while the forces of the inter-mediate strut levels experience higher exceedance rations, and the strut force intensity at the top and the bottom strut levels are conforming well with the reference APD.

### 3.3.4   Effect of excavation width($B$)

Varying excavation widths appear to have no significant effect on the strut force as long as the excavation width is about 3 times the excavation depth. A constant $B/T$ ratio of 1 is maintained for these studies for the purpose of eliminating possible interference from the clay thickness effect. However, for a narrower excavation, more passive resistance below the excavation level is mobilized due to the interaction from both sides of the wall, which results in reduced strut forces, especially as the lower strut levels. It appears that, when the $B/H$ ratio is below 1, the reference APD is able to compute strut forces satisfactorily.

### 3.3.5 Effect of thickness to hard stratum($T$)

The thickness of soft clay below the excavation level is seen to impose strong effect on the strut force. When the $T/B$ ratio drops below 1, the restraining effect from the presence of hard stratum at shallow depth reduces the strut force. A maximum of 80% reduction is noted at the lowest strut level as the $T/B$ ratio scales down from 1 to 0.25. Conversely, as the $T/B$ ratio increases above 1, the effect of clay thickness becomes negligible. It is, therefore, postulated that, for a braced excavation with a $T/B$ ratio more than 1, the layer of soil below a depth of $1B$ from the excavation level could be neglected from the strut force analysis.

### 3.3.6 Effect of preloading

To stiffen the support system, higher preload has been used in an attempt to reduce wall deflections and around settlements. The wall may be pushed back under this load. But one should be careful to avoid excessive preload, which can cause passive failure of the soil behind the wall, and large bending moment can be induced in the wall. Singapore Post Center used 100% preload. Most MRT stations along the NEL required a minimum of 50%. Some (O'Rourke, 1974) consider little benefit in introducing the additional load. The way of preloading includes the following three types, as shown in Figures 3.11 to 3.13, showing the effectiveness of preloading using different methods.

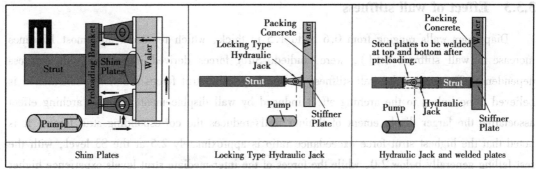

**Figure 3.11　Three types of applying preload**

In general, strut forces reduce under increasing preloads, but with a diminishing efficiency. The forces at the top and bottom strut levels again agree well the reference APD; whereas forces at the intermediate strut levels are seen to exceed by as much as 100%.

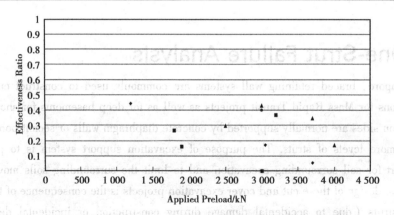

**Figure 3.12　Effectiveness of preloading using shim plate method**

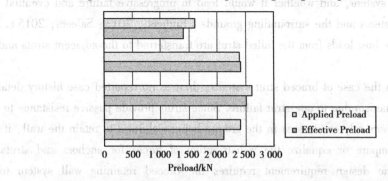

**Figure 3.13　Effectiveness of preloading using locking type hydraulic jacks**

### 3.3.7　Effect of temperature

Struts will expand or contract according to $\Delta L = \alpha \Delta t L$

where

　$L$ = strut length;

　$\Delta t$ = change in strut temperature;

　$A$ = thermal coefficient of expansion, $1.2 \times 10^{-5}/°C$ for steel.

　If fully restrained (i.e. $\Delta L = 0$)

$$\Delta P_{temp} = \Delta \sigma A = \left(\frac{\Delta L}{L}\right) EA = \alpha \Delta TEA \tag{3.7}$$

The actual $\Delta P_{field}$ measured in the field are much less than $\Delta P_{temp}$ because of wall yielding.

　Degree of restrained $(\%) = \dfrac{\Delta P_{field}}{\Delta P_{temp}} \times 100\%$

$$\Delta P = k\alpha \Delta TEA \tag{3.8}$$

A temperature change of $\pm 10\ °C$ is expected. $k = 1.0$ is for a fully restrained strut where both ends are prevented to expand freely. In the absence of rigorous analysis, $k = 0.6$ is recommended for flexible sheet pile walls and $k = 0.8$ for stiff wall with stiff soil condition. Temperature effects are normally added to the predicted strut loads after the analysis is completed.

# 3.4　One-Strut Failure Analysis

In Singapore, braced retaining wall systems are commonly used to construct cut and cover tunnels/stations for Mass Rapid Transit projects as well as for deep basements for shopping malls. The excavation sides are normally supported by concrete diaphragm walls or secant bored pile walls with two or more levels of struts. The purpose of excavation support system is to provide rigid lateral support for soil surrounding excavation and to limit the surrounding soils movement. One concern in the design of these cut and cover excavation projects is the consequence of the failure of one or two struts (due to accidental damage during construction or incidental design/quality problems) in bracing system, and whether it would lead to progressive failure and eventual total collapse of bracing systems and the surrounding grounds (Endicott, 2013; Saleem, 2015). This part aims to investigate how loads from the failed strut are transferred to the adjacent struts and the whole support system.

Unfortunately, in the case of braced strut systems, there is no reported case history detailing the load transfer mechanism due to one-strut failure. Since struts provide passive resistance to wall movement, while anchors rely on stresses in the ground being mobilized to retain the wall, it may not be feasible to compare or equalize the redistribution of forces for anchors and struts. In Singapore, part of the design requirement requires the braced retaining wall system to be structurally safe, robust and has sufficient redundancy to avoid catastrophic collapse. In the conventional approach for one-strut failure using 2D analysis, the entire level of the failing strut is removed and thus the forces can only be distributed vertically. This generally leads to a more conservative design with heavier strut sections. Thus, 3D analysis of one-strut failure is essential to provide more realistic understanding of the force/stress transfer behavior of the braced excavation system. This part describes the use of both 2D and 3D FE analyses to assess the impact of the failure of one strut on the remaining struts.

## 3.4.1　Numerical schemes

Figure 3.14 shows a typical cross-section and plan view for the cases considered. The parameters shown in the figure include $L$ = excavation length, $B$ = excavation width, $D$ = wall penetration depth, $T$ = clay thickness below the final excavation level (FEL), $S_H$ = horizontal strut spacing, $S_V$ = vertical strut spacing and $H_e$ = depth of final excavation. Vertical retaining walls along the excavation boundary were installed together with a five-level strut and waling system.

For 2D analyses, a half mesh was used due to geometrical symmetry. A very fine mesh size was used for 2D analysis to improve the accuracy of FE calculations. For 3D analyses, a medium mesh size in the horizontal direction and medium coarse mesh size in vertical direction were used to reach a balance between processing time and accuracy.

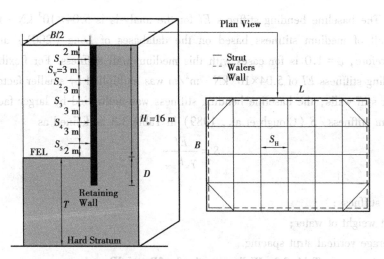

**Figure 3.14    Cross-section and plan view of the model for braced excavation**

For the 2D simulations, fourth-order 15-node triangular elements, which are considered to be very accurate elements, were used to model the soil while the interface elements have 5 integration points. In 3D PLAXIS, the interface elements have 9-point Gauss integration with three translational degrees of freedom for each node. This is described in greater detail in Van Langen (1991). For the 2D analysis, the retaining wall is simulated using 5-node elastic plate elements. The elastic behaviour is defined by the following parameters: $EA$(normal stiffness), $EI$(bending stiffness, m), and Poisson's ratio. For the 3D analysis, the wall is simulated using 8-node quadrilateral plate elements with six degrees of freedom per node. For brevity, only a typical 3D half mesh is shown in Figure 3.15, comprising of 15,679 nodes and 4,980 15-noded wedge elements.

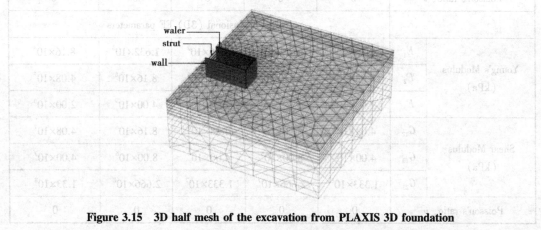

**Figure 3.15    3D half mesh of the excavation from PLAXIS 3D foundation**

Three wall types, with five different stiffness values, were considered for each soil type as listed in Table 3.3. Based on the approach adopted by Finno et al. (2007), the wall thickness of 0.42 m was set to an arbitrary (constant) value so that the moment of inertia I and area $A$ were kept constant, and only the wall elastic modulus $E$ was varied. A wall stiffness coefficient $\alpha$ was introduced to represent walls with different rigidities (Bryson and Zapata-Medina, 2012) as shown in Table 3.3. The baseline bending stiffness $EI$ for the analysis is $5.04\times10^5$ kN $\cdot$ m$^2$/m, which refers to a wall of medium stiffness based on the databases of Long (2001) and Moormann (2004). Therefore, $\alpha = 1.0$ is for cases with this medium wall stiffness. For flexible walls, the baseline bending stiffness $EI$ of $5.04\times10^5$ kN $\cdot$ m$^2$/m was multiplied by smaller factors of 0.1 and 0.2, while for stiff walls, the baseline bending stiffness was multiplied by larger factors of 2 and 10. The system stiffness, $S$ (Clough et al., 1989) in Table 3.3 is defined as

$$S = \frac{EI}{\gamma_w h_{avg}} \qquad (3.9)$$

where

$EI$ = wall stiffness;

$\gamma_w$ = unit weight of water;

$h_{avg}$ = average vertical strut spacing.

**Table 3.3  Wall properties for 2D and 3D analyses**

| Parameters | | Wall types | | | | |
|---|---|---|---|---|---|---|
| | | flexible | | medium | stiff | |
| $\alpha$ | | Plane strain (2D) FE parameters | | | | |
| | | 0.1 | 0.2 | 1.0 | 2.0 | 10 |
| System stiffness, $S$ | | 32 | 64 | 320 | 320 | 3 200 |
| Wall stiffness, $EI$(kN $\cdot$ m$^2$/m) | | $5.04\times10^4$ | $1.008\times10^5$ | $5.04\times10^5$ | $1.008\times10^6$ | $5.04\times10^6$ |
| Compressive stiffness, $EA$(kN/m) | | $3.427\times10^6$ | $6.854\times10^6$ | $3.427\times10^7$ | $6.854\times10^7$ | $3.427\times10^8$ |
| Poisson's ratio, $\nu$ | | 0 | 0 | 0 | 0 | 0 |
| | | Three-dimensional (3D) FE parameters | | | | |
| Young's Modulus (kPa) | $E_1$ | $8.16\times10^6$ | $1.632\times10^7$ | $8.16\times10^7$ | $1.632\times10^8$ | $8.16\times10^8$ |
| | $E_2$ | $4.08\times10^5$ | $8.16\times10^5$ | $4.08\times10^6$ | $8.16\times10^6$ | $4.08\times10^7$ |
| | $E_3$ | $2.00\times10^8$ | $4.00\times10^8$ | $2.00\times10^9$ | $4.00\times10^9$ | $2.00\times10^6$ |
| Shear Modulus (kPa) | $G_{12}$ | $4.08\times10^5$ | $8.16\times10^5$ | $4.08\times10^6$ | $8.16\times10^6$ | $4.08\times10^7$ |
| | $G_{13}$ | $4.00\times10^5$ | $8.00\times10^5$ | $4.00\times10^6$ | $8.00\times10^6$ | $4.00\times10^7$ |
| | $G_{23}$ | $1.333\times10^6$ | $2.666\times10^7$ | $1.333\times10^7$ | $2.666\times10^8$ | $1.33\times10^8$ |
| Poisson's ratio, $\nu$ | | 0 | 0 | 0 | 0 | 0 |

The struts were simulated using node-to-node anchor elements in 2D analyses. For 3D analyses, the struts and walers were modeled as beam elements, which have six degree of freedom per node, and the struts were placed horizontally at a spacing of 4 or 5 meters (for different case studies) in two directions to form a frame net. The walings were used to connect the excavation wall and the struts. The material properties are tabulated in Table 3.4.

**Table 3.4  Properties of waling system**

| Parameter | | Struts | Walers |
|---|---|---|---|
| Unit weight $\gamma(\text{kN/m}^3)$ | | 78.5 | 78.5 |
| Cross section area $A(\text{m}^2)$ | | 0.007 367 | 0.008 682 |
| Young's Modulus $E(\text{kPa})$ | | 2.1E8 | 2.1E8 |
| Moment of inertia $(\text{m}^4)$ | $I_3$ | 5.073E-5 | 1.045E-4 |
| | $I_2$ | 5.073E-5 | 3.668E-4 |
| | $I_{23}$ | 0 | 0 |

A typical staged construction simulation is shown in Table 3.5. The original ground water table was assumed to be 2 m below the ground surface in the retained soil. The water table inside the excavation was progressively lowered with the excavation of the soil during each phase.

**Table 3.5  Typical construction sequence for 2D analyses**

| Phases | Construction Details |
|---|---|
| Phase 1 | Install the excavation wall |
| Phase 2 | Excavate to 2 m below ground surface |
| Phase 3 | Excavate to 3 m |
| Phase 4 | Install strut system at 2 m below ground surface |
| Phase 5 | Excavate to 5 m below ground surface |
| Phase 6 | Excavate to 6 m below ground surface |
| Phase 7 | Install strut system at 5 m below ground surface |
| Phase 8 | Excavate to 8 m below ground surface |
| Phase 9 | Excavate to 9 m below ground surface |
| Phase 10 | Install strut system at 8 m below ground surface |
| Phase 11 | Excavate to 11 m below ground surface |
| Phase 12 | Excavate to 12 m below ground surface |
| Phase 13 | Install strut system at 11 m below ground surface |
| Phase 14 | Excavate to 14 m below ground surface |
| Phase 15 | Excavate to 15 m below ground surface |
| Phase 16 | Install strut system at 14 m below ground surface |
| Phase 17 | Excavate to 16 m below ground surface |

The properties of three different types of clays which were considered in this parametric study are similar to the properties assumed by Bryson and Zapata-Medina (2012) and are tabulated in Table 3.6. The soils are assumed to follow the Hardening Soil (HS) model. The three soil types are soft clay, medium clay and stiff clay. The clays are real soils whose properties have been extensively reported in the literature. The properties of the soft clay with average $c_u = 20$ kPa are based on the Upper Blodgett soft clay reported by Finno et al. (2002). The medium clay with average $c_u = 45$ kPa are based on the Taipei silty clay found at the Taipei National Enterprise Center (TEC) project (Ou et al., 1998). The Gault clay at Cambridge (Ng, 1992; Ng and Yan, 1998) with average $c_u = 125$ kPa was used as the model for the stiff clay.

**Table 3.6  Input HS soil parameters of three clays**

| Parameter | Unit | A: soft clay (Chicago clay) | B: medium clay (Taipei silty clay) | C: stiff clay (Gault clay) |
|---|---|---|---|---|
| $\gamma_{unsat}$ | kN/m$^3$ | 18.1 | 18.1 | 20 |
| $\gamma_{sat}$ | kN/m$^3$ | 18.1 | 18.1 | 20 |
| $E_{50}^{ref}$ | kN/m$^2$ | 2 350 | 6 550 | 14 847 |
| $E_{oed}^{ref}$ | kN/m$^2$ | 2 350 | 6 550 | 14 847 |
| $E_{ur}^{ref}$ | kN/m$^2$ | 7 050 | 19 650 | 44 540 |
| $C$ | kN/m$^2$ | 0.05 | 0.05 | 0.05 |
| $\varphi$ | ° | 24.1 | 29 | 33 |
| $\Psi$ | ° | 0 | 0 | 0 |
| $\nu_{ur}$ | [−] | 0.2 | 0.2 | 0.2 |
| $p^{ref}$ | kN/m$^2$ | 100 | 100 | 100 |
| $M$ | [−] | 1.0 | 1.0 | 1.0 |
| $K_0^{nc}$ | [−] | 0.59 | 0.55 | 1.5 |
| $R_f$ | [−] | 0.7 | 0.95 | 0.96 |
| $R_{inter}$ | [−] | 1 | 1 | 1 |

## 3.4.2  One-strut failure analysis: two hypothetical cases

In order to investigate the influence of the one-strut failure on the adjacent struts, the strut forces before and after strut failure are examined. By tabulating the load transfer percentage, the influence of the strut failure can be demonstrated. Assume $N_{pre}$ is the load on the strut before strut failure; $N_{post}$ is the load on the strut after strut failure and $N_{fail}$ is the load on the failed strut before failure. Then the load transfer percentage is defined as Eq. 3.10.

$$\text{Load Transfer}(\%) = \frac{N_{post} - N_{pre}}{N_{fail}} \times 100\% \qquad (3.10)$$

## 1) One-Strut Failure for Soft Clay: $H_e = 12$ m with 3-Levels of Struts

The load transfer percentages from 2D analyses are tabulated in Table 3.7, and the 3D results are shown in Table 3.8. The 2D results indicate that failure of $S_3$ leads to considerable increase in the load for the 2nd level $S_2$ strut. On the other hand, the 3D results indicate that the load of the failed strut is not only transferred vertically upward, but also to the adjacent horizontal and diagonal struts as shown in Figure 3.16, where larger arrows denote the larger magnitude of the transferred loads and the red circles denote the struts with more than 10% load increase after the one-strut failure. It should be noted from Table 3.8 that the load transfer percentage values along the left and right sides of the failed strut (i.e., 9.7 and 9.2 for soft clay, a=0.1, $S_2$ in Table 3.8, 23.3 and 21.1 for soft clay, a=0.1, $S_3$ in Table 3.8) are not equal considering that the numerical model is not symmetrical about the failed ($x$=0) strut. In addition, for the 3D analyses, the magnitude of the load transfer percentages is much smaller, up to approximately 20% for flexible walls and 50% for medium walls compared with the results from the 2D analyses. Therefore, compared to 3D results, the 2D analyses overestimate the possible consequence of the one-strut failure for soft clay, as it ignores the restraining effects of the adjacent horizontal struts.

**Table 3.7　Load transfer percentages from 2D analyses for the one-strut failure of $S_3$(soft clay)**

| Struts | Load Transfer/% | |
|:---:|:---:|:---:|
| | $\alpha = 0.1$ | $\alpha = 1.0$ |
| $S_1$ | −37.7 | −30.4 |
| $S_2$ | 122.2 | 120.3 |
| $S_3$ | Failed strut | |

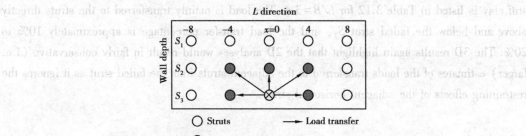

Figure 3.16　Influence zone of the one-strut failure of $S_3$(soft clay)

**Table 3.8　Load transfer percentage from 3D analyses for the one-strut failure of $S_3(L/B=2.2)$**

| Struts | Load Transfer(%) | | | | |
|---|---|---|---|---|---|
| | $x=-8$ | $x=-4$ | $x=0$ | $x=4$ | $x=8$ |
| soft clay, $\alpha=0.1$ | | | | | |
| $S_1$ | −0.1 | −2.5 | −9.2 | −2.5 | 0.2 |
| $S_2$ | −0.5 | 9.7 | 15.4 | 9.2 | −2.2 |
| $S_3$ | 2.7 | 23.3 | Failed strut | 21.2 | −0.1 |
| soft clay, $\alpha=1.0$ | | | | | |
| $S_1$ | −5.7 | −4.4 | −3.3 | −3.7 | −3.0 |
| $S_2$ | 1.5 | 11.9 | 53.6 | 11.2 | −0.7 |
| $S_3$ | 4.2 | 24.4 | Failed strut | 22.5 | 3.3 |

## 2) One-Strut Failure for Medium and Stiff Clays: $H_e = 16$ m with 5-Levels of Struts

The load transfer percentages of the one-strut failure from 2D analyses for medium and stiff walls are shown in Table 3.9. For medium clay, most of the load of the failed strut is transferred to the struts immediately above and below the failed strut, with approximately 30% to 50% of the load transferred to the upper strut and the remainder carried by the lower strut. For stiff clay, the failure of strut $S_3$ leads to the force redistribution for the $S_2$, $S_4$ and $S_5$ struts. The $S_2$ and $S_4$ struts each carry approximately 30% of the load of the failed strut, and $S_5$ carries approximately 10% to 20%. When the wall is also stiff, the top strut is able to carry approximately 10% of the load. For the 3D analyses, the percentage of the load transfer to adjacent struts of the relevant struts for medium clay with stiff walls are tabulated in Table 3.10 for $L/B=2.2$ and Table 3.11 for $L/B=3.4$, respectively. The tables show that the struts affected most are located directly above or below the failed strut $S_4$ or diagonally across, as plotted in Figure 3.17. The influence of the one-strut failure from 3D analyses for the stiff clay is shown in Figure 3.17. The load transfer percentages for stiff clay is listed in Table 3.12 for $L/B=3.4$. The load is mainly transferred to the struts directly above and below the failed strut $S_3$, and the load transfer percentage is approximately 10% to 20%. The 3D results again highlight that the 2D analyses would result in fairly conservative (i.e. larger) estimates of the loads transferred to the adjacent struts from the failed strut as it ignores the restraining effects of the adjacent horizontal struts.

**Table 3.9  Load transfer percentage from 2D analyses（medium and stiff clays）**

| clay type | Struts | Load Transfer（%） | | |
|---|---|---|---|---|
| | | $\alpha = 0.1$ | $\alpha = 1.0$ | $\alpha = 10$ |
| medium clay | $S_1$ | −0.9 | −0.9 | −0.4 |
| | $S_2$ | −5.2 | 11.2 | 15.5 |
| | $S_3$ | 46.0 | 41.7 | 31.1 |
| | $S_4$ | Strut Failure | | |
| | $S_5$ | 53.4 | 47.1 | 46.3 |
| stiff clay | $S_1$ | −0.8 | 2.4 | 15.4 |
| | $S_2$ | 27.6 | 30.2 | 23.1 |
| | $S_3$ | Strut Failure | | |
| | $S_4$ | 29.3 | 29.5 | 22.1 |
| | $S_5$ | 10.8 | 11.9 | 14.6 |

**Table 3.10  Load transfer percentage from 3D analyses for the one-strut failure of $S_4$ for medium clay（$L/B = 2.2$）**

| Struts | Load Transfer（%） | | | | |
|---|---|---|---|---|---|
| | $x = -8$ | $x = -4$ | $x = 0$ | $x = 4$ | $x = 8$ |
| $S_1$ | −1.1 | 0.2 | 1.6 | 0.2 | −1.0 |
| $S_2$ | −0.3 | 3.1 | 7.9 | 3.2 | −0.3 |
| $S_3$ | 0.5 | 5.5 | 15.3 | 5.5 | 0.5 |
| $S_4$ | 1.2 | 7.2 | Failed strut | 7.2 | 1.3 |
| $S_5$ | 2.8 | 9.0 | 19.5 | 9.0 | 2.7 |

**Table 3.11  Load transfer percentage from 3D analyses for the one-strut failure of $S_4$ for medium clay（$L/B = 3.4$）**

| Struts | Load Transfer（%） | | | | |
|---|---|---|---|---|---|
| | $x = -8$ | $x = -4$ | $x = 0$ | $x = 4$ | $x = 8$ |
| $S_1$ | −0.3 | 0.8 | 2.0 | 0.5 | −0.7 |
| $S_2$ | −0.1 | 3.4 | 8.0 | 3.2 | −0.2 |
| $S_3$ | −0.1 | 5.3 | 15.1 | 5.4 | 0.4 |
| $S_4$ | 0.2 | 6.5 | Failed strut | 6.5 | 0.7 |
| $S_5$ | 1.4 | 7.5 | 17.9 | 7.4 | 1.5 |

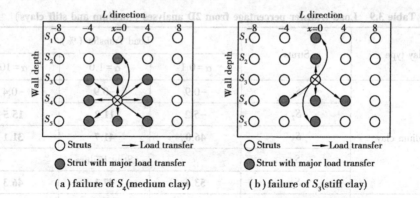

(a) failure of $S_4$(medium clay)        (b) failure of $S_3$(stiff clay)

**Figure 3.17   Influence zone of the failure of strut**

**Table 3.12   Load transfer percentage from 3D analyses**

**for the one-strut failure of $S_3$ for stiff clay ($L/B=3.4$)**

| Struts | Load Transfer (%) | | | | |
|---|---|---|---|---|---|
| | $x=-8$ | $x=-4$ | $x=0$ | $x=4$ | $x=8$ |
| stiff clay, $\alpha=0.1$ | | | | | |
| $S_1$ | −0.5 | −3.7 | −5.1 | 0.0 | 0.0 |
| $S_2$ | 0.1 | 4.8 | **12.6** | 5.8 | 1.6 |
| $S_3$ | 1.4 | 4.6 | Failed strut | 5.4 | 2.6 |
| $S_4$ | 3.4 | 6.8 | **12.3** | 6.6 | 7.2 |
| $S_5$ | 4.5 | 5.0 | 4.7 | 4.8 | 3.8 |
| stiff clay, $\alpha=1.0$ | | | | | |
| $S_1$ | −0.2 | 0.6 | 0.6 | 0.8 | 0.1 |
| $S_2$ | 1.2 | 4.6 | **17.8** | 4.9 | 1.6 |
| $S_3$ | 1.3 | 4.1 | Failed strut | 4.4 | 1.8 |
| $S_4$ | 3.2 | 6.0 | **17.3** | 6.1 | 3.0 |
| $S_5$ | 4.6 | 5.9 | 6.6 | 5.8 | 4.2 |
| stiff clay, $\alpha=10$ | | | | | |
| $S_1$ | −0.3 | 3.0 | 5.9 | 3.1 | −0.1 |
| $S_2$ | 0.7 | 4.4 | **12.1** | 4.5 | 0.9 |
| $S_3$ | 1.6 | 5.0 | Failed strut | 5.1 | 1.7 |
| $S_4$ | 3.0 | 4.4 | **11.1** | 5.9 | 3.0 |
| $S_5$ | 4.7 | 6.6 | 9.3 | 6.5 | 4.4 |

# 4 Retaining Wall and Bending Moment

## 4.1 Introduction

Stress analysis is necessary for the design of structural components, and deformation analysis aims at diagnosing the wall deflections and soil movements caused by excavation to protect adjacent properties. What will be introduced in this chapter is about wall bending moment. Bending moment and the shear is relevant to the choice of the appropriate type and dimension of retaining walls, and sometimes to the design of reinforcements. The differences in the profiles of the wall maximum bending moments as the excavation depth increases and the system response was governed by the movements in the underlying clay (Whittle and Hashash, 1993). The stress analysis methods for excavation include the simplified method and the numerical method, and the latter can be further classified into the beam on elastic foundation method and the finite element method, all of which will be introduced in this chapter.

## 4.2 Wall Types

### 4.2.1 Soldier piles

Types of steel for soldier piles include the rail pile, the steel H-pile (or $W$ section) and the steel I-pile (or $S$ section). The rail pile and the steel H-pile are more commonly used than the steel I-pile. It is optional to place laggings between soldier piles. Whether to place them or not depends on the in situ soil properties and strength characteristics. As Figure 4.1 illustrates, the construction procedure for soldier piles can be described as follows:

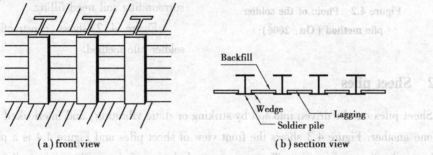

(a) front view　　　　(b) section view

**Figure 4.1　Soldier piles**

①Strike soldier piles into soil. In non-urban areas, it will be all right to strike them into soil directly. In urban areas, however, static vibrating installation would be a better way to have soldier piles penetrated into the soil. If encountering a hard soil layer, pre-bore the soil.

②Place laggings as excavation proceeds. Then backfill the voids between soldier piles and laggings.

③Install horizontal struts in proper places during the excavation process.

④Excavation completed, begin constructing the inner walls of the basement. Then remove struts level by level and construct floor slabs.

⑤Complete the base.

⑥Pull out the piles.

The merits of soldier piles include:

a. Easier and faster construction with lower cost.

b. Piles can be easily pulled out.

c. Less ground disturbance is caused when pulling out the piles, compared to pulling out sheet piles.

d. The pile tip can be strengthened with special materials used in gravel soils.

e. Soldier piles are reusable.

The drawbacks of soldier piles include:

a. Sealing is difficult. In sandy soils with high groundwater level, some dewatering measures may be necessary.

**Figure 4.2   Photo of the soldier pile method (Ou, 2006)**

b. Installing soldier piles by striking will cause much noise and vibration. The latter will render sandy soils below the foundation denser in such a way that uneven settlement of the adjacent buildings may occur.

c. Backfilling is necessary if soldier piles are driven using pre-boring. Deficiency in backfilling will cause bad effects in the vicinity.

d. The voids between retaining walls and surrounding soil need filling.

Figure 4. 2 shows construction with the soldier pile method.

## 4.2.2   Sheet piles

Sheet piles can be driven into soil by striking or static vibrating, and interlocked or connected with one another. Figure 4.3 shows the front view of sheet piles and Figure 4.4 is a photo showing the sheet piles in an excavation. There are several shapes of sheet pile sections. Some commonly used include the U-section, Z-section, and the line-section, as shown in Figure 4.5. If the interlocking is well done, sheet piles can be quite efficient in water sealing. If not, leaking may occur at the joints. In clayey soils having low permeability, sheet piles do not necessarily require perfect connection to prevent leaking. On the other hand, if they are used in sandy soils with high permeability, any breach in sheet piles may well cause leaking. If leaks occur, sand in back of the retaining walls will very possibly flow out, which may cause settlement in turn. If leakages are too great, the excavation might be endangered. The construction method for sheet piles can be described as follows:

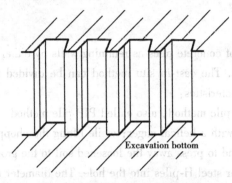

Figure 4.3   Steel sheet pile method

Figure 4.4   Photo of the steel sheet pile method (Ou, 2006)

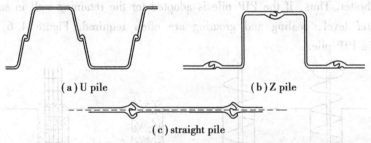

(a) U pile        (b) Z pile

(c) straight pile

Figure 4.5   Sections of steel sheet piles

①Drive sheet piles into soil by striking or static vibrating.

②Proceed to the first stage of excavation.

③Place wales in proper places and install horizontal struts.

④Proceed to the next stage of excavation.

⑤Repeat procedures 3 and 4 till the designed depth.

⑥Complete excavation and begin to build the foundation of the building.

⑦Build the inner walls of the basement. Dismantle the struts level by level and build the floor slabs accordingly.

⑧Complete the basement.

⑨Dismantle the sheet piles.

The merits of the steel sheet piles method are as follows:

a. It is highly watertight.

b. It is reusable.

    c.It has higher stiffness than soldier piles.

    The drawbacks of the steel sheet piles method are as follows:

    a.Lower stiffness than column piles or diaphragm walls.

    b.Susceptible to settlement during striking or dismantling in a sandy ground.

    c.Not easy to strike piles into hard soils.

    d.A lot of noise is caused during striking.

    e.Leaks cannot be completely avoided but sealing and grouting are probably necessary if leaks occur.

## 4.2.3 Column piles

    The column piles method is to construct rows of concrete piles as retaining walls by either the cast-in situ pile method or the precast pile method. The cast-in situ method can be divided into three subtypes according to their construction characteristics:

    ①Packed in place piles. The packed in place pile method, also called PIP pile method, can be described as follows: dig to the designed depth with a helical auger; while lifting the chopping bit gently, fill in prepacked mortar from the front end to press away the loosened soil to the ground surface; after grouting is finished, put steel cages or steel H-piles into the hole. The diameter of a PIP pile is around 30-60 cm. It often happens that PIP piles are not capable of being installed completely vertically, so connections are not always watertight, and connection voids often cause leaks of groundwater. Thus, if the PIP pile is adopted for the retaining wall in sandy soils with high groundwater level, sealing and grouting are often required. Figure 4.6 illustrates the construction of a PIP pile.

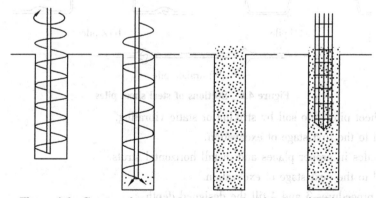

**Figure 4.6  Construction procedure of a packed in place (PIP) pile**

    ②Concrete piles. The construction of concrete piles can be described as follows: drill a hole to the designed depth by machine, put the steel cages into it, and fill it with concrete using Tremie tubes. The reverse circulation drill method (also called the reverse method), which is to employ stabilizing fluid to stabilize the hole wall during drilling, is the most commonly used construction method for concrete piles. It is also feasible to build following the all casing method, which is to drill with simultaneous casing-installment to protect the hole wall. Since the wall is protected by casings, stabilizing fluid is not required. The cost of the all casing method is rather high.

Nevertheless, it can be easily applied to cobble-gravel layers or soils with seepage; whereas the reverse method cannot. The diameters of the concrete piles are around 60-200 cm.

③Mixed piles. Mixed piles are also called MIP piles (mixed in place piles) or SMW (soil mixed wall). The method is to employ a special chopping bit to drill a hole with the concrete mortar sent out from the front of the bit to be mixed with soil. When the designed depth is reached, lift the bit a little, keeping swirling and grouting simultaneously, and let mortar mix with soil thoroughly. After pulling out the drilling rod, put steel cages or H-piles into the hole if necessary. Figure 4.7 illustrates the construction process for a mixed pile. Figure 4.8 shows MIP piles with H steels.

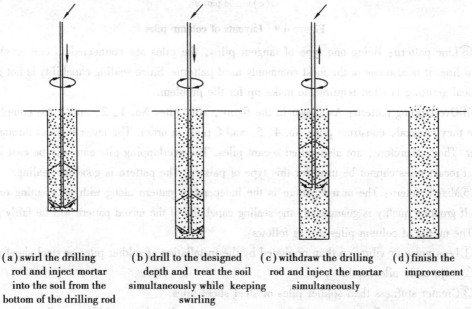

(a) swirl the drilling rod and inject mortar into the soil from the bottom of the drilling rod   (b) drill to the designed depth and treat the soil simultaneously while keeping swirling   (c) withdraw the drilling rod and inject the mortar simultaneously   (d) finish the improvement

**Figure 4.7   Construction procedure of a mixed in place (MIP) pile**

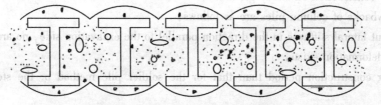

**Figure 4.8   Soil mixed wall (SMW)**

As illustrated in Figure 4.9, the patterns of column piles include the independent pattern, the S pattern, the line pattern, the overlapping pattern, and the mixed pattern. They can be described as follows:

①Independent pattern: As shown in the figure, sealing is impossible with the independent pattern. Thus, excavation in permeable ground with high groundwater level requires dewatering in advance. The pattern, nevertheless, is especially fitting for soil with high strength, such as gravel soils.

②S pattern: The second row of piles fits into gaps in the first row, which have been arranged in the independent pattern. This pattern is also called the tangent piles. The method is relatively simple in construction. On the other hand, the sealing capability is weak because of the less orderly layout. Grouting is necessary for sealing.

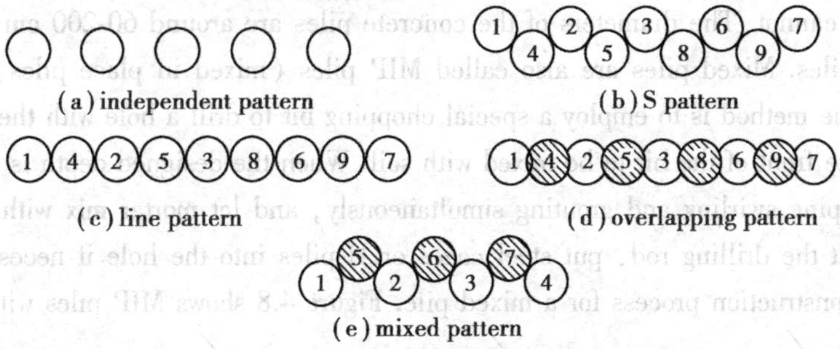

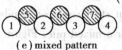

(a) independent pattern　　　　　　　(b) S pattern

(c) line pattern　　　　　　　(d) overlapping pattern

(e) mixed pattern

**Figure 4.9　Layouts of column piles**

③Line pattern: Being one type of tangent piles, the piles are connected to one another to form a line. It is also one of the most commonly used patterns. Since sealing capability is not good, chemical grouting is often required to make up for the problem.

④Overlapping pattern: As shown in the figure, after piles No. 1, 2, and 3 are completed, before they congeal, construct piles No. 4, 5, and 6 piles in order. The latter will cut through the former. They, therefore, are also called secant piles. The overlapping pile can only be cast in the field. Precast piles cannot be used for this type of pattern. The pattern is good for sealing.

⑤Mixed pattern: The mixed pattern is the independent pattern along with jet grouting or MIP piles. If grouting quality is guaranteed, the sealing capability of the mixed pattern will be fairly good.

The merits of column piles are as follows:

①Less noise or vibration than produced by the installation of soldier piles or steel sheet piles.

②Adjustable pile depth.

③Greater stiffness than soldier piles or steel sheet piles.

④When equipped with a special bit, they are also applicable to cobble-gravelly soils.

⑤Easier construction on sandy ground.

The drawbacks of column piles are as follows:

①Without lateral stiffness in the direction parallel to the excavation side, no arching effect to prevent wall deformation exists.

②Longer construction period than that for the soldier pile method or the steel sheet pile method.

③Lower stiffness than diaphragm walls.

④Highly susceptible to construction deficiency.

## 4.2.4　Diaphragm walls

Diaphragm walls are also called slurry walls. Since first adopted in Italy in the 1950s, they have been widely used around the world. With technological advances, more and more new methods and construction equipment have been developed. The basement wall (BW) method and Impresa Construzioni Opere Specializzate (ICOS) method, designed separately by a Japanese company and an Italian company are commonly used in some Asian countries. The Masago Hydraulic Long bucket (MHL) method, taking advantage of a bailing bucket to excavate the trenches of the diaphragm wall, are also used in many countries. The teeth of the steel bailing bucket can clutch soils and rocks and store them inside the bucket. Then, the full bucket is lifted out of the trench, and soil and rocks inside are bailed out. Thus, stabilizing fluid need not be pumped out and mud separation equipment is saved. The method is easy in operation. The span of the bailing bucket is about 2.5-3.3 m.

The first stage of the constructing of diaphragm walls is to divide the whole length into several panels according to the construction conditions. The construction procedure of each panel is as follows: the construction of guided walls, the excavation of trenches, placing steel cages, and concrete casting, as shown in Figure 4.10. After excavating the trench, mud in the trench must be cleared from the trench. Concrete casting, the last stage of diaphragm wall panel construction, adopts the Tremie pipe to pour concrete into the trench and form a diaphragm wall panel. The construction procedure of a diaphragm wall panel, including excavating trenches, placing connection pipes (depending on the method), placing steel cages, and concrete casting.

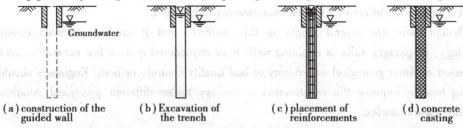

(a) construction of the guided wall    (b) Excavation of the trench    (c) placement of reinforcements    (d) concrete casting

**Figure 4.10　Construction procedure of a diaphragm wall panel**

The joints between panels of a diaphragm wall have to be carefully treated so that they are watertight and are able to transmit bending moments and shear forces. There are many types of joints of diaphragm walls with no standard form. Some patent methods are available.

The connection pipe method and the end-plate method are the commonly used joint methods. The former is highly watertight; on the other hand, its capability of transmitting bending moments and shear forces is not good enough. If the diaphragm walls are confined to the tempo-rary use of soil retaining, the connection pipe method can serve the purpose well. If they are meant for permanent structures, the end-plate method is recommended for its better capability to transmit bending moments and shear forces. As shown in Figure 4.11, when the trench excavation of the primary panel is finished, put a connection pipe into the trench, then place the steel cage with a crane and pour concrete into the trench. Once the concrete has been poured, dismantle the connection pipe in two or three hours and proceed to the construction of the secondary panel.

To ensure the quality of diaphragm walls, many other construction details need to be taken

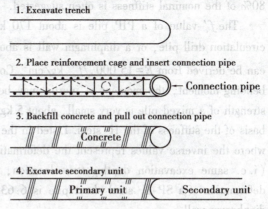

1. Excavate trench

2. Place reinforcement cage and insert connection pipe

Connection pipe

3. Backfill concrete and pull out connection pipe

Concrete

4. Excavate secondary unit

Primary unit    Secondary unit

**Figure 4.11　Joint of diaphragm walls:**
**the connection pipe method**

care of. For the construction details of diaphragm walls, readers are advised to read Xanthakos (1994).

The merits of diaphragm wall methods include:

①Low vibration, low noise, high rigidity, and relatively small wall deformation.

②Adjustable thickness and depth of the wall.

③Good sealing capability.

④May be used as a permanent structure.

⑤The diaphragm wall and foundation slabs form a unity, so that the former can serve as pile foundations.

The drawbacks of the diaphragm wall method are as follows:

①Massive equipment is required, long construction period, and great cost.

②The peripheral equipment (e.g. the sediment pool) occupies a large space.

③This method is not applicable to cobble-gravelly grounds.

④It is difficult to construct when encountering quick sand.

Though there are several merits in this method, and it involves matured construction technology, diaphragm walls as retaining walls have engendered quite a few excavation accidents. The reason is either geological uncertainty or bad quality control, or both. Engineers should keep studying how to improve the construction technology under different geological conditions by consulting case histories.

## 4.2.5 Selection of the retaining system

The selection of retaining walls has to consider the excavation depth, geological conditions, groundwater conditions, adjacent building conditions, the site size, the construction period, and the budget, etc. Table 4.1 offers the application ranges of various retaining walls for design or construction reference. Table 4.2 illustrates the nominal stiffness per unit length where the value of moment of inertia is not reduced. A stress or deformation analysis, however, has to consider the decrease of stiffness of a soldier pile or a steel sheet pile because of repetitive use. A discount of 80% of the nominal stiffness is often suggested.

The $f_c'$-value of a PIP pile is about 170 kg/cm$^2$ while that of a concrete pile, a reverse circulation drill pile, or a diaphragm wall is about 210 kg/cm$^2$ or more. Their Young's modulus can be derived from $E = 15\,000\sqrt{f_c'}$ kg/cm$^2$. Considering the cracks in the retaining wall due to bending moment, the moment of inertia may be reduced by 30%-50%. Since the compressive strength of a mixed pile is very small, about 5 kg/cm$^2$, its stiffness can be estimated solely on the basis of the stiffness of the H- steel. Listed in the sixth column of Table 4.2 are the stiffness ratios where the inverse values represent the deformation ratios under the same construction condition (i. e. same excavation depth, strut location, and strut stiffness, etc.). For example, the deformation of a SP-Ⅲ steel sheet pile is 6.63 (5.3/0.8) times as great as a 50 cm thick diaphragm wall.

**Table 4.1 Application conditions for retaining walls**

| Wall type | Soil type | | | Sealing and stiffness | | Construction conditions | | | | | Construction period | Budget |
|---|---|---|---|---|---|---|---|---|---|---|---|---|
| | Soft clay | Sand | Gravel soil | Sealing | stiffness | Noise and vibration | Treatment of dump mud | Surface settlement | Underground obstruction | Excavation depth | | |
| Soldier pile | × | ○ | ○¹ | × | × | ×² | ◎ | × | ○ | × | ◎ | ◎ |
| Steel sheet pile | ○ | ◎ | × | ○ | × | ×² | ◎ | × | ○ | × | ◎ | ◎ |
| PIP pile | ◎ | ○ | × | ○ | ◎ | ◎ | × | ○ | × | ○ | × | ○ |
| Reinforced concrete column pile | ◎ | ◎ | × | ◎ | ◎ | ◎ | × | ◎ | × | ◎ | × | × |
| MIP pile | ○ | ○ | × | ◎ | ○ | ◎ | ○ | ○ | × | ○ | ○ | ○ |
| Diaphragm wall | ◎ | ◎ | ○ | ◎ | ◎ | ◎ | × | ○ | × | ◎ | × | × |

Notes: ◎: good  ○: acceptable  ×: not good

1. Should be applied along with special drill and striking device.

2. If driven into soil by static vibrating process, noise and vibration can be reduced.

**Table 4.2 Nominal stiffness (before reduction)**

| Retaining wall | | $E(\text{kg/cm}^2)$ | $I(\text{cm}^4/\text{m})$ | $EI(\text{kN} \cdot \text{m}^2/\text{m})$ | Stiffness ratio |
|---|---|---|---|---|---|
| Method | Type and dimension | | | | |
| Soldier pile[1] | H300×300×10×15 | $2.04\times10^6$ | 20 400 | 4 160 | 1.0 |
| | H350×350×12×19 | $2.04\times10^6$ | 40 300 | 8 220 | 2.0 |
| Steel sheet pile[2] | SP-III | $2.04\times10^6$ | 16 400 | 3 350 | 0.8 |
| | SP-IV | $2.04\times10^6$ | 31 900 | 6 500 | 1.6 |

continued

| Retaining wall | | $E(\mathrm{kg/cm^2})$ | $I(\mathrm{cm^4/m})$ | $EI(\mathrm{kN \cdot m^2/m})$ | Stiffness ratio |
|---|---|---|---|---|---|
| Method | Type and dimension | | | | |
| Column pile[3] | 30 cm(diameter) | $2.10\times10^5$ | 132 500 | 2 780 | 0.7 |
| Column pile[3] | 80 cm(diameter) | $2.10\times10^5$ | 2 513 300 | 52 780 | 12.7 |
| MIP pile[4] | SMW method H400×200×8×13 | $2.04\times10^6$ | 59 250 | 12 090 | 2.9 |
| Diaphragm wall[5] | 50 cm thick | $2.10\times10^5$ | 1 041 700 | 21 900 | 5.3 |
| Diaphragm wall[5] | 100 cm thick | $2.10\times10^5$ | 8 333 300 | 175 000 | 42.0 |

Notes:

1. The distance between H-steels is 1.00 m. The stiffness of the soldier pile can be reduced by 20% in analysis.

2. The stiffness of the sheet pile can be reduced by 20% in analysis.

3. $f'_c$ for PIP piles is about 170 kg/cm² ; $f'_c$ for reinforced concrete column piles is about 280 kg/cm² ; in the table $f'_c$ is assumed to be 210 kg/cm² , which can be reduced by 30%-50% in analysis.

4. $f'_c$ for MIP piles is about 5 kg/cm² , that is, the stiffness of the MIP can be ignored and only the stiffness of H-piles is to be considered. The distance between H-piles in the table is 40 cm.

5. $f'_c$ is assumed to be 210 kg/cm² . The stiffness can be reduced by 30%-50% in analysis.

# 4.3　Stress Analysis Method

## 4.3.1　Simplified methods

Generally speaking, simplified methods employ the monitoring results of excavation case histories and then sort them into the stress and deformation characteristics of retaining walls and soils. The characteristics are useful not only to help understand the actual excavation behavior but to offer information for excavation-induced stress and deformation analyses.

The strut-retaining system of an excavation is, in nature, a highly static indeterminate structure and thereby is difficult to analyze by hand calculation unless the loading pattern, boundary conditions, and analysis method are simplified. The stress analysis methods that will be introduced in this chapter are induced from accumulated experience or observations of designers and are more applicable to common excavations. As for special excavations (e.g. large scale or great depth), the numerical method introduced in the next part, is recommended instead, considering that the simplified method is lacking in solid theoretical support.

For cantilevered walls, the design of a cantilevered wall is based on the fixed earth support method, that is, the embedded part of the wall is assumed to be fixed at a certain depth. Thus, in the limiting state, the active earth pressure above the excavation surface can develop fully; whereas the passive and active earth pressures near the fixed point cannot. On the other hand, under working load, though the active earth pressure above the excavation surface may still fully develop, the passive as well as the active earth pressures near the fixed point are not fully developed. It is therefore difficult to estimate the distribution of earth pressure on the retaining wall. To be conservative and simplify the procedure of computing the stress of cantilevered walls, we may assume the earth pressure is in the limiting state.

With the earth pressure on the wall known, we can then adopt the simplified gross pressure method for the analysis of the stress of a cantilevered wall (Padfield and Mair, 1984), as shown in Figure 4.12(a). The earth pressure distribution must not include the safety factors, or the distribution of earth pressures will be distorted. Though the depth of the wall used for analysis, as shown in Figure 4.12(a), is not as great as designed [Figure 4.12(b)], the maximum bending moment of the wall thus derived is the closest to the real value, as shown in Figure 4.12(b). If sheet piles or soldier piles are to be adopted, we can design according to the maximum bending moment without knowing the distribution of bending moments. If the retaining wall is an RC wall (such as diaphragm walls or column piles), the bending moment derived from the simplified method has to be modified. For a cantilevered wall, the real distribution of earth pressures under working load is difficult to obtain. To be conservative and simplify the calculation, the earth pressure in the limiting state can be adopted for the stress analysis of a cantilevered wall. Both the gross pressure method and the net pressure method are applicable to compute shear and bending moment of a cantilevered wall.

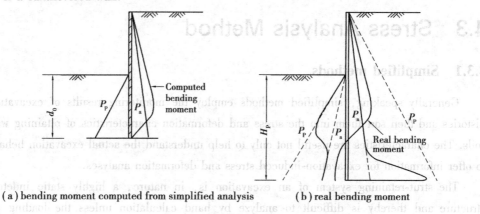

(a) bending moment computed from simplified analysis        (b) real bending moment

**Figure 4.12    Computation of bending moment of cantilever walls**

The simplified methods are mostly empirical formulas or diagrams based on the monitoring results of excavations. Since an excavation can be viewed as a full scale test, with similar geological conditions and excavation methods, simplified methods based on past experience are feasible to make reasonable predictions.

## 4.3.2    Beam on Elastic Foundation Method

The finite element method and the beam on elastic foundation method are two commonly used numerical methods. Theories concerning the finite element method are quite complicated and some of them are not fully developed. The finite element analysis normally requires enormous preprocessor and postprocessor time, computation time, and analysts have to be well equipped with comprehensive geotechnical knowledge and experience. Thus, the method is not widely adopted in the analysis and design of excavations. The beam on elastic foundation method, on the other hand, is simpler in its analysis model. With succinct input parameters, it does not take much time for processing and therefore is favored by most engineers. Nevertheless, the simplicity of the beam on elastic foundation method requires more delicacy and prudence when dealing with complicated excavation problems so that mistakes can be avoided.

In foundation engineering, the soil-structure interaction problem is often simulated as a series of springs to simplify analysis. Among them, Winkler's model (Winkler, 1867) is most widely applied.

As shown in Figure 4.13, the basic assumption of the Winkler model is given that the foundation is a structure with stiffness and soils are of elastic foundation, their interaction can be simulated as a series of individual springs. The spring constant is the ratio of stress ($p$) to displacement ($\delta$), which can be expressed as follows:

$$k_s = \frac{p}{\delta} \tag{4.1}$$

where the constant $k_s$ is called the coefficient of subgrade reaction, the modulus of subgrade reaction, or the soil spring constant, the unit of which is $(\text{force}) \times (\text{dimension})^{-3}$. The strength of the Winkler model is that it greatly simplifies analysis, for it assumes the elements are individually acting without interaction.

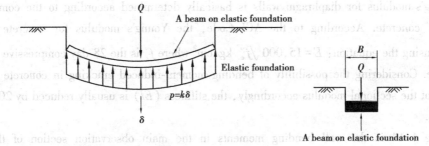

**Figure 4.13  Winkler's model**

The beam on elastic foundation analysis of an excavation assumes the retaining wall to be a beam on an elastic foundation, which is simulated as a series of soil springs and the earth pressures on both sides of the wall before excavation is taken to be the at-rest earth pressure ($Ko$-condition). After excavation is started, unloading induced by excavation will cause unbalanced forces between the two sides of the wall and make the wall deform. The amount of the unbalanced force is the difference between the at-rest earth pressures on the two sides of the wall. When the wall is kept unmoved, acted on by the unbalanced forces, the beam is displaced, and will change the distribution of earth pressures. The earth pressure from outside is decreased to $p_o - k_h\delta$ ($k_h$ is the horizontal coefficient of subgrade reaction and $\delta$ is the lateral displacement of the wall) with the increase of displacement. The minimum lateral earth pressure is the active earth pressure. The earth pressure from inside is increased to $p_o - k_h\delta$ due to the inward displacement of the wall. When soil springs develop up to the passive condition, the soil reaction on the passive side ceases increasing and stays at the passive earth pressure. This state is called the plastic state. When the reaction forces of soil springs are smaller than the passive earth pressure at a point, this is called the elastic state.

As discussed in Section 4.2.1, soldier piles commonly used in excavations are H steels and rail piles. For the dimensions and related properties of H steels and rail piles, please refer to books on steel structures or steel structure design manuals. The nominal Young's modulus for solider piles is $2.04 \times 10^6$ kg/cm$^2$. Theoretically, the stiffness ($EI$) does not need reduction in analysis. Considering the repeated use of solider piles, which decreases their stiffness as a result; therefore, the nominal Young's modulus is usually reduced by 20%.

The nominal Young's modulus for sheetpiles is also $2.04 \times 10^6$ kg/cm$^2$. Some people consider that sheetpiles are not rigidly jointed together and advise the nominal moment of inertia per unit width to be reduced by 40%. The author, however, does not think it necessary to take the question of joining into consideration since it is an analysis on the basis of plane strain, that is, only the vertical stiffness is to be considered. Therefore, the nominal stiffness is recommended for use. Considering the repeated use of sheetpiles, however, the stiffness can be assumed to be 80% of the nominal value in analysis. When analyzing the three dimensional behaviors of sheetpiles, on the other hand, the joining should be considered and a suitable reduction factor for the horizontal stiffness should be taken into account.

Young's modulus for diaphragm walls is basically determined according to the compressive strength of concrete. According to the ACI Code, the Young's modulus for concrete can be estimated using the equation: $E = 15,000 \sqrt{f_c'}$ kg/cm$^2$ where $f_c'$ is the 28-day compressive strength of concrete. Considering the possibility of bending moment-induced cracking in concrete and the reduction of the sectional modulus accordingly, the stiffness ($EI$) is usually reduced by 20%-40% in analysis.

Figure 4. 14 is the wall bending moments in the main observation section of the TEC excavation. The solid line refers to the bending moments obtained from the rebar strain meters embedded inside the diaphragm wall. Because the cracking of the diaphragm wall will influence the measurement results of the rebar strain meter, the computed bending moments contain the effect of cracking of the diaphragm wall. The dotted line in the figure represents the bending moment computed from the deformation curve of the diaphragm wall. The computation is as follows: First, compute the radius of the curvature by differentiating twice the multinomial function simulating the deformation curve. Assuming $E$ is not reduced, the moment at a certain depth of the diaphragm wall can be obtained using the equation: $M = EI/r$, where $r$ is the radius of curvature. The bending moment computed from the wall deformation curve excludes the effect of the diaphragm wall cracking. The ratios of the bending moments from rebar strain meters to those from wall deformation curves are then the reduction factors ($R$), as shown in Figure 4.15. From the figure, we can see that the reduction factors at different depths are different. Basically, the reduction factors at the top and bottom of the wall are close to 1.0 and that near the excavation bottom is as low as 0.5. In analysis, we can then assign different reduction factors for different depths of the diaphragm wall.

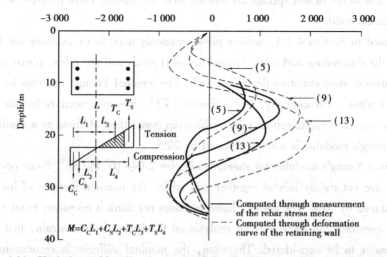

**Figure 4.14   Variations of bending moments of the diaphragm wall in the TEC excavation**

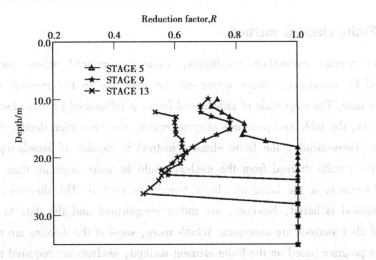

**Figure 4.15　Reduction factors of bending moments for the diaphragm wall in the TEC excavation**

As elucidated in Section 4.2.3, column piles can be distinguished into packed-in-place (PIP) piles, reinforced concrete piles, and mixed piles. Reinforced concrete piles can further be classified into reverse circulation drill piles and all casing piles. The value of $f_c'$ for PIP piles is about 170 kg/cm$^2$ and that for reinforced concrete piles is about 280 kg/cm$^2$. In analysis, it can be reduced by 30%-50%. Which value of the reduction factor is to be taken, however, depends on the construction quality of the column pile. Besides, if considering the different degrees of cracking of concrete at different depths of the column pile, we can assign different reduction factors at different depths of the column pile, which is similar to the approach for diaphragm walls.

The value of $f_c'$ for mixed piles is about 5 kg/cm$^2$. With rather low strength, the stiffness of mixed piles can be ignored and we can, instead, only consider the stiffness of the H steel (or W section) within mixed piles.

Affected by bending moment, the concrete of a diaphragm wall may crack and its section modulus may be thereby decreased. Thus, the section modulus in analysis is usually reduced by 20%-40%. According to field measurements, however, the bending moments of the diaphragm wall vary with depth and the section modulus should be reduced discriminately. Basically, the top and bottom of the diaphragm wall are the places where the reduction rate should be the lowest; whereas the wall near the excavation bottom requires the highest reduction rate. In analysis, we can assign different reduction rates for different depths. Basically, Young's moduli of sheetpiles and soldier piles are not affected by the cracking or bending moment and their section moduli are not supposed to be reduced either. If considering the repeated use of sheetpiles and soldier piles, however, reduction is still necessary. The stiffness is usually reduced by 20%.

Due to a relatively simple analytic model, simple input parameters and quick computation time, the beam on elastic foundation method has been widely used in the analysis and design of deep excavations. Because of its easier model, however, it is not readily adopted for simulations of more complicated excavations.

### 4.3.3 Finite element method

Under normal excavation conditions, excavation-induced stress and deformation are engendered by unbalanced forces acting on the wall due to the removal of soils within the excavation zone. The magnitude of unbalanced forces is influenced by many factors: the conditions of soil layers, the table and pressures of groundwater, the excavation depth, the excavation width and so on. Theoretically, the finite element method is capable of simulating these factors and therefore the results derived from the method would be more accurate than those derived from simplified methods or the beam on elastic foundation method. The theories on which the finite element method is based, however, are rather complicated and the data to be processed both before and after analysis are enormous. What's more, some of the theories are not fully developed. To apply a program based on the finite element method, analysts are required to be well equipped with comprehensive geotechnical knowledge and experience. All this adds confusion and trouble for analysts.

Considering the complexity of the finite element method and that any small neglect is likely to lead to wrong results, the results of the finite element method should be examined by other methods, for example, the simplified methods, to ensure the reasonability of the results.

The yielding stress of the retaining wall and struts is usually very high and can be analyzed using linear elastic models. That is, Young's modulus and Poisson's ratio are assumed to be constants. Either plane strain elements or beam elements can be used for the analysis of retaining walls. If beam elements are adopted, the bending moment of the retaining wall can be found at each node of the elements. If plane strain elements are adopted instead, the bending moment of the retaining wall can be computed through the stresses at the integration points within the element. For example, a $Q_8$ plane strain element has 9 integration points. The relative positions of the integration points are as shown in Figure 4.16(a). The finite element method can directly obtain the stresses at these 9 integration points. Figure 4.16(b) shows the typical distribution of stresses due to bending in the central section of the retaining wall. The stress includes the axial stress caused by the self-weight of concrete and the stress induced by the bending moment. Taking a unit width for computation, the wall bending moment ($M_{wall}$) and the flexural stress ($\delta_{max}$) at the outer fiber of the wall have the following relation:

$$M_{wall} = \frac{\sigma_{max} t^2}{6} \tag{4.2}$$

where

$t$ = thickness of the wall.

If the stresses at No.6 and 8 integration points are used to compute the wall bending moment, the above $\delta_{max}$ can be computed by the following equation:

$$\sigma_{max} = \frac{\sigma_8 - \sigma_6}{2 \times 0.775} \tag{4.3}$$

where $\sigma_8$ is the horizontal stress at No. 8 integration point and $\sigma_6$ is the horizontal stress at No. 6 integration point.

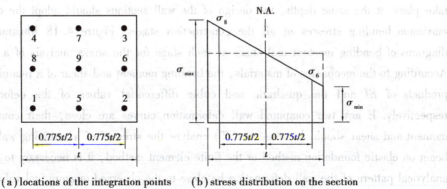

(a) locations of the integration points    (b) stress distribution on the section

**Figure 4.16   Computation of the bending moment of the retaining wall**

If the computer outputs do not provide the stresses at the integration points, or only provide the stresses at the center of the element, the retaining wall should then be divided into two lines (see Figure 4.17) and then we should compute bending moments through the stresses at the centers of the elements using the method above.

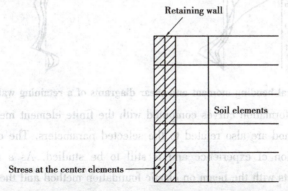

**Figure 4.17   Double elements used for the retaining wall**

# 4.4  Design of Retaining Walls

The penetration depth of a retaining wall directly affects the stability of the excavation. Concerning the design of the penetration depth of a retaining wall, please refer to the contents of stability analysis. This section focuses on designs for sections and dimensions of different types of retaining walls. Table 4.2 lists the comparative values of nominal stiffness for different retaining walls for the references in preliminary design.

To design the sections and dimensions of a retaining wall, we should first carry out the stress analysis. Three methods can be adopted for the stress analysis of a retaining wall: the assumed support method, the beam on elastic foundation method, and the finite element method, which have been introduced in Section 4.3. The actual construction procedures should be simulated in a stress analysis. Because the maximum bending stress occurring at each construction stage does not take place at the same depth, the design of the wall sections should adopt the envelope of the maximum bending stresses of all the construction stages. Figure 4.18 illustrates the typical diagrams of bending moment and shear at each stage for the stress analysis of a retaining wall. According to the mechanics of materials, the bending moment and shear of a retaining wall are the products of $EI$ and the quadratic and cubic differential values of the deformation curve, respectively. If any two computed wall deformation curves are close, their computed bending moment and shear should be close too. To analyze the stresses of a retaining wall by using the beam on elastic foundation method or the finite element method, it is necessary to ensure that the analytical pattern of the wall deformation be close to the observed or empirical values.

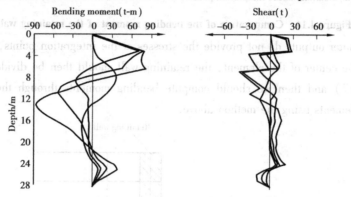

**Figure 4.18　Typical bending moment and shear diagrams of a retaining wall by stress analysis**

Basically, the deformation curves computed with the finite element method and the beam on elastic foundation method are also related to the selected parameters. The correct choosing of the parameters is a question of experience and is still to be studied. As a result, to discuss the differences of the results with the beam on elastic foundation method and the finite element method respectively is meaningless. The assumed support method and its determination of the assumed support are too rough. Besides, the method cannot simulate the construction process completely. Thus, the analytic results by using the assumed support method and those by using the finite element method or beam on elastic foundation method may be different. The results of the assumed support method may be only applicable to small scaled excavations.

### 4.4.1　Soldier piles

The commonly used types of soldier piles in excavations are the H steel, I steel, and rail piles. Concerning the dimensions and properties, the books on steel structures or the AISC Specification can be referred to. Types of rail pile are usually classified in terms of weight per length (kg/m). The rail pile, having a smaller section and is therefore easier to be driven into soils, is mostly used in hard soils or cobble-gravelly soils.

The dimensions of soldier piles and the distance between them are determined based on the results of stress analysis. Then take the maximum bending moment ($M_{max}$) from the typical bending moment envelope (Figure 4.18). According to the ASD method, we can obtain the section modulus of the soldier pile as

$$S = \frac{M_{max}}{\lambda \sigma_a} \tag{4.4}$$

where

$\sigma_a$ = allowable stress of the steel;

$\lambda$ = short-term magnified factor of the allowable stress, which can be found from the country building codes.

The dimensions and spans of rail piles can thus be selected according to the computed section modulus. Basically, under a certain stress, the longer the span is, the larger the required dimension of the soldier pile and the thicker the lagging should be. On the contrary, the shorter the distance is, the smaller the required dimension of the soldier pile and the thinner the lagging could be. The numbers of soldier piles are then increased. To compute the thickness of the laggings, we usually assume the lagging to be the simply supported beam on the soldier piles. The computed thickness of the lagging often comes out larger than the commonly used laggings in general excavations, which are around 3-4 cm thick. Considering the lateral earth pressure on the back of the wall is not necessarily uniformly acting on the laggings, sometimes it is centering on soldier piles, which are of higher rigidity, and sometimes the pressure is less than expected due to the effect of soil arching. The 3-4 cm thick lagging is often adopted if the excavation is shallow.

## 4.4.2 Sheet piles

The dimensions of a sheet pile are determined on the basis of the results of the stress analysis. According to the envelope of bending moments, take the maximum bending moment $M_{max}$ and compute the section modulus using Eq. 4.2, which is then used to find the dimension of the sheet pile by consulting the related steel manual.

## 4.4.3 Column piles

Column piles used in excavations include the PIP pile, reinforced concrete pile, and the mixed pile. The reinforced concrete pile can be further divided into the reversed pile and the all casing pile. Basically, the stiffness of the reinforced concrete pile is the highest, that of the PIP pile the second, and that of the mixing pile the smallest. For more about the characteristics, strengths and shortcomings, and construction of column piles, see Section 4.2.3.

Column piles bear the axial load and flexural load simultaneously. Therefore, their behavior is similar to that of the reinforced concrete columns. No matter which type of column pile is used, it is necessary to transform the flexural rigidity per pile into that per unit width in a plane strain analysis. The thus obtained bending moment and shear envelopes are then used for the design of reinforced concrete columns. Since the design of the columns is extremely complicated, to save space, this book is not going to further explain this subject. Reader can refer to the design chart of reinforced concrete columns or the American Concrete Institute (ACI) code.

## 4.4.4  Diaphragm walls

The design of a diaphragm wall includes specifying the wall thickness and the reinforcements. The thickness is usually determined according to the results of the stress analysis, the deformation analysis, and the feasibility of detailing of concrete reinforcements. According to the experience of excavations in Taipei, the thickness of a diaphragm wall can be assumed to be 5% $H_e$ ($H_e$ is the excavation depth) in the preliminary design.

The design of reinforcements usually follows the widely used strength design method (the LRFD method). The major items of design include the vertical main reinforcement, the horizontal main reinforcement, the shear reinforcement, and the lap splice length and development length of the reinforcement. The design of reinforcements of a diaphragm wall is based on the bending moment and shear envelope obtained from the stress analysis. For the detailed design process, please refer to the ACI code.

# 5　Ground Movements

## 5.1　Introduction

Lateral wall deformations and ground surface settlements represent the performance of excavation support systems. These are closely related to the stiffness of the supporting system, the soil and groundwater conditions, the earth and water pressures, and the construction procedures. The decision to set particular wall deflection and ground movement limits can be of significant economic importance. The setting of appropriate limits should be considered carefully and clearly communicated.

Figure 5.1 shows the general deflection behavior of the wall in response to the excavation presented by Clough and O'Rourke (1990). Figure 5.1(a) shows that at early phases of the excavation, when the first level of lateral support has to be installed, the wall will deform as a cantilever. Settlements during this phase may be represented by a triangular distribution having the maximum value very near to the wall. As the excavation activities advance to deeper elevations, horizontal supports are installed restraining upper wall movements. At this phase, deep inward movements of the wall occur [Figure 5.1(b)]. The combination of cantilever and deep inward movements results in the cumulative wall and ground surface displacements shown in Figure 5.1(c). Clough and O'Rourke (1990) stated that if deep inward movements are the predominant form of wall deformation, the settlements tend to be bounded by a trapezoidal displacement profile as in the case with deep excavations in soft to medium clay; and if cantilever movements predominate, as can occur for excavations in sands and stiff to very hard clay, then settlements tend to follow a triangular pattern. Similar findings were presented by Ou et al. (1993) and Hsieh and Ou (1998) who, based on observed movements of case histories in clay, proposed the spandrel and concave settlement profiles. It has to be noted that Figure 5.1 only describes the general wall deflection behavior in response to the excavation and neglects important factors such as soil conditions, wall installation methods, and excavation support system stiffness, which have been shown to influence the magnitude and shape of both lateral wall movements and ground settlements.

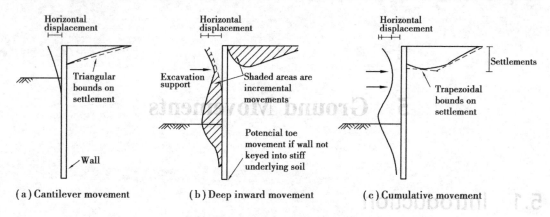

Figure 5.1   Typical profiles of movement for braced and tieback walls ( After Clough and O' Rourke, 1990)

# 5.2   Sources of Ground Movements

Ground movements arise from:

a. wall installation (Section 5.2.1).

b. excavation in front of the wall (Section 5.2.2).

c. groundwater flow resulting in loss of ground and consolidation caused by changes in water pressures due to seepage through and/or around the wall (Section 5.2.3).

d. other site-specific sources of movements, such as:

● construction of large diameter bored piles within or around the proposed excavation.

● ground improvement installations such as grouting ( compensation or permeation ) in or around the site.

● installation of excavation access ramps.

● miscellaneous shallow excavations ( e.g. for drains ).

● removal of temporary sheet piles.

● installation of temporary anchors, etc.

Each of these are discussed in detail in the following sections. Many of the site-specific causes are not predictive analysis or assessment, and the designer should carefully consider the impact of these when making site-specific editions of ground movements. Such impacts will be additional to the generic causes as follows.

## 5.2.1   Wall installation

The construction of a diaphragm wall first partitions the wall into several panels. The construction process of each panel includes guided trench excavation, guided wall construction, trench excavation (for diaphragm wall), and reinforcement placement and concrete casting, the depth of a guided trench is generally about 2-3 m, sometimes 5 m. Before concreting guided walls, guided trenches, not strutted, are open ditches. The maximum settlement induced by excavation of the guided trench occurs at the verge of the trench. The settlement decreases with the distance from the trench. Considering that both measurement of and literature on this field are almost nonexistent and that no significant settlement occurs during this stage (Woo, 1992), this chapter will not delve into the subject.

　　The stress condition of soil in the vicinity of trenches during diaphragm wall construction is rather complicated. Take the construction of a single panel of a diaphragm wall for example. To keep the trench wall from falling, it is necessary to fill the panel with stabilizing fluid during the excavation process of the trench panel. Under normal construction conditions, excavating a trench panel filled with bentonite will cause the stress state of the soil around the trench panel to change from the original $K_0$ to the balanced state of the fluid pressure. However, the fluid pressure is normally not equal to the original earth and water pressures in the trench panel, but is usually smaller. The trench excavation will decrease the total lateral stress of the soil within a specific range around the trench, and thereby produce lateral movement of the soil in the vicinity of the trench. Ground settlement is thus produced. During concrete casting, the lateral pressure in the panel during this stage should be greater than the fluid pressure during the stage of excavation because the unit weight of concrete is greater than that of stabilizing fluid. Therefore, the lateral movement caused at the previous stage will be pushed back and decrease in amount while the amount of ground settlement does not change significantly.

　　The soil deformation behavior caused by trench excavation is not the same as that caused by main excavation. The reasons for the differences are the differences in excavation geometric shapes and the strutting methods. The ratio of the depth of a trench panel to its width and that of the depth to length are both much larger than those in main excavations. What's more, there is the influence of stabilizing fluid employed to counteract the lateral earth pressure and to ensure the stability of trench walls. Nevertheless, in spite of the differences in geometric shapes and construction techniques, the excavation of a trench panel is also a type which can produce deformations, though few and with small influence range. The shape of ground surface settlement is basically similar to that induced by main excavation.

　　Besides, because the retaining wall is the combined whole of many connected diaphragm wall panels, settlement will be accumulated panel by panel and the final deformation gets more serious accordingly. Though the problems of ground displacement induced by the construction of diaphragm walls have gradually drawn attention from engineers, there are few study results available due to the complexity of the construction process and the fact that monitoring results are almost nonexistent. In the 1980s, there were some in situ monitoring projects and some results were obtained. Nevertheless, most of them were confined to the deformation induced by the construction process of a single panel; for example, those carried out in Oslo and in Singapore. As for the monitoring of the final deformation after the completion of a whole retaining wall, there is almost no literature. According to the monitoring results of the rapid transit system in Hong Kong (Cowland and Thorley, 1985), after the completion of the diaphragm walls and before the main excavation, the accumulated deformation can be 40%-50% of the total deformation after the completion of the main excavation. Clough and O'Rourke (1990) found that the ratio of the maximum settlement induced by the construction of diaphragm walls to the depth of the trench is 0.15%, according to many in situ monitoring results, as shown in Figure 5.2. We can see that soil settlement in the vicinity of diaphragm wall panels, induced by their construction, is significant and that caution is strictly required to protect adjacent properties.

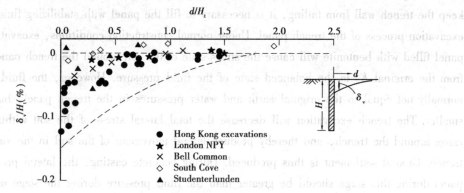

**Figure 5.2  Envelope of ground surface settlements induced by trench excavations ( Clough and O'Rourke, 1990)**

## 5.2.2　Excavation in front of wall

For any given applied loading conditions ( actions) behind an embedded retaining wall, ground movements arising from excavation in front of the wall are influenced by:

　　a. stress changes due to wall installation and excavation.

　　b. excavation geometry.

　　c. ground stiffness and strength.

　　d. changes in groundwater conditions.

　　e. type and stiffness of the wall and its support system.

　　f. construction sequence and methods.

　　g. quality of workmanship.

Although any one of these factors may control the overall movement of a supported excavation, the interaction between these factors in the actural three dimensions (3D) makes the problem even more complicated. For this reason and the fact that it is impossible to quantify many of these factors, structural modelling and analysis of excavation support systems seldom results in reliable prediction of movement. So, the estimation of ground movements associated with supported excavations is generally based on a combination of empirical and analytical methods in conjunction with the application of sound judgment and experience.

However, before considering in detail how such estimates can be made quantitatively, it is appropriate to first consider each of these factors in more detail.

## 1) Stress changes due to wall installation and excavation

The complexity of stress changes at four locations beneath, adjacent to and remote from an excavation in an overconsolidated clay supported by a diaphragm wall during excavation is illustrated in Box 5.1.

As can be seen from the stress path in Box 5.1, the proximity of element $B$ to the failure envelope is the most significant factor that influences the horizontal movement of the soil below excavation level. Yield will be small, and both heave and horizontal movements will also be small if stress path 2 to 4 is well within the passive failure envelope. If the effective stress points for element $B$ are close to the failure envelope, this is indicative of significant yield and local passive failure resulting in relatively high magnitude lateral movements.

Box 5.1    Stress paths for soil elements near an excavation retained

by a cast in situ embedded wall in overconsolidated clay

It is assumed that pore water pressures are initially hydrostatic below an in situ groundwater level. After excavation, steady-state seepage eventually develops from the initial groundwater level behind the wall to a groundwater level at formation level in front.

The geological stress history of the clay comprises deposition followed by the removal of overburden, resulting in an overconsolidated material with $\sigma_h' > \sigma_v'$ in situ. This is represented on the indicative stress path 0'-0 in the figure. There may also be reloading by superficial deposits, which is not shown with figure.

During wall installation, both elements $A$ and $B$ will be subjected to a reduction in lateral total stress as the excavation is made under support fluid, followed by an increase in lateral total stress as the concrete is poured (assuming uncased bores or panels). Field measurements (Symons and Carder, 1992) and centrifuge model tests (Powrie and Kantartzi, 1996) show that pore water pressures will fall during excavation under bentonite and recover to approximately their in situ values during concreting. Indicative effective stress paths are shown in the figure(0-1 under bentonite, 1-2 during concreting).

During excavation in front of the wall, the wall is likely to move forward into the excavation resulting in a reduction in horizontal total stress for element $A$ behind the wall. This will cause a reduction in pore water pressure behind the wall. In an overconsolidated clay, shear following yield will also generate negative excess pore water pressures (2-3 in the figure). The long-term steady-state seepage pore water pressure behind the wall is less than the initial in situ hydrostatic value, but probably greater than the pore water pressure immediately following excavation so that the pore water pressure will probably increase in the long term as steady state conditions are approached (3-4). Overall (2-4) element $A$ will experience a reduction in pore water pressure and a reduction in horizontal total stress at constant vertical total stress. These changed in pore water pressures and boundary stresses will result in an increase in vertical effective stress and a decrease in horizontal effective stress, bringing element $A$ towards the active condition.

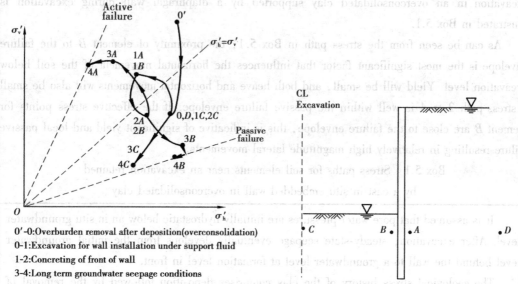

0′-0:Overburden removal after deposition(overconsolidation)
0-1:Excavation for wall installation under support fluid
1-2:Concreting of front of wall
3-4:Long term groundwater seepage conditions

During excavation, element $B$ will experience a large reduction in vertical total stress, which will result in a large reduction in pore water pressure. Movement of the wall below formation level into the soil in front tend to increase the horizontal total stress. These changes are likely to result in an increase in horizontal effective stress and a reduction in vertical effective stress during excavation (2-3 in the figure). In the long term, pore water pressure will increase again as steady state seepage develops, reducing both vertical and horizontal effective stress and bringing the soil towards passive failure (3-4).

These changes in stress from the initial stage prior to excavation are summarized in the table:

| | Element $A$ | Element $B$ | Element $C$ | Element $D^1$ |
|---|---|---|---|---|
| Vertical total stress during excavation | Constant | Decreases | Decreases | Unchanged |
| Horizontal total stress during excavation | Decreases | Decreases due to unloading. Increase due to wall movement | Decreases | Unchanged |
| Pore water pressure during excavation | Decreases | Decreases | Decreases | See note[2] |
| Pore water pressure in the long term | Probably increases | Increases | Decreases | See note[2] |

continued

| | Element A | Element B | Element C | Element D[1] |
|---|---|---|---|---|
| Undrained shear strength in the long term | Probably increases | Decreases | Decreases | Unchanged |
| Strain during excavation | Vertical compression | Vertical extension | Vertical extension | Unchanged |
| Strain in the long term | Vertical compression | Vertical extension | Vertical extension | Unchanged |

Notes: 1. Assumed to be located sufficiently remotely from the wall so as not to be affected by changes in soil stress due to excavation in front of the wall.

2. Depends on ground permeability.

## 2) Excavation geometry

The geometrical shape and area in plan and the excavation depth all have a critical influence on the magnitude and distribution of ground movements around an excavation in any given ground conditions. In many situations, embedded retaining walls form a closed box, or a more complex shape. In such cases, the distribution of movement will be complex and its magnitude is difficult to estimate without the benefit of comparable experience. The corners of an excavation tend to restrict movement.

Analyses undertaken to date allowing for D effects are site specific and the extrapolation of the results of such studies for general application is limited only to reasonably comparable situations. Further work is necessary before reliable correlations can be established for general application. However, St John (1975) found that plane strain and axi-symmetric analyses gave similar vertical movements. Horizontal movements from the axi-symmetric analysis were some 50 percent of those computed in the plane strain model. Simic and French (1998) found that steel quantities in diaphragm wall reinforcement could be reduced by about 25 percent at the corners of the excavation they studied. Temporary prop loads measured across the corners of the Mayfair car park excavation in London indicate a 40 percent reduction within a horizontal distance from the corner equal to the excavation depth, compared to the central props where plane strain conditions predominated (Richards et al., 1999). Being significantly economical are possible.

In addition, the effects of the existence of a corner on the deflection behavior of an excavation wall as represented by the 3D response of excavations have been reported in the literature by a number of researchers, namely, Ou et al. (1996, 2000), Chew et al. (1997), Lee et al. (1998), Finno and Roboski (2005), Roboski and Finno (2006), Finno et al. (2007) For a typical rectangular shaped excavation, Ou et al. (1996) defined the plane strain ratio (PSR) as the maximum lateral wall displacement computed from the results of 3D analysis normalized by that derived from a plane strain analysis. Finno et al. (2007) extended this work and based on field and parametric numerical studies, concluded that the PSR for excavations in clays was strongly influenced by the ratio of the length of the excavation in relation to its width ($L/B$) and the ratio of the length of wall to the excavation depth($L/H$). Their findings are presented in Figure 5.3 and show that PSR is less than 1 for $L/B$ and $L/H$ ratios of less than 2 and 6 respectively. In general, greater corner effects are observed for relatively deep excavations (as evidenced by small $L/H$ ratios). That is, for any given geometry, higher reductions in movements occur near the corners of deep excavations as compared to shallower excavations in similar soil conditions with similar support systems. When $L/H$ is larger than 6, plane strain and 3D analyses yield similar wall movements along the perimeter wall at the center of the excavation.

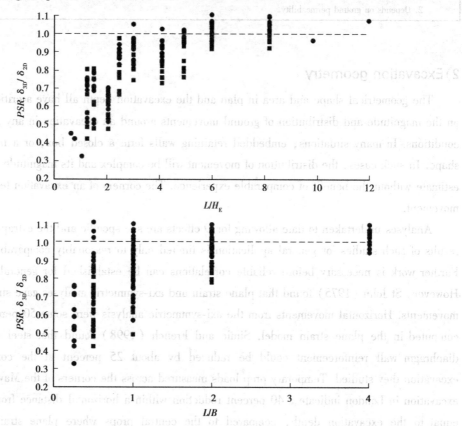

Figure 5.3  Relationship between PSR and excavation geometry (Finno, 2007)

## 3) Ground stiffness and strength

Overall ground movements (heave beneath the excavation, wall deflections and ground movements around the excavation) are strongly influenced by the stiffness of the ground and in weaker ground soil strength. Movements in competent ground conditions, such as dense coarse-grained soils, stiff clays and weak rocks are generally much smaller than in less competent ground conditions, such as soft and firm clays and loose sands. Soft and firm clays can be susceptible to large movements even at small excavation depths, owing to the potential for undrained bearing capacity failure at the toe of the wall. The requirement that the difference in vertical stress in the soil on either side of the wall at the toe should not exceed the bearing capacity of the soil gives a theoretical upper limit to the excavation depth, irrespective of the nature of the support system provided.

## 4) Changes in groundwater condition

If the retaining wall does not provide a hydraulic cut-off at depth, a steady-state groundwater seepage condition will develop involving flow around the toe of an impermeable retaining wall and subsequently upwards to the excavation formation level. Unless recharged from another source, such groundwater flow will result in drawdown behind the wall (and related decreased pore water pressures) and reduced pore water pressures within the excavation compared to the initial hydrostatic condition. Establishment of a steady-state groundwater flow regime may result in both vertical ground movements, owing to the consolidation or swelling of fine-grained soils, and horizontal ground movements associated with the lateral movement of the wall into the excavation.

More sources of ground movements potentially arising from groundwater flow are considered in Section 5.2.3.

## 5) Type and stiffness of the wall and its support system

The type of wall selected and the excavation construction method are major influences on the magnitudes of wall and ground movements. It is well known that stiffer walls attract larger bending moments than more flexible walls with the same ground and support conditions (Rowe et al., 1952). Wall flexure redistributes the stresses imposed by the soil away from the linear, limit

equilibrium ideal, which reduces the structural stress resultants, but at the expense of larger wall and ground movements. For a propped or anchored embedded wall of given overall height $H$ and flexural stiffness $EI$, bending effects are most significant when the wall is supported at the top. In general, wall deformation occurs partly due to rigid body rotation (in the case of a propped or anchored wall, about the position of the prop or anchor), and partly due to bending (Figure 5.4).

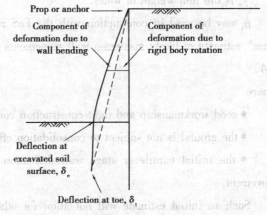

**Figure 5.4  Component of wall displacements and definition of a stiff wall**

The relative importance of wall movement resulting from bending and rigid body rotation may be characterized by means of a dimensionless flexibility number $R$, given by:

$$R = G^* H^4 / EI \tag{5.1}$$

where

$G^*$ is the rate of increase in soil shear modulus with depth in $kN/m^3$.

$H$ is the total height of the wall in m.

$EI$ is the flexural stiffness of the wall, in $kN \cdot m^2/m$.

Diakoumi et al. (2013) show that, depending on the soil strength and groundwater conditions, wall flexibility effects will reduce bending moments in an embedded wall propped at the crest to below those calculated using an EC7-1 type limit equilibrium linear stress distribution at values of $\log_{10}R$ between 1.5 and 2.5 ($R$ between 32 and 320) (see Figure 5.5).

For multi-propped walls, Addenbrooke et al. (2000) define displacement flexibility (in $kN/m^4$) $\Delta$ as:

$$\Delta = EI/h^5 \tag{5.2}$$

where

$h$ is the average vertical prop spacing of a multi-propped support system.

By keeping $\Delta$ constant, a designer can consider different wall types and associated propping systems for the same absolute displacement. This can be useful when comparing value engineer alternatives to satisfy particular deformation performance requirements alternatively for multi-propped walls, Clough et al. (1989) and Clough and O Rourke (1990) defined dimensionless system stiffness $\rho_s$ as

$$\rho_s = EI/(\gamma_w h^4) \tag{5.3}$$

where

$\gamma_w$ is the unit weight of water.

$\rho_s$ may be used in conjunction with the FoS against base failure to provide an initial "first Pass" estimate of likely maximum wall movements for an excavation of given depth (see Figure 5.4)

where

● good workmanship and tight construction control have been employed.

● the ground is not subject to consolidation effects.

● the initial cantilever stage wall deflection is a relatively small component of the total movement.

Such an initial estimate will not allow for other significant factors (previously discussed), which may strongly influence wall deflection. It is evident from Figure 5.5 that in this circumstance and with good quality workmanship and well-controlled construction practice, wall deformation due to excavation should be relatively insensitive to the magnitude of system stiffness. This means that with adequate propping, flexible walls (e.g. sheet pile walls) embedded in stiff clays and other adopted without significant increase in ground movements.

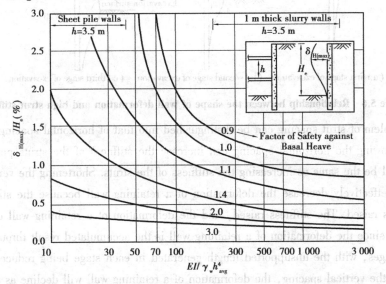

**Figure 5.5　Maximum lateral wall movements vs. system stiffness (After Clough et al., 1989)**

As for multi-propped excavation, the prop or strut properties also have significant influence on wall movement. As shown in Figure 5.6(a), with the start of the first stage of excavation, wall movement will be produced and form a cantilever shape. The second stage of excavation starts after the installation of the first level of struts. If the stiffness of the struts is high enough, the compression of the struts will be rather small, so that the retaining wall will rotate about the contact point between the struts and the wall, and wall deformation is thus generated. The maximum wall deformation will occur near the excavation surface as shown in Figure 5.6(b). With the completion of the second level of struts, the third stage of excavation starts. Suppose the stiffness of the second level of struts is also strong enough, the retaining wall will continue rotating about the contact point with the second level of the struts, and wall deflection is produced again. The location of the maximum deformation will be near the excavation surface [Figure 5.6(c)]. If the soil below the excavation surface is soft soil, the resisting force to prevent the retaining wall from pushing in will be weak and the location of the maximum deformation will be mostly below the excavation surface. Inferred from the same extrapolation, excavation in stiff soils (such as sand) will mostly produce the maximum deformation above the excavation surface. Actually, the locations of the maximum deformations are found near the excavation surfaces in most of the excavations in Taipei (Ou et al., 1993).

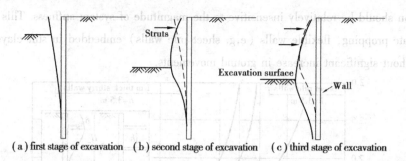

(a) first stage of excavation    (b) second stage of excavation    (c) third stage of excavation

**Figure 5.6    Relationship between the shape of wall deformation and high strut stiffness**

The problem of strut spacing can be distinguished into that of horizontal spacing and vertical spacing. Narrowing the horizontal spacing can increase the stiffness of the struts per unit width. The effect will be the same as increasing the stiffness of the struts. Shortening the vertical spacing of struts can effectively decrease the deformation of a retaining wall because the stiffness of the strut system is raised. The stiffness raised, and the deformation of a retaining wall declines. Put another way, since the deformation of a retaining wall is the accumulated result throughout all the excavation stages, with the unsupported length generated in each stage being reduced due to the shortening of the vertical spacing, the deformation of a retaining wall will decline as a result. The "unsupported length" refers to the distance between the lowest level of strut and the excavation surface.

When applying the braced excavation method (or the anchored excavation method), preload is often exerted onto struts. Suppose the struts are placed at shallower levels. Thus, under normal conditions (with the preload not too small), the preload will be capable of pushing the retaining wall out. If the struts are placed at deeper levels, with the earth pressure growing with the depth, the preload of struts will not be able to push the wall outward easily (Ou et al., 1998). Actually, whether the preload is capable of causing a retaining wall to move, preload is always helpful to reduce the displacement of a retaining wall or the ground settlement.

## 6) Construction sequence and methods

The choice of wall type and its installation, and the overall excavation construction sequence and duration significantly influence ground movements around an excavation for any given ground and groundwater conditions; excavation geometry and dimensions construction sequence (e.g. top-down versus bottom-up) appropriate for particular circumstances is provided in Gaba et al. (2017) where the respective advantages and disadvantages of each approach are discussed in detail. It is generally assumed that the top-down construction method helps to minimize ground movements because this applies early stiff support (typically via a structural slab) near the top of the wall ahead of any significant excavation in front of the wall. However, the regularity of support provided by subsequent floors at each storey height of the structure may not necessarily provide support to the wall at optimum levels (e.g. where external surcharge loading may be applied adjacent to the excavation from existing foundations, or where lateral support to the wall may be referable at locations other than where the structural slabs are located due to, for example, the need to minimize the vertical height from the penultimate support to excavation formation level where weak strata exist immediately below final formation level). CGS (2006) reports that movements of anchored walls can be less than those that are supported struts for the following reasons:

①Anchors can be stressed to a proportion of the design load to the completed excavation before excavation below the anchors level.

②Typically, little excavation occurs below the anchor level to enable anchor installation equipment to operate.

③The connection between struts, waler beams and the wall is often imperfect and without significantly pre-loading the struts; compression of these connections can lead to additional deformation.

④When completing the structures within the excavation by the bottom-up method, struts are typically removed but anchors can be left in place for longer duration, if required.

## 7) Quality of workmanship

Not surprisingly, unprofessional or inappropriate attitudes or failure to effectively apply risk management and mitigation measures on site in routine good construction practice, or not to adhere to recognized good workmanship standards can lead to uncontrolled ground movements and even local failure and progressive collapse. In this regard, according to Puller (2003), there are numerous examples of poor site practices such as:

①late installation of supports.

②unplanned over-excavation.

③poor pile driving and caisson construction.

④water seepage and loss of ground through holes for tie backs and at joints or sheet pile interlocks or diaphragm wall joints.

⑤remoulding and undercutting of clay berms.

⑥excessive surcharge loads from spoil heaps and construction equipment.

⑦lack of rigidity and tightness of shores and braces.

⑧control of ground movements.

Ground movements are best minimized and controlled at source. In this regard, measures that can be adopted by varying the construction sequence and methods while maintaining a good standard of workmanship. Measures that can be adopted to minimize ground movements around and beneath an excavation are summarized here:

• Good workmanship is essential. Supports should be installed tight to the wall. The prop, and any packing between the prop and waling, should not rely on friction or adhesion between the prop end and waling to hold it in place.

● The wall should have adequate embedment in competent (stiff) strata for satisfactory vertical and lateral stability.

● Minimize the first-stage excavation and install the first (stiff) support as early as possible in the construction sequence. With a high stiffness prop installed early during excavation, the horizontal wall movement is not likely to be measured at the wall top, but a depth of some 0.7 to 0.9 times the maximum excavation depth (Carder, 1995).

● Minimize the extent of the dig beyond the proposed support levels.

● Minimize delays to the construction of the wall and its support system.

● Prevent deterioration of lateral support from a clay berm by blinding it or covering it with a waterproof membrane to maintain the berms natural moisture content.

● Avoid unplanned over-excavation.

● Minimize removal of fines during dewatering and loss of fines through wall joints.

● Minimize drawdown outside excavation.

### 5.2.3 Movements due to water

Figure 5.7 shows some potential water flow situations that can result in ground movements, water movements in and around an excavation can occur:

● Through flaws in an impervious wall;

● By flow through a wall (e.g. contiguous bored pile wall, leakage through sheet pile interlock and king post wall lagging);

● By flow under the wall;

● By flow along boundaries between soils of different permeability;

● By flow along the wall if the wall penetrates an underlying aquifer;

● By dewatering.

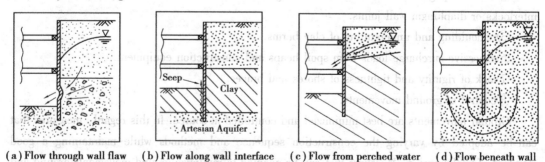

(a) Flow through wall flaw    (b) Flow along wall interface    (c) Flow from perched water    (d) Flow beneath wall

**Figure 5.7    Movements potentially associated with water flow**

If piezometric pressures in an aquifer underlying an excavation are not properly reduced, heaving of the base can occur, leading to loss of passive restraint to the wall. Piping may also occur in coarse grained soils. A case study illustrating base instability in a cofferdam excavation is discussed by Preene et al. (2016) Rowe (1986) noted a case study of piping failure at the base of an excavation in interbedded sand and clay horizons due to water leakage through sheet pile interlocks along confined horizons. This illustrates that dangerous situations can arise when excavating below the water table in fine-grained soils (sands and silts). If leaks develop at joints in the wall, then loss of fines can very quickly lead to catastrophic ground movements in this high-risk situation.

Consolidation settlements may also occur in fine-grained soils due to a combination of these effects.

# 5.3  Ground Movement Predictions Adjacent to Excavations

This section will introduce empirical formulas to predict ground surface settlement and the characteristics of soil movement. Though many empirical formulas have been proposed, only four of the most well-known ones among them are to be discussed for reference and application without a comprehensive introduction to all of the available formulas.

## 5.3.1  Peck's method

Peck (1969a) was the first, based on field observations, to propose a method to predict excavation-induced ground surface settlement. He mainly employed the monitoring results of case histories in Chicago and Oslo and established the relation curves between the ground surface settlement ($\delta_v$) and the distance from the wall (d) for different types of soil, as shown in Figure 5.8. The method classifies soil into three types according to the characteristics of soil:

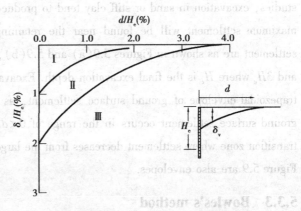

**Figure 5.8  Peck's method (1969a) for estimating ground surface settlement**

Type I : Sand and soft to stiff clay, average workmanship.

Type II : Very soft to soft clay.

1. Limited depth of clay below the excavation bottom.

2. Significant depth of clay below the excavation bottom but $N_b<N_{cb}$.

Type Ⅲ: Very soft to soft clay to a significant depth below the excavation bottom and $N_b \geqslant N_{cb}$.

where $N_b$, the stability number of soil, is defined as $\gamma H_e / s_u$, where $\gamma$ is the unit weight of the soil, $H_e$ is the excavation depth, $s_u$ is the undrained shear strength of soil, and $N_{cb}$ is the critical stability number against basal heave.

Since Peck's method took the monitoring results of case histories before 1969, most of which employed steel sheet piles or soldier piles with laggings as the retaining wall, quite different and more advanced design and construction methods (e.g. the diaphragm wall method which offers higher stiffness) have been employed in excavation projects in recent years, and the relation curves proposed by Peck are not necessarily applicable to all excavations. Basically, the curves derived from Peck's method are envelopes. Because Peck's method is the first to derive an empirical formula to predict the ground surface settlement induced by excavation and is simple for application, it is still used by some engineers.

## 5.3.2 Clough and O'Rourke's method

Clough and O'Rourke (1990) proposed various types of envelopes of excavation-induced ground surface settlements for different soils on the basis of case studies. According to their studies, excavation in sand or stiff clay tend to produce triangular ground surface settlement. The maximum settlement will be found near the retaining wall. The envelopes of ground surface settlement are as shown in Figures 5.9(a) and 5.9(b), whose influence ranges are separately $2H_e$ and $3H_e$ where $H_e$ is the final excavation depth. Excavation in soft to medium clay will produce a trapezoidal envelope of ground surface settlement, as shown in Figure 5.9(c). The maximum ground surface settlement occurs in the range of $0 < d/H_e < 0.75$ while $0.75 < d/H_e < 2.0$ is the transition zone where settlement decreases from the largest to almost none. Basically, the curves in Figure 5.9 are also envelopes.

## 5.3.3 Bowles's method

Bowles (1988) suggested a procedure to estimate excavation-induced ground surface settlements, which can be described as follows (Figure 5.10):

①Compute the lateral displacement of the wall using the finite element method or the beam on elastic foundation method.

②Compute the area of the lateral wall deflection (act).

③Estimate the influence range of ground surface settlement (D) following Capse's method (1966):

$$D = (H_e + H_d) \tan(45° - \varphi/2)$$

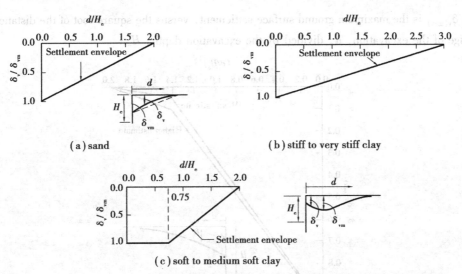

(a) sand  (b) stiff to very stiff clay

(c) soft to medium soft clay

**Figure 5.9  Clough and O'Rourke's method (1990) for estimating ground surface settlement**

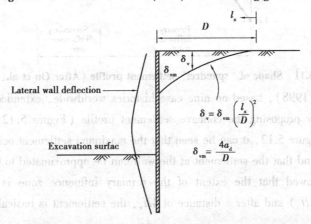

$$\delta_v = \delta_{vm} \left( \frac{l_x}{D} \right)^2$$

$$\delta_{vm} = \frac{4a_d}{D}$$

**Figure 5.10  Bowles's method (1986) for estimating ground surface settlement**

where $H_e$ = the excavation depth, $H_d = B$ if $\varphi = 0$ and $H_d = 0.5 B \tan(45° + \varphi/2)$ if $\varphi \geqslant 0$, $B$ = the excavation width.

④Maximum ground surface settlement $\delta_{vm} = 4a_d/D$.

⑤The ground surface settlement $\delta_v$ is assumed to be parabolic $\delta_v = \delta_{vm}(l_x/D)^2$.

### 5.3.4  Ou and Hsieh method

Ou et al. (1993) proposed a procedure to estimate excavation-induced ground settlement profile normal to the excavation support wall. Their work was based on observation of 10 case histories in soft soils. From these data, they developed a trilinear settlement profile (Figure 5.11) called spandrel-type settlement, which presents the maximum settlement very near to the wall. The spandrel type of settlement profile occurs if a large amount of wall deflection occurs at the first phase of excavation when cantilever conditions exist and the wall deflection is relatively small due to subsequent excavation. The data presented in Figure 5.11 is normalized settlement, $\delta_v/\delta_{v(max)}$, where $\delta_{v(max)}$ is the maximum ground surface settlement, versus the square root of the distance from the edge of the excavation, $d$, divided by the excavation depth, $H_e$.

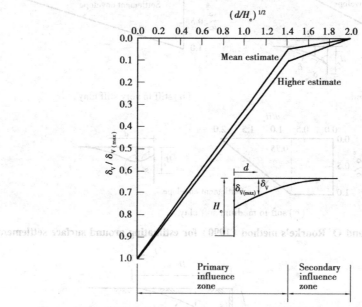

**Figure 5.11   Shape of "spandrel" settlement profile (After Ou et al., 1993)**

Hsieh and Ou (1998), based on nine case histories worldwide, extended the work done by Ou et al. (1993) by proposing the concave settlement profile (Figure 5.12) induced by deep excavations. From Figure 5.12, it can be seen that the maximum settlement occurs at a distance of $H_e/2$ from the wall and that the settlement at the wall can be approximated to $0.5\delta_{v(max)}$. The case history data also showed that the extent of the primary influence zone is approximately two excavation depths ($2H_e$) and after a distance of $4H_e$, the settlement is basically negligible.

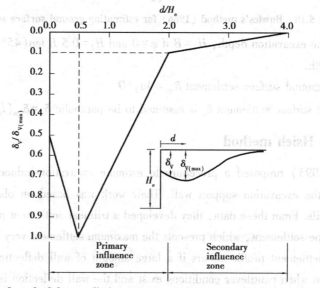

**Figure 5.12   Proposed method for predicting concave settlement profile (After Hsieh and Ou, 1998)**

Hsieh and Ou (1998) also established the relationship of cantilever area and deep inward area of wall deflection, similar to the one proposed by O'Rourke (1981), as the first approximation to predict the type of settlement profile. They suggested the following procedures for predicting the settlement profile:

①predict lateral deformations using finite element or beam on elastic foundation methods;

②determine the type of settlement profile by calculating the areas of the cantilever and inward bulging of the wall displacement profile;

③estimate the maximum ground surface settlement as $\delta_{v(max)} \approx 0.5\,\delta_{H(max)}$ to $1.0\,\delta_{H(max)}$;

④plot the surface settlement profile using Figure 5.11 for spandrel settlement profile or Figure 5.12 for concave settlement profile.

## 5.3.5  Newly proposed methods

Finno and Roboski (2005), and Roboski and Finno (2006) proposed parallel distributions of settlement and lateral ground movement for deep excavations in soft to medium clays. The parallel distribution profiles were based on optical survey data obtained around a 12.8 m-deep excavation in Chicago supported by a flexible sheet pile wall and three levels of anchors. They found that when using the complementary error function (erfc), just geometry and maximum movement parameters are necessary for defining the parallel distributions of ground movement. The complementary erfc function is defined as

$$\delta(x) = \delta_{max}\left\{1 - \frac{1}{2}erfc\left[\frac{2.8\left[x + L\left(0.015 + 0.035\ln\frac{H_e}{L}\right)\right]}{0.5L - L\left(0.015 + 0.035\ln\frac{H_e}{L}\right)}\right]\right\}  \quad (5.4)$$

where

$\delta_{max}$ can be either maximum settlement or maximum lateral movement;

$L$ is the length of the excavation;

$H_e$ is the height of the excavation as presented in Figure 5.13.

Although Equation 6.4 was derived from observations of flexible wall excavations, it has been reported by Roboski and Finno (2006) that it can predict with reasonable agreement the ground movement profiles for stiffer walls.

Special attention is needed in excavations where there are larger diameter utility pipes, buildings with stiff floor systems, buildings supported on deep foundations, and deep foundations between the building and the excavation because they provide restraint for the movements and will consequently affect their distribution. Roboski and Finno (2006) concluded that the complementary error function is applicable to excavations where the induced ground movements can develop with little restraint.

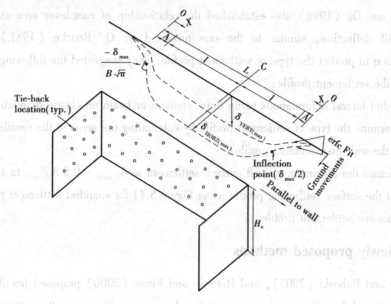

**Figure 5.13** **Derived Fitting Parameters for the Complementary Error Function.** $\delta_{\text{VERT}}$,

settlement; $\delta_{\text{HORZ}}$, lateral movement (After Roboski and Finno, 2006)

Recently, a series of finite element analyses were carried out by Zhang and Goh (2013), and Goh et al. (2017) to investigate the influences of soil properties, wall stiffness, excavation length, excavation depth, clay thickness at the base of the excavation and wall embedment depth, on the maximum wall deflection induced by braced-excavation in clay. Some advanced computational learning tools such as multivariate adaptive regression splines (MARS) and logarithmic regression model were used to estimate the simple wall deflection equation. As for 2D cases, a simple logarithmic regression model was developed for estimating $\delta_{\text{Hm}}$ in terms of $\alpha$, $c_{\text{u}}$, $H_{\text{e}}$, $D$, $L/B$ and $D/B$, as shown in Eq. 5.5. Meanwhile, the optimal MARS models was also developed based on a total of 1,120 cases to determine the maximum diaphragm wall deflection; the ANOVA parameter relative importance assessment indicates that the two variables which contribute most to the diaphragm wall deflection are $h$ (excavation depth) and $B$ (excavation width). The BFs (basic functions) and their corresponding equations for the optimal MARS model is tabulated in Table 5.1 and the interpretable MARS model is given by Eq. 5.6.

$$\delta_{\text{Hm}}(\text{mm}) = \alpha^{-0.061\,7} c_{\text{u}}^{-3.089\,9} H_{\text{e}} D^{4.599\,7} \left(\frac{L}{B}\right)^{0.302\,4} \left(\frac{D}{B}\right)^{-4.801\,9} \tag{5.5}$$

where

$\alpha$ represents walls with different rigidities, a smaller $\alpha = 0.1$ represents flexible walls and larger $\alpha$ represents stiff walls;

$c_{\text{u}}$ is the undrained shear strength of the soil;

$H_{\text{e}}$ is the excavation depth;

$D$ is the wall penetration depth;

$L$ and $B$ are excavation length and width respectively.

**Table 5.1  Basis functions and corresponding equations of**
**MARS model for diaphragm wall deflection prediction**

| Basis function | Equation | Basis function | Equation |
|---|---|---|---|
| BF1 | $\mathrm{Max}(0,\ln\ (EI/\gamma_{\mathrm{w}}h_{\mathrm{avg}}^{4})-7.313)$ | BF12 | $\mathrm{Max}(0,30-T)$ |
| BF2 | $\mathrm{Max}(0,7.313-\ln\ (EI/\gamma_{\mathrm{w}}h_{\mathrm{avg}}^{4}))$ | BF13 | $BF6*\mathrm{Max}(0,\ln(EI/\gamma_{\mathrm{w}}h_{\mathrm{avg}}^{4})-7.313)$ |
| BF3 | $\mathrm{Max}\left(0,\dfrac{E_{50}}{c_{\mathrm{u}}}-200\right)$ | BF14 | $BF6*\mathrm{Max}(0,7.313-\ln(EI/\gamma_{\mathrm{w}}h_{\mathrm{avg}}^{4}))$ |
| BF4 | $\mathrm{Max}\left(0,200-\dfrac{E_{50}}{c_{\mathrm{u}}}\right)$ | BF15 | $BF7*\mathrm{Max}(0,\ln\ (EI/\gamma_{\mathrm{w}}h_{\mathrm{avg}}^{4})-8.176)$ |
| BF5 | $\mathrm{Max}(0,\gamma-17)$ | BF16 | $BF7*\mathrm{Max}(0,8.176-\ln\ (EI/\gamma_{\mathrm{w}}h_{\mathrm{avg}}^{4}))$ |
| BF6 | $\mathrm{Max}(0,17-\gamma)$ | BF17 | $BF10*\mathrm{Max}(0,\ln\ (EI/\gamma_{\mathrm{w}}h_{\mathrm{avg}}^{4})-7.313)$ |
| BF7 | $\mathrm{Max}\left(0,\dfrac{c_{\mathrm{u}}}{\sigma_{\mathrm{v}}'}-0.25\right)$ | BF18 | $BF10*\mathrm{Max}(0,7.313-\ln\ (EI/\gamma_{\mathrm{w}}h_{\mathrm{avg}}^{4}))$ |
| BF8 | $\mathrm{Max}\left(0,0.25-\dfrac{c_{\mathrm{u}}}{\sigma_{\mathrm{v}}'}\right)$ | BF19 | $BF10*\mathrm{Max}(0,T-30)$ |
| BF9 | $\mathrm{Max}(0,h-17)$ | BF20 | $BF10*\mathrm{Max}(0,30-T)$ |
| BF10 | $\mathrm{Max}(0,17-h)$ | BF21 | $\mathrm{Max}(0,B-40)$ |
| BF11 | $\mathrm{Max}(0,T-30)$ | BF22 | $\mathrm{Max}(0,40-B)$ |

Note: $EI$ is the wall stiffness; $h_{\mathrm{avg}}$ is the average vertical distance; $h$ is excavation depth; $B$ is excavation width; $\gamma$ is soil unit weight; $T$ is soft clay thickness; $\dfrac{E_{50}}{c_{\mathrm{u}}}$ is soil stiffness; $\dfrac{c_{\mathrm{u}}}{\sigma_{\mathrm{v}}'}$ is soil shear strength ratio.

$$\begin{aligned}
\delta_{h0} = {} & 165 - 50.889 \times BF1 + 66.598 \times BF2 - 0.191\,4 \times BF3 + 0.495\,6 \times BF5 + \\
& 19.135 \times BF6 - 326.34 \times BF7 + 815.69 BF8 + 4.998\,1 \times BF9 - 6.189\,1 \times BF10 + \\
& 7.4897 \times BF11 - 7.0073 \times BF12 - 13.712 \times BF13 + 24.131 \times BF14 + \\
& 540.93 \times BF15 - 331.28 \times BF16 + 2.771\,6 \times BF17 - 4.582\,1 \times BF18 - \\
& 1.1808 \times BF19 + 0.861\,2 \times BF20 + 0.511\,4 \times BF21 - 1.547\,4 \times BF22 \quad (5.6)
\end{aligned}$$

## 5.3.6  Relation between $\delta_{\mathrm{H(max)}}$ and $\delta_{\mathrm{V(max)}}$

In general, the maximum ground surface settlement $\delta_{\mathrm{V(max)}}$, can be estimated by referring to the value of the maximum wall deflection $\delta_{\mathrm{H(max)}}$. From the data of case histories reported by Mana and Clough (1981), Ou et al. (1993), and Hsieh and Ou (1998), the relationship between maximum wall deflection and maximum ground surface settlement can be obtained as

$$\delta_{V(max)} \approx 0.5\delta_{H(max)} \ to \ 1.0 \ \delta_{H(max)} \tag{5.7}$$

Long (2001) analyzed the effect of different support types on $\delta_{vmax}$ and $\delta_{hmax}$. It was found that there is a variable response in the ratio of $\delta_{vmax}$ and $\delta_{hmax}$ between different support types and ground conditions. The findings of Long (2001) are summarized in Table 5.2. This shows that, as the soil conditions deteriorate, the ratio of $\delta_{vmax}/\delta_{hmax}$ changes from less than 1 to greater than 1. Long (2001) did not quantify the relationship between support stiffness and ground movements, despite identifying the possibility that existed. Long (2001) suggested that there is a linear relationship between $H$ and $\delta_{vmax}$ and $\delta_{hmax}$ once classification of different ground conditions has been made, though this was not proven in terms of statistical significance. He also identified that there is an increase in $\delta_{hmax}$ with a decreasing FoS against basal heave, confirming the conclusions of the earlier study by Clough and O'Rourke (1990).

**Table 5.2　Findings of Long (2001) for magnitudes of ground movement based on split database**

| Dataset | $\delta_{hmax}/H(\%)$ | $\delta_{vmax}/H(\%)$ | Description of data |
|---------|-----------------------|-----------------------|---------------------|
| 1 | 0.18 | 0.14 | $h<0.6H$ |
| 2 | 0.38 | 0.48 | $h>0.6H$, FoS>3 |
| 3 | 0.82 | 1.1 | $h>0.6H$, FoS<3 |

Based on the cases by Long (2001) using the same variables. In total, 389 case studies have been compiled, with 235 from the database constructed by Long (2001), and 154 cases added by Holmes et al. 2019, dating from 2002 to 2016. The database is attached in the Appendix I. It should be noted that only the propped or anchored excavation case studies from the database of Long (2001) were included.

# 5.4　Damage to Buildings

This section considers the key principles of building damage assessment; it does not cover assessment of damage to utilities, which require very specific considerations.

Cracks are the main indicators of damage to a building. Cracks can be caused by effects such as temperature variations, moisture content changes (shrinkage), chemical reactions etc. or by deformations of the building.

Buildings located near excavations can experience several types of deformation (e. g. "sagging" or "hogging", rigid body tilt) and can be damaged depending on their construction type, stiffness, openings and joints. The 3D behaviour of a deep excavation can reduce or increase the amount of damage suffered by nearby buildings. Where the subsidence contours are oblique to the structure, cracking occurs in the walls and floor, accompanied by diagonal cracks. When assessing actual or potential building damage, it is important to distinguish between the different deformation types.

Damage due to deformations is related to the curvature of the building. More curvature is indicative of higher strains and more damage. The most likely deformation types are "sagging" and "hogging" as described by Burland and Wroth (1974). The sides of the building settle more than the average in the "hogging" type of deformation; whereas in "sagging" the central part of the building settles most (Figure 5.14).

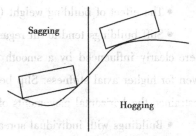

**Figure 5.14 Sagging and hogging deformation modes**

Rigid body tilt should not contribute to the stresses and strains in the building and causes indirect damage due to gravity forces on structural elements (such as walls). However, Leonards (1975) reported that for framed structures founded on isolated footings, tilt contributes to stress and strain in the frame, unless each footing tilts or rotates through the same angle as the overall structure, which is highly unlikely.

In the previous sections, all excavation-induced ground movements are considered as "greenfield" displacements, i. e. ignoring the presence of any surrounding buildings or developments. However, it is known that the presence of the building and its interface with the ground also influences the profile of the settlement and the way the displacements are transferred to the building.

Potts and Addenbrooke (1997) investigated the effect of building stiffness on tunnelling-induced displacements by undertaking a parametric study using finite element methods with a non-linear elastic-plastic soil model. The building was represented as an equivalent beam having bending and axial stiffness $EI$ and $EA$ (where $E$ is the Young's modulus of elasticity, $A$ is the cross-sectional area and $I$ is the moment of inertia of the beam). They defined bending stiffness $\rho^*$ and axial stiffness $\alpha^*$ as

$$\rho^* = EI/E_sH^4 \tag{5.8}$$
$$\alpha^* = EA/E_s \tag{5.9}$$

where

$H$ is the half-width of the beam;

$E_s$ is representative soil stiffness.

Design curves were established for the likely modification to the Greenfield settlement profile caused by a surface structure. Franzius et al. (2004) extended the work of Potts and Addenbrooke by including the effect of building weight. Franzius et al. (2006) then discussed this further by varying the soil-structure interface. These effects were also investigated by EL Shafie (2008) who performed centrifuge tests on model buildings subject to excavation induced ground movements. EL Shafie (2008) concluded the following:

• The effect of building weight (up to 40 kPa) is small.

• Stiff buildings tend to tilt regardless of the soil-structure interface. Horizontal displacements were clearly influenced by a smooth interface, leaving the greenfield soil displacements intact, even for higher axial stiffness. Slip between the building and the soil occurred. Rough inter faces restrained the horizontal movements of the building.

• Buildings with individual spread footings experienced large differential settlements resulting in significant distortions and tensile strains concentrating at the weaker parts of the building.

Cording et al. (2010) proposed that geometry, age of construction and condition, including previous deterioration and finishes, all affect the response of the building to ground movements. So, an understanding of the structural characteristics of the building will help in determining potential distortion and damage.

A three-stage approach should be adopted for assessing potential damage to buildings near excavations supported by embedded retaining walls (see Figure 5.15).

**Stage 1:** Ground movements behind the retaining wall should be estimated as described in Section 6.3 assuming greenfield conditions, i.e. ignoring the presence of the building or utility and the ground above foundation level. Contours of ground surface movements should be drawn and a zone of influence established based on specified settlement and distortion criteria. All structures and utilities within the zone of influence should be identified.

**Stage 2:** A condition survey should be carried out on all structures and utilities within the zone of influence before starting work on site. The structure or utility should be assumed to follow the ground (i.e. it has negligible stiffness), so the distortions and consequently the strains in the structure or utility can be calculated. In some circumstances, a "wave" of settlement may be experienced by the structure as the settlement develops over time. The method of damage assessment should adopt the limiting tensile strain approach. See also Table 5.3, Figure 5.16 and Box 5.2.

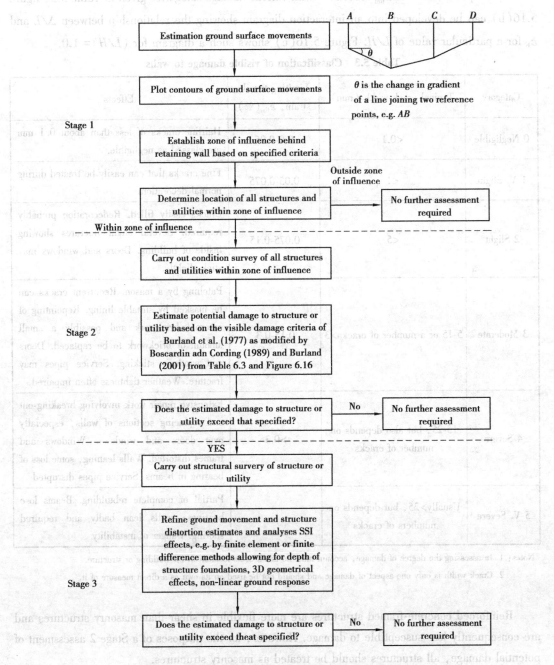

θ is the change in gradient
of a line joining two reference
points, e.g. *AB*

Figure 5.15 Procedure for building damage assessment

Note:

By adopting values of $\varepsilon_{lim}$ associated with various damage categories given in Table 5.3, figure 5.16(b) can be developed into an interaction diagram showing the relationship between $\Delta/L$ and $\varepsilon_h$ for a particular value of $L/H$. Figure 5.16(c) shows such a diagram for $(L/H) = 1.0$.

**Table 5.3  Classification of visible damage to walls**

| Category | Typical crack width (mm) | Limiting tensile strain, $\varepsilon_{lim}(\%)$ | Effects |
|---|---|---|---|
| 0 Negligible | <0.1 | 0.0-0.05 | Hairline cracks of less than about 0.1 mm are classed as negligible. |
| 1 V. Slight | <1 | 0.05-0.075 | Fine cracks that can easily be treated during normal decoration. |
| 2 Slight | <5 | 0.075-0.15 | Cracks easily filled. Redecoration probably required. Several slight fractures showing inside of building. Doors and windows may stick slightly. |
| 3 Moderate | 5-15 or a number of cracks>3 | 0.15-0.3 | Patching by a mason. Recurrent cracks can be masked by suitable lining. Repointing of external brickwork and possibly a small amount of brickwork to be replaced. Doors and windows sticking. Service pipes may fracture. Weather tightness often impaired |
| 4 Severe | 15-25, but also depends on number of cracks | >0.3 | Extensive repair work involving breaking-out and replacing sections of walls, especially over doors and windows. Windows and frames distorted. Walls leaning, some loss of bearing in beams. Service pipes disrupted |
| 5 V. Severe | Usually>25, but depends on numbers of cracks | | Partial or complete rebuilding. Beams lose bearings; walls lean badly and required shoring. Danger of instability |

Notes: 1. In assessing the degree of damage, account must be taken of its location in the building or structure.

2. Crack width is only one aspect of damage and should not be used on its own as a direct measure of it.

Reinforced concrete-framed structures are more flexible in shear than masonry structures and are consequently less susceptible to damage. However, for the purposes of a Stage 2 assessment of potential damage, all structures should be treated as masonry structures.

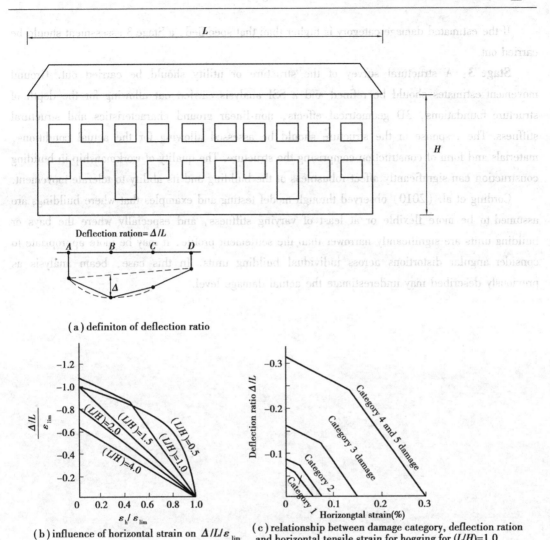

(a) definiton of deflection ratio

(b) influence of horizontal strain on $\Delta/L/\varepsilon_{lim}$

(c) relationship between damage category, deflection ration and horizontal tensile strain for hogging for $(L/H)=1.0$

**Figure 5.16  Relationship between damage category, deflection ratio and horizontal tensile strain (after Burland, 2001)**

Box 5.2  Procedure for stage 2 damage category assessment

The following steps should be undertaken in making a Stage 2 assessment of the damage to a structure:

1. Establish $L$ and $H$ for the structure (see Figure 5.16(a) for definitions of $L$ and $H$.

2. Determine $(L/H)$.

3. Determine relationship between $(\Delta/L)$ and $\varepsilon_h$, for the required $(L/H)$ from Figure 5.16 (b) for $\varepsilon_{lim}$ values from Table 5.3.

4. Estimate vertical and horizontal ground surface movements in the vicinity of the structure.

5. Determine $(\Delta/L)$ and $\varepsilon_h (=\varepsilon_h/L)$ where $\varepsilon_h$ is the horizontal movement.

If the estimated damage category is higher than that specified, a Stage 3 assessment should be carried out.

**Stage 3**: A structural survey of the structure or utility should be carried out. Ground movement estimates should be refined and a SSI analysis carried out allowing for the depth of structure foundations, 3D geometrical effects, non-linear ground characteristics and structural stiffness. The response of the structure should be assessed allowing for the actual conditions, materials and form of construction comprising the structure. The quality of workmanship in building construction can significantly affect robustness of the building and its ability to tolerate movement.

Cording et al. (2010) observed through model testing and examples that where buildings are assumed to be more flexible or at least of varying stiffness, and especially where the bays or building units are significantly narrower than the settlement profile, it may be more appropriate to consider angular distortions across individual building units. In this case, beam analysis as previously described may underestimate the actual damage level.

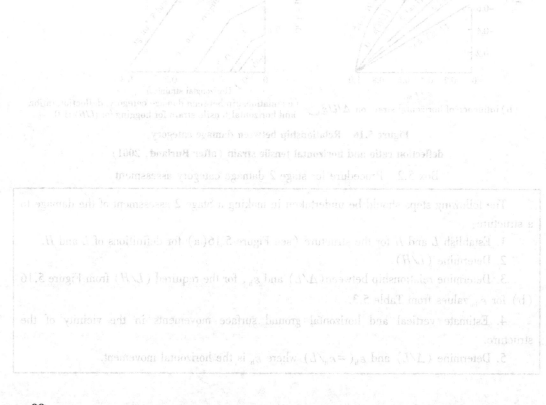

# 6　Finite Element Method

## 6.1　Introduction

As explained in Chapter 5, under normal excavation conditions, excavation-induced stress and deformation are engendered by unbalanced forces acting on the wall due to the removal of soils within the excavation zone. The magnitude of unbalanced forces is influenced by many factors: the conditions of soil layers, the table and pressures of groundwater, the excavation depth, the excavation width and so on. Theoretically, the finite element method is capable of simulating these factors and therefore the results derived from the method would be more accurate than those derived from simplified methods or the beam on elastic foundation method. The theories on which the finite element method is based, however, are rather complicated and the data to be processed both before and after analysis are enormous. What's more, some of the theories are not fully developed. To apply a program based on the finite element method, analysts are required to be well equipped with comprehensive geotechnical knowledge and experience. All this adds confusion and trouble for analysts. Considering the complexity of the finite element method and that any small neglect is likely to lead to wrong results, the results of the finite element method should be examined by other methods, the simplified methods, for example, to ensure the reasonability of the results.

Besides, some researchers write the governing equation in the form of an explicit finite difference equation and solve it by dynamic relaxation. This method solves the velocity and movement through the movement equation by assigning a damping value close to the critical damping. The strain rate is then obtained from velocity and used to solve the new stress increment. The process continues till the unbalanced forces are in equilibrium or the system reaches a steady state. The main theory on which the finite difference method is based is not the same as that of the finite element method. However, other theories, such as constitutive laws of soil, drained or undrained behaviors, determination of soil parameters, simulations of excavation are identical with the finite element method.

In terms of stress type (see Section 6.6.1), the finite element method can be split into the total stress analysis and the effective stress analysis methods. Concerning the analysis of drained conditions (see Section 6.6.2), the finite element method can also be divided into the undrained analysis method, drained analysis method, and the partially drained analysis method. With the finite element method, the undrained behaviors of clayey soils can be analyzed by total stress analysis, effective stress analysis, and coupled analysis.

Although many of the theoretical descriptions in this chapter are complicated, they are necessary knowledge for excavation analyses using the finite element method, and are also applicable to other problems concerning geotechnical engineering. Readers are advised to run programs according to the contents of this chapter to examine the correctness of analyses.

Although this chapter tries to elucidate the application of the finite element method on deep excavation in details, there remains much to be explored. Readers interested in the method can see the related references. Besides, considering the complexity of the analysis process of the finite element method and the comprehensive knowledge required by analysts, this chapter is recommended for readers above the graduate level.

## 6.2 Basic Principles

This section explains the basic principles of the finite element method. In an excavation, find a section, the central section in most cases, whose behaviors meet the plane strain condition as shown in Figure 6.1(a). Take the profile of this section and divide the soils and structures within the excavation influence range into many meshes, each of which is called an element [see Figure 6.1(b)]. According to the properties of the material of each element, establish its stress-strain relation, which is called the constitutive law. The constitutive law of an isotropic material can be expressed as follows:

$$\{\sigma\} = [C]\{\varepsilon\} \tag{6.1}$$

where

$\{\sigma\}$ = stress matrix. The sign $\{\}$ refers to a column matrix;

$\{\varepsilon\}$ = stram matrix;

$[C]$ = stress-strain relational matrix.

Under the condition of plane strain, the matrices are

$$\{\sigma\} = \begin{Bmatrix} \sigma_{xx} \\ \sigma_{yy} \\ \tau_{xy} \end{Bmatrix} \tag{6.2}$$

$$\{\sigma\} = \begin{Bmatrix} \varepsilon_{xx} \\ \varepsilon_{yy} \\ \gamma_{xy} \end{Bmatrix} \tag{6.3}$$

$$[C] = \frac{E}{(1+v)(1-2v)} \begin{bmatrix} (1-v) & v & 0 \\ v & (1-v) & 0 \\ 0 & 0 & \frac{(1-2v)}{2} \end{bmatrix} \tag{6.4}$$

E and $v$ in Eq. 6.4 are Young's modulus and Poisson's ratio, respectively.

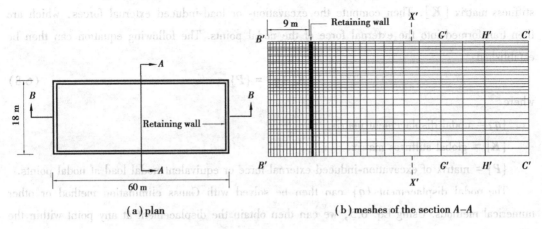

(a) plan

(b) meshes of the section A–A

**Figure 6.1  Finite element analysis of an excavation**

As shown in Figure 6.2, the relation between the displacement at any point within the element and that of the nodal point of the element can be expressed as follows:

$$\{u\} = [f]\{q\} \tag{6.5}$$

where $[f]$ = displacement shape function.

According to the theory of elasticity, the strain and displacement at a point within the element have a relation which can be expressed as follows:

$$\{\varepsilon\} = [d]\{u\} = [d][f]\{q\} = [B]\{q\} \tag{6.6}$$

where $[d]$ = linear partial differential operator, such as $\dfrac{\partial}{\partial x}, \dfrac{\partial}{\partial y}$; $[B] = [d][f]$ = relational matrix between the strain and the nodal displacement.

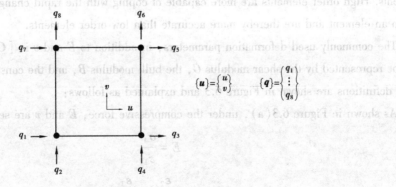

$$\{u\} = \begin{Bmatrix} u \\ v \end{Bmatrix} \qquad \{q\} = \begin{Bmatrix} q_1 \\ \vdots \\ q_8 \end{Bmatrix}$$

**Figure 6.2  Four-node element**

According to the principle of virtual work, we can derive the stiffness matrix of the element to be

$$[K_e] = \int_V [B]^T [C][B] \, dV \tag{6.7}$$

After establishing the relational matrices for all the elements, combine them into the global stiffness matrix $[K]$. Then compute the excavation- or load-induced external forces, which are then transformed into the external force of the nodal points. The following equation can then be established:

$$[K]\{q\} = \{P\} \tag{6.8}$$

where

$\{q\}$ = nodal displacement matrix.

$[K]$ = global stiffness matrix.

$\{P\}$ = matrix of excavation-induced external force or equivalent nodal load at nodal points.

The nodal displacement $\{q\}$ can then be solved with Gauss elimination method or other numerical methods. Using Eq. 6.5, we can then obtain the displacement at any point within the element. By means of Eq. 6.6, the strain at any point within the element can be obtained. Lastly, use Eq. 6.1 for obtaining the stresses at any point within the element. As a result, we can obtain the deformation, the stress, the strain, the bending moment of the retaining wall, the ground surface settlement, and the movement of the excavation bottom.

When the displacement shape function $[f]$ is quadratic, differentiated by the partial differencial operator $[d]$, the matrix $[B]$ becomes linear, which shows that the strain within the element changes linearly. The element within which the strain changes linearly is called the low order element. Otherwise, those within which strains do not change linearly are called high order elements.

Since the order of the shape functions of high order elements is higher than those of the low order elements, the number of nodes of high order elements is larger than that of low order elements. High order elements are more capable of coping with the rapid change of stress or strain within an element and are thereby more accurate than low order elements.

The commonly used deformation parameters, in addition to $E$ and $v$ in $[C]$ in Eq. 6.4, can also be represented by the shear modulus $G$, the bulk modulus $B$, and the constrained modulus $M$. Their definitions are shown in Figure 6.3 and explained as follows:

As shown in Figure 6.3(a), under the compressive force, $E$ and $v$ are separately defined as

$$E = \frac{\sigma_1}{\varepsilon_1} \tag{6.9}$$

$$v = -\frac{\varepsilon_2}{\varepsilon_1} = -\frac{\varepsilon_3}{\varepsilon_1} \tag{6.10}$$

Figure 6.3(b) shows the strains $\varepsilon_1, \varepsilon_2$ and $\varepsilon_3$ produced under the action of the stresses, $\sigma_1$, $\sigma_2$ and $\sigma_3$. Thus, the bulk modulus is

$$B = \frac{\sigma_{avg}}{\varepsilon_v} = \frac{\sigma_1 + \sigma_2 + \sigma_3}{3\varepsilon_v} \tag{6.11}$$

where

$$\varepsilon_v = \text{volumetric strain} = \frac{\Delta V}{V} \approx \varepsilon_1 + \varepsilon_2 + \varepsilon_3;$$

$\Delta V$ = change of the volume;

$V$ = volume.

Figure 6.3(c) shows the shear strain produced under the action of the shear stress, $\tau$. Thus, the shear modulus is

$$G = \frac{\tau}{\gamma} \tag{6.12}$$

where $\gamma$ = shear strain.

Figure 6.3(d) displays the axial load, $\sigma_1$, on the material while the lateral strain is restrained. That is, the lateral strain is 0. The constrained modulus is

$$M = \frac{\sigma_1}{\varepsilon_1} \tag{6.13}$$

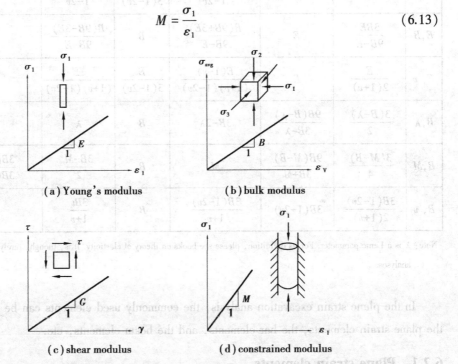

(a) Young's modulus　(b) bulk modulus

(c) shear modulus　(d) constrained modulus

Figure 6.3　Definitions of various deformation moduli

According to the theory of elasticity, the relationship between the deformation parameters $E$, $v$, $G$, $B$, and $M$ can be derived, as shown in Table 6.1. That is, $[C]$ in Eq. 6.4 can also be expressed by any two parameters among $E$, $v$, $G$, $B$, and $M$. Thereby, the matrix $[C]$ in Eq. 6.4 is also often expressed in $G$ and $B$ as follows:

$$[C] = \begin{bmatrix} (3B+4G)/3 & (3B-2G)/3 & 0 \\ (3B-2G)/3 & (3B+4G)/3 & 0 \\ 0 & 0 & G \end{bmatrix} \tag{6.14}$$

**Table 6.1   Relations between elastic deformation parameters (After Chen and Saleeb, 1982)**

|        | $G$                   | $E$                       | $M$                          | $B$                    | $\lambda$                  | $v$                         |
|--------|-----------------------|---------------------------|------------------------------|------------------------|----------------------------|-----------------------------|
| $G,E$  | $B$                   | $E$                       | $\dfrac{G(4G-E)}{3G-E}$      | $\dfrac{GE}{9G-3E}$    | $\dfrac{G(E-2G)}{3G-E}$    | $\dfrac{E-2G}{2G}$          |
| $G,M$  | $G$                   | $\dfrac{G(3M-4G)}{M-G}$   | $M$                          | $M-\dfrac{4G}{3}$      | $M-2G$                     | $\dfrac{M-2G}{2(M-G)}$      |
| $G,B$  | $G$                   | $\dfrac{9GB}{3B+G}$       | $B+\dfrac{4G}{3}$            | $B$                    | $B-\dfrac{2G}{3}$          | $\dfrac{3B-2G}{2(3B+G)}$    |
| $G,\lambda$ | $G$              | $\dfrac{G(3\lambda+2G)}{\lambda+G}$ | $\lambda+2G$       | $\lambda+\dfrac{2G}{3}$ | $\lambda$                 | $\dfrac{\lambda}{2(\lambda+G)}$ |
| $G,v$  | $G$                   | $2G(1+v)$                 | $\dfrac{2G(1-v)}{1-2v}$      | $\dfrac{2G(1+v)}{3(1-2v)}$ | $\dfrac{2Gv}{1-2v}$    | $v$                         |
| $E,B$  | $\dfrac{3BE}{9B-E}$   | $E$                       | $\dfrac{B(9B+3E)}{9B-E}$     | $B$                    | $\dfrac{B(9B-3E)}{9B-E}$   | $\dfrac{3B-E}{6B}$          |
| $E,v$  | $\dfrac{E}{2(1+v)}$   | $E$                       | $\dfrac{E(1-v)}{(1+v)(1-2v)}$ | $\dfrac{E}{3(1-2v)}$  | $\dfrac{vE}{(1+v)(1-2v)}$  | $v$                         |
| $B,\lambda$ | $\dfrac{3(B-\lambda)}{2}$ | $\dfrac{9B(B-\lambda)}{3B-\lambda}$ | $3B-2\lambda$ | $B$               | $\lambda$                  | $\dfrac{\lambda}{3B-\lambda}$ |
| $B,M$  | $\dfrac{3(M-B)}{4}$   | $\dfrac{9B(M-B)}{3B+M}$   | $M$                          | $B$                    | $\dfrac{3B-M}{2}$          | $\dfrac{3B(2M-1)+M}{3B(2M+1)-M}$ |
| $B,v$  | $\dfrac{3B(1-2v)}{2(1+v)}$ | $3B(1-2v)$            | $\dfrac{3B(1-2v)}{1+v}$      | $B$                    | $\dfrac{3Bv}{1+v}$         | $v$                         |

Note: $\lambda$ is a Lame parameter. For its definition, please see books on theory of elasticity. It is, though, rarely used in geotechnical analyses.

In the plane strain excavation analysis, the commonly used elements can be categorized into the plane strain elements, the bar elements, and the beam elements, etc.

## 6.2.1   Plane strain elements

In terms of shape, plane strain elements are usually categorized into triangular elements and quadrilateral elements. As shown in Figure 6.4, the commonly used triangular elements are the constant strain triangular elements (CST elements), also called the T3 elements (Triangular with 3 nodes) where the strain variation is constant, and the linear strain elements (LST elements), also called the T6 elements (Triangular with 6 nodes) where the strain varies linearly. The commonly used quadrilateral elements are Q4 elements and Q8 elements. The former consist of 4 nodes for each and strains change linearly. They are thereby called low order elements. The latter have 8 nodes for each and the strain changes non-linearly. They thus belong to high order elements, as shown in Figure 6.5.

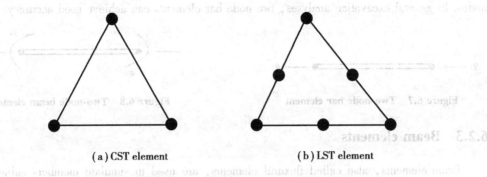

(a) CST element     (b) LST element

**Figure 6.4　Triangular elements**

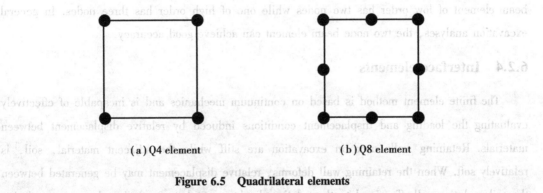

(a) Q4 element     (b) Q8 element

**Figure 6.5　Quadrilateral elements**

In general excavation analysis, the plane strain elements are used to simulate soils and structural materials. High order plane strain elements, such as a Q8 element, are more recommended, considering the accuracy of analysis. Usually, the accuracy of a Q8 element is better than that of four Q4 elements, as shown in Figure 6.6.

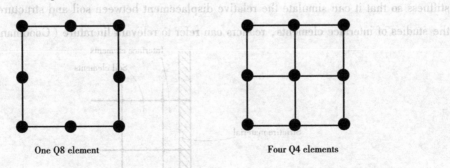

One Q8 element     Four Q4 elements

**Figure 6.6　Comparison of accuracy between a Q8 element and four Q4 elements**

## 6.2.2　Bar elements

Bar elements, also called truss elements, are used to simulate struts, anchors, or other members bearing only axial stress, as shown in Figure 6.7. Each node of a bar element has only one degree of freedom. A bar element of low order has two nodes while one of high order has three nodes. In general excavation analyses, two node bar elements can achieve good accuracy.

$q_1 \longrightarrow \bullet\!\!\!=\!\!\!=\!\!\!=\!\!\!\bullet \longrightarrow q_2$

$q_1 \longrightarrow \quad q_2 \quad q_3$

**Figure 6.7　Two-node bar element**　　　　　**Figure 6.8　Two-node beam element**

### 6.2.3　Beam elements

Beam elements, also called flexural elements, are used to simulate members subjected to moment, as shown in Figure 6.8. Each node of a beam element has two degrees of freedom. A beam element of low order has two nodes while one of high order has three nodes. In general excavation analyses, the two node beam element can achieve good accuracy.

### 6.2.4　Interface elements

The finite element method is based on continuum mechanics and is incapable of effectively evaluating the loading and displacement conditions induced by relative displacement between materials. Retaining walls used in excavation are stiff while the adjacent material, soil, is relatively soft. When the retaining wall deforms, relative displacement may be generated between the soil and the wall. To simulate the relative displacement between soil and structures during excavation, interface elements are sometimes used in analysis.

As shown in Figure 6.9, an interface element is an element connecting structures and soil, with or without thickness, which has a quite large normal stiffness but relatively small shear stiffness so that it can simulate the relative displacement between soil and structures. Concerning the studies of interface elements, readers can refer to relevant literature (Goodman et al., 1968;

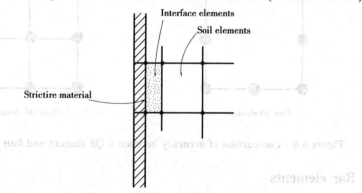

**Figure 6.9　Interface element**

Pande and Sharma, 1979; Sachdeva and Ramakrishnan, 1981; Desai and Nagaraj, 1988; Sharman and Desai, 1992).

Though interface elements can rationally simulate the relative displacement between soil and structures, since extra parameters, which are not easily obtained from conventional soil tests, will be introduced, and numerical instability during analysis often occurs, they have to be used especially carefully. If interface elements are not to be adopted, the soil in the vicinity of the structure can be considered to divide into fine elements. When the retaining wall is deformed, these fine soil elements can easily attain the plastic state, which will then produce larger deformation.

This section only succinctly elucidates the basic principles of the finite element method. As for other inferring processes and important related theories, please refer to related publications.

# 6.3 Determination of Initial Stresses

In the non-linear finite element analysis, the magnitude of the initial stresses is one of the crucial factors to the analysis result. The initial stresses have to be determined at the beginning of analysis. There are two methods for the computation of the initial stresses: the direct input method and the gravity generation method. They are introduced as follows:

## 6.3.1 Direct input method

As shown in Figure 6.10, compute the vertical stress ($\sigma_y$) and the horizontal stress ($\sigma_x$) at a point in the ground in a free field surface:

$$\sigma_y = \gamma_m h_1 + \gamma_{sat} h_2 \tag{6.15}$$
$$\sigma'_y = \sigma_y - u \tag{6.16}$$
$$\sigma'_x = K_0 \sigma'_y \tag{6.17}$$
$$\sigma_x = \sigma'_x + u \tag{6.18}$$

where

$\gamma_m$ and $\gamma_{sat}$ are the moisture unit weight and the saturated unit weight of soil, respectively. $u$ is the porewater pressure;

$K_0$ is the coefficient of at-rest earth pressure;

$h_1$ and $h_2$ are the depth of the groundwater level and the depth between the point and the groundwater level, respectively.

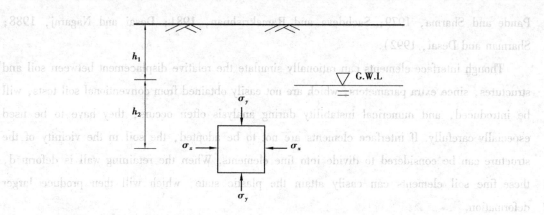

**Figure 6.10  Computation of the initial stresses**

As shown in Figure 6.11, the initial stresses in sloping ground with no groundwater is

$$\sigma'_y = \gamma_m h \tag{6.19}$$

$$\sigma'_x = K_0 \sigma'_y \tag{6.20}$$

$$\tau_{xy} = \frac{1}{2} \gamma_m h \sin \alpha \tag{6.21}$$

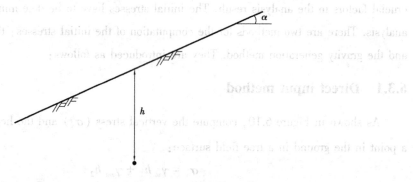

**Figure 6.11  The stresses in sloping ground**

If there is groundwater seepage, the porewater pressure can be estimated using the flow net method and then $\sigma'_y$ and $\sigma'_x$ can be determined accordingly. The shear stress $\tau_{xy} = 1/2\gamma'h \sin \alpha$.

## 6.3.2  Gravity generation method

Assume the boundaries of the excavation profile are all rollers ( this is a temporary assumption to compute the initial stresses. When starting analysis, it should be set to what it should be). Assign suitable $E$ and $v$ to each element. Considering the lateral strain of each element in the initial state is 0, according to the theories of elasticity, we can infer that $v=K_0/(1+K_0)$. $E$ can be an arbitrary large number. Then have the body forces act throughout the whole area and use the finite element method to solve the initial stresses for each element.

Considering that the values of $E$ and $v$ are not easily determined, a modification can be made by determining the stresses for each element with direct input method first and then by having the body forces act throughout all elements. The stresses of an element can be transformed into the equivalent internal nodal forces and the body force can be transformed into the equivalent external nodal force. The difference between the internal nodal force and the external nodal force can be expressed as follows:

$$\{R\} = \int_V [B]^T \{\sigma\} \, dV - \int_V [f]^T \{\gamma\} \, dV \qquad (6.22)$$

where

$\{R\}$ = residual force matrix;

$[B]$ = strain-displacement matrix;

$\{\sigma\}$ = initial stress matrix;

$[f]$ = displacement shape function matrix;

$\{\gamma\}$ = unit weight of soil.

The stresses of the elements are solved repeatedly with the residual forces acting at the nodes until the residual forces $\{R\}$ equal to 0. The stresses are then the initial stresses.

# 6.4 Modeling of an Excavation Process

The conception of simulating excavations using the finite element method can be illustrated by Figure 6.12. As shown in Figure 6.12(a), before excavation, the initial stresses of the soil are in equilibrium. Once excavation is started, the stresses on the excavation surface and the wall above the excavation surface shall become 0. Thus, an excavation simulation can be described as follows:

- Compute the initial stress state on the designed excavation surface [Figure 6.12(b)].
- According to the stress state obtained above, compute the equivalent nodal force ($P_{eq}$).
- Load forces of $-P_{eq}$ on the excavation surface.
- Reduce the stiffness of the soil that is to be excavated to 0 (or to a very low value).
- Compute the displacement, stress, and strain in the unexcavated area caused by $-P_{eq}$.

Following Brown and Booker's study (1985), the equivalent nodal force on the excavation surface in Step 2 can be computed as follows:

$$\{P_{eq}\} = \int_{V_i} [B]^T \{\sigma_{i-1}\} \, dV - \int_{V_i} [f]^T \{\gamma\} \, dV \qquad (6.23)$$

where

$P_{eq}$ = the equivalent nodal force on the excavation surface;

$i$ = the present excavation stage;

$i - 1$ = the previous excavation stage;

$V_i$ = the volume in the present excavation stage;

$\sigma_{i-1}$ = the stress in the previous excavation stage.

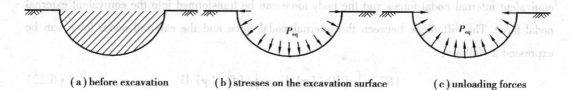

(a) before excavation     (b) stresses on the excavation surface     (c) unloading forces

**Figure 6.12  Simulation of excavation**

The first item in the above equation represents integration of the nodal forces derived from the stresses at the integration points in the previous excavation stage (the stresses before excavation) over the volume in the ith stage (the volume after excavation). The second item represents integration of the nodal forces derived from gravity over the volume in the ith stage (the volume after excavation). Subtract the latter from the former, and we can then obtain the equivalent nodal forces on the excavation surface. If dewatering is carried out in the excavation zone, due to the change of the water pressure before and after excavation, a stress increment $\{U\}$ will be generated. The stress increment can be transformed into the equivalent nodal force. Eq. 6.24 can then be rewritten into the following:

$$\{P_{eq}\} = \int_{V_i} [B]^T \{\sigma_{i-1}\} \, dV - \int_{V_i} [f]^T \{\gamma\} \, dV + \int_{V_i} [B]^{'} \{U\} \, dV \qquad (6.24)$$

where $U$ = change in the water pressure due to dewatering.

The equivalent nodal forces can thus be obtained from Eq. 6.24, The changes of stresses and strains for soil elements can be obtained by loading the equivalent nodal forces on the excavation area and perform finite element analysis.

# 6.5  Mesh Generation

The shape of the element used in an analysis can strongly affect the results obtained. The following are the commonly adopted rules for mesh generation:

## 6.5.1  Shape of the element

Elements used in finite element analysis should avoid irregular shapes. It is better to be as regular as possible, because elements in irregular shapes will cause numerical instability or inaccuracy of numerical analysis. Whether an element is in a good shape can be evaluated by its aspect ratio. The aspect ratio is the ratio of the length to the width of an element ($L/B$), as shown in Figure 6.13. The closer to 1.0 is the aspect ratio, the better is the shape. That is, the square or an equilateral triangle is the best choice. Elements with angles of 90° (quadrangles) or 60°

(triangles) are also good elements. Since neither squares nor equilateral triangles are easily found, elements with an aspect ratio within the range $1.0 \leqslant \dfrac{L}{B} \leqslant 2.0\text{-}2.5$ can be viewed as good ones.

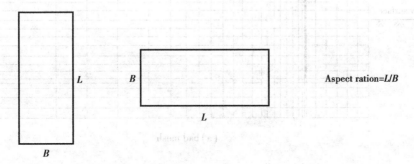

**Figure 6.13  Definition of the aspect ratio**

The shape of an element will influence the analytical accuracy of the element and the surrounding elements. It is therefore necessary to place good elements in crucial areas. In less crucial areas, some elements not so good can be placed. For example, if the retaining wall is an important object of analysis, good elements should be placed in its surroundings. On the other hand, if the boundary areas are not important areas, some elements not that good can be placed there.

## 6.5.2  Density of mesh

In principle, the mesh in the area of stress concentration, of rapid strain changing, the crucial areas, and the object zones should be finer. The retaining wall is a rigid structure and soil is comparatively a soft material. The mesh in the transition zone between the wall and the surrounding soils should, therefore, be as fine as possible, since a larger stress gradient will be generated there. The farther from the wall, the lower the density of mesh can be.

Since the unloading forces caused by excavation act directly on the excavation bottom in the excavation zone, the density of mesh in the excavation zone will greatly affect the analysis results. Thus, the density of mesh in the excavation zone should be as fine as possible, as shown in Figure 6.14 (Ou et al., 1996).

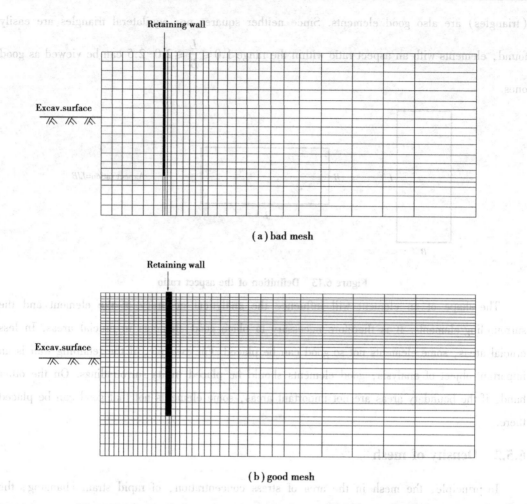

(a) bad mesh

(b) good mesh

**Figure 6.14   The finite element meshes used in the analysis of excavation**

## 6.5.3   Boundary conditions

If considering the symmetry of an excavation and taking a half for analysis as shown in Figure 6.1, the symmetric boundary [line $B'$-$B'$ in Figure 6.1(b)] should be equipped with rollers to restrain the lateral displacement and allow vertical displacement. According to Ou and Shiau (1998), to analyze movement in an excavation, rollers will be more efficient than hinges placed on line $C'$-$C'$ [Figure 6.1(b)]. Also, the rollers should be placed at a distance more than three excavation depths from the retaining wall for the analysis of wall deformation. To analyze ground settlement, the rollers should be placed at a distance of more than four excavation depths from the retaining wall. In principle, the farther the boundary is to the retaining wall, here designated as $C'$-$C'$, the better the analysis results are, though it takes more computation time. On the base, either rollers or hinges can be placed. In general, the hinges or rollers should be placed on hard soils or several meters below the retaining wall bottom.

The location of the boundary of mesh can also be determined from convergence study of the finite element analysis. As shown in Figure 6.1(b), assume the boundary is line $G'$-$G'$, and then carry out the analysis. Extend the boundary to line $H'$-$H'$ and perform the analysis again. If the two analyses come out similar in stress, strain, or displacement along, for example, $X'$-$X'$, it means that the boundary can be set at $G'$-$G'$ or $H'$-$H'$. Otherwise, the boundary should be moved to $C'$-$C'$.

# 6.6  Excavation Analysis Method

## 6.6.1  Total stress analysis and effective stress analysis

The total stress analysis generally refers to analyses where the total stress is used and the input parameters are total stress parameters such as $s_u$ and $\varphi = 0$ (for saturated clay), or $c_T$ and $\varphi_T$(for unsaturated clay). The stresses thus obtained are then the total stress. The effective stress analysis refers to the analyses where the effective stress is used and the input parameters are effective parameters such as $c'$ and $\varphi'$. The stresses obtained from the analysis are then the effective stress.

In the beam on elastic foundation method introduced in the previous chapter, the analyses can be split into drained analysis, undrained analysis, and partially drained analysis. The corresponding type of stress used in these methods is the effective stress, the total stress, and the total stress, respectively. Therefore, the undrained behavior can only be analyzed by the total stress method.

In the finite element method, drained analysis should be done with the effective stress while undrained analysis can be carried out with either the total stress or effective stress.

For example, if the hyperbolic model is expressed by the effective stress, $c'$ and $\varphi'$ being the input parameters, $\{\Delta\sigma\}$ in Eq. 6.1 will then imply the effective stress. The method is an effective stress analysis. If the hyperbolic model is expressed by the total stress, $\{\Delta\sigma\}$ in Eq. 6.1 will then represent the total stress because $E_t$ and $v$ in $[C]$ are obtained from the hyperbolic model expressed by the total stress. The method is a total stress analysis.

Similarly, if the yielding function in the elastoplastic model is expressed by the effective stress, $\{\Delta\sigma\}$ will then imply the effective stress. If the yielding function in the elastoplastic model is expressed by the total stress, $\{\Delta\sigma\}$ will then imply the total stress.

## 6.6.2  Drained analysis, undrained analysis, and partially drained analysis

Drained analysis refers to methods where the excess pore water pressure is assumed to be completely dissipated in analysis (i.e. $u_e = 0$) and the soil goes through a volume change accordingly. Drained analyses are mainly applied to granular soils and the long-term behavior of clayey soils. Thus, effective stress analysis and effective stress parameters should be adopted. Undrained analysis refers to methods where the excess pore water is not dissipated at all in analysis ($u_e \neq 0$). If the soil is in the saturated state, the volume change would be none.

In the hyperbolic model, the undrained analysis for short-term behavior of saturated clay has to adopt the total stress analysis. That is, the analysis should adopt $s_u$ and assume $\varphi = 0$. The undrained analysis of unsaturated clay directly adopts the results of the triaxial $UU$ test, $c_T$ and $\varphi_T$. For granular soils, it is only the effective stress analysis that has to be adopted. That is, $c'$ and $\varphi'$ are the required parameters.

In the elastoplastic model, the undrained analysis for short-term behavior of saturated clay can adopt either the effective stress analysis or the total stress analysis. If adopting the effective stress analysis, the yielding function has to be capable of representing the fact that the soil strength increases with the increase of the average effective stress ($p'$). The yielding function can be the Mohr-Coulomb model (input parameters $c'$ and $\varphi'$), the Drucker-Prager model, the Cam-clay model, etc. The input parameters are all the effective stress parameters and the stiffness of pore water also has to be fed. If adopting the total stress analysis, the yielding function should be capable of simulating the fact that the soil strength does not increase with the increase of the average total stress ($p$). The yielding function can be the Mohr-Coulomb model ($s_u$, $\varphi = 0$), the von Mises model, the Tresca model, etc. The input parameters are all the total stress parameters.

In some cases, the behaviors of clay are neither completely drained nor perfectly undrained. Rather, they are in between. They are partially drained behaviors. The analysis of the partially drained behavior of clay can be carried out through undrained analysis, which takes advantage of the known excess pore water at each stage. To obtain the pore water pressure at a certain stage, one of the methods is to install a piezometer to measure the pore water pressure, which may partially dissipate at a construction stage. Then use the equation, $\sigma' = \sigma - u$, to derive the effective stress of the soils. Given the effective stress of soils, according to the normalized behavior of soils ($s_u / \sigma'_v =$ constant), the undrained shear strength of the soils can then be obtained. Then carry out an undrained analysis of the excavation for each stage to obtain the partially drained behavior of soils. Another method is to obtain the undrained shear strengths for each stage by means of laboratory or in situ shear strength tests first, which are then used for undrained analysis to analyze the partially drained behavior.

### 6.6.3　Coupled analysis

In the method just mentioned, analysis is restricted to either drained or undrained analysis. However, real soil behavior is often time related, with porewater pressure response and rate of loading. To account for such behavior, it is necessary to couple the continuity of equation or general consolidation equation with the constitutive and equilibrium equations. The method is then called the coupled analysis, which is subsumed in effective stress analysis, that is, the effective stress parameters are required in analysis. In addition to the parameters of soil models, the coupled analysis also requires the coefficient of permeability and loading time. The coupled method uses displacement and porewater pressure as unknowns and therefore results in both displacement and porewater pressure degrees of freedom at element nodes.

As a result, it can compute the displacements and stresses of a soil element as well as the porewater pressures based on the effective stress. Figure 6.15 shows a commonly used element for coupled analysis. The element has eight displacement nodes, four of which are simultaneously porewater pressure nodes.

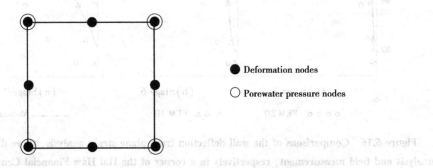

**Figure 6.15  A commonly used element for the coupled analysis**
**(eight deformation nodes and four porewater pressure nodes)**

If using the coupled method to analyze completely undrained behavior, we could simply set the time of loading within a very short time. For example, assume there is a site to be excavated 4 m deep. The coupled method simply sets the 4 m excavation to be finished within a very short time when considering the undrained condition. If considering the drained condition, we can set the excavation time for a very long period. When considering partially drained behavior, set the excavation time in between (the actual excavation time).

## 6.6.4  Planc strain analysis and three-dimensional analysis

Though in general engineering practice, plane strain analysis is capable of obtaining a rational result, the wall deflection and ground settlement on the section of the shorter side [B-B section in Figure 6.1(a)] or that near the corner are three-dimensional behaviors and plane strain analysis would overestimate the deformation or settlement. Three-dimensional analysis can solve the problems. Figure 6.16 shows wall deformation at the comer in the Haihaw Financial Center excavation from plane strain analysis, three-dimensional analysis and field measurement. As shown in the Figure, plane strain analysis will overestimate; whereas three-dimensional analysis will obtain rational results. Besides, the wall deformation and ground surface settlement in excavations with soil improvement or property protection measures such as counterfort walls or cross walls also constitute three-dimensional behavior. In such cases, only three-dimensional analysis can obtain rational results.

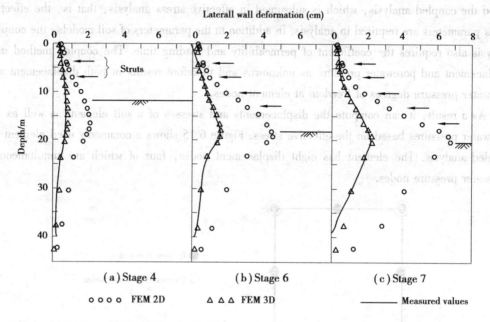

(a) Stage 4          (b) Stage 6          (c) Stage 7

○ ○ ○ ○  FEM 2D        △ △ △  FEM 3D        ——— Measured values

**Figure 6.16  Comparisons of the wall deflection from plane strain analysis, three dimensional analysis and field measurement, respectively in a corner of the Hai Haw Financial Center excavation**

In three dimensional analysis, the stress state at a point has six components: $\sigma_{xx}$, $\sigma_{yy}$, $\sigma_{zz}$, $\tau_{xy}$, $\tau_{yz}$, $\tau_{zx}$. The theories of finite element analysis, the soil models, and others are all similar to those for the plane strain analysis. In practice, the element of the retaining wall and soil used in the three-dimensional analysis is usually an 8-noded low order hexahedron element, which is called a $H8$ (8-noded hexahedron) element, the solid element, or brick element, as shown in Figure 6.17(a). A 20-node high order element [Figure 6.17(b)], though more accurate, is usually not adopted because it requires too much computer memory and is very time consuming in computation.

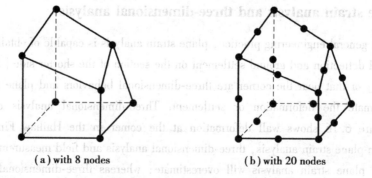

(a) with 8 nodes                    (b) with 20 nodes

**Figure 6.17  Three dimensional elements**

# 6.7 Example: Excavation in Sand

PLAXIS is a finite element package intended for the two-dimensional or three-dimensional analysis of deformation, stability and groundwater flow in geotechnical engineering. Geotechnical applications require advanced constitutive models for the simulation of the non-linear, time-dependent and anisotropic behaviour of soils and/or rock. In addition, since soil is a multi-phase material, special procedures are required to deal with pore pressures and (partial) saturation in the soil. Although the modelling of the soil itself is an important issue, many geotechnical projects involve the modelling of structures and the interaction between the structures and the soil. PLAXIS is equipped with features to deal with various aspects of complex geotechnical structures.

This part describes the construction of an excavation pit in soft clay and sand layers by using the software PLAXIS 3D. The pit is a relatively small excavation of 12 by 20 m, excavated to a depth of 6.5 m below the surface. Struts, waling and ground anchors are used to prevent the pit from collapsing. After the full excavation, an additional surface load is added on one side of the excavation.

The proposed geometry for this exercise is 80 m wide and 50 m long, as shown in Figure 6.18. The excavation pit is placed in the center of the geometry. Figure 6.19 shows a cross section of the excavation pit with the soil layers. The clay layer is considered to be impermeable.

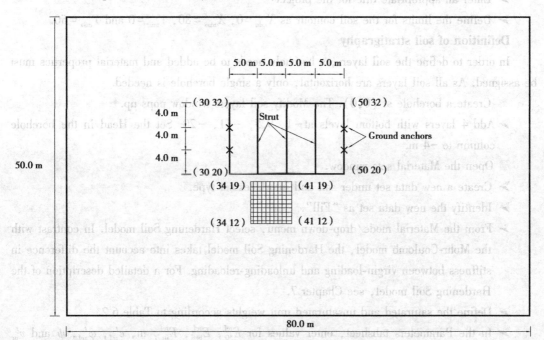

**Figure 6.18   Top view of the excavation pit**

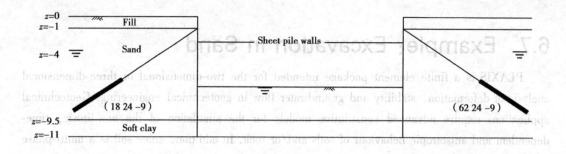

**Figure 6.19    Cross section of the excavation pit with the soil layers**

## 6.7.1    Geometry

To create the geometry model, follow these steps:

**Project properties**

➤ Start a new project

➤ Enter an appropriate title for the project

➤ Define the limits for the soil contour as $X_{min}=0$, $X_{max}=80$, $Y_{min}=0$ and $Y_{max}=50$.

**Definition of soil stratigraphy**

In order to define the soil layers, a borehole needs to be added and material properties must be assigned. As all soil layers are horizontal, only a single borehole is needed.

&#x2795; Create a borehole at $(0,0)$. The Modify soil layers window pops up.

➤ Add 4 layers with bottom levels at $-1$, $9.5$, $-11$, $-20$. Set the Head in the borehole column to $-4$ m.

&#x25a4; Open the Material sets window.

➤ Create a new data set under soil and interfaces set type.

➤ Identify the new data set as "Fill".

➤ From the Material mode/drop-down menu, select Hardening Soil model. In contrast with the Mohr-Coulomb model, the Hardening Soil model takes into account the difference in stiffness between virgin-loading and unloading-reloading. For a detailed description of the Hardening Soil model, see Chapter 7.

➤ Define the saturated and unsaturated unit weights according to Table 6.2.

➤ In the Parameters tabsheet, enter values for $E_{50}^{ref}$, $E_{oed}^{ref}$, $E_{ur}^{ref}$, m, $c'_{ref}$, $\varphi'_{ref}$, $\psi$ and $v'_{ur}$ according to Table 6.2. Note that Poisson's ratio is an advanced parameter.

➤ As no consolidation will be considered in this exercise, the permeability of the will not influence the results. Therefore, the default values can be kept in the parameters tabsheet.

**Table 6.2   Material properties for the soil layers**

| Parameter | Name | Fill | Sand | Soft clay | Unit |
|---|---|---|---|---|---|
| General | | | | | |
| Material model | Model | Hardening Soil model | Hardening Soil model | Hardening Soil model | — |
| Drainage type | Type | Drained | Drained | Undrained | — |
| Unit weight above phreatic level | $\gamma_{unsat}$ | 16.0 | 17.0 | 16.0 | kN/m$^3$ |
| Unit weight below phreatic level | $\gamma_{sat}$ | 20.0 | 20.0 | 17.0 | kN/m$^3$ |
| Parameters | | | | | |
| Secant stiffness for CD triaxial test | $E_{50}^{ref}$ | 2.2×10$^4$ | 4.3×10$^4$ | 2.0×10$^3$ | kN/m$^3$ |
| Tangent oedometer stiffness | $E_{oed}^{ref}$ | 2.2×10$^4$ | 2.2×10$^4$ | 2.0×10$^3$ | kN/m$^3$ |
| Unloading/reloading stiffness | $E_{ur}^{ref}$ | 6.6×10$^4$ | 1.29×10$^5$ | 1.0×10$^4$ | kN/m$^3$ |
| Power for stress level dependency of stiffness | m | 0.5 | 0.5 | 1.0 | — |
| Cohesion | $c_{ref}'$ | 1 | 1 | 5 | kN/m$^2$ |
| Friction angle | $\varphi'$ | 30.0 | 34.0 | 25.0 | ° |
| Dilatancy angle | $\psi$ | 0.0 | 4.0 | 0.0 | ° |
| Poisson's ratio | $v_{ur}'$ | 0.2 | 0.2 | 0.2 | — |
| Interfaces | | | | | |
| Interface strength | — | Manual | Manual | Manual | — |
| Interface reduction factor | $R_{inter}$ | 0.65 | 0.7 | 0.5 | — |
| Initial | | | | | |
| $K_0$ determination | — | Automatic | Automatic | Automatic | — |
| Lateral earth pressure coefficient | $K_0$ | 0.500 0 | 0.440 8 | 0.741 1 | — |
| Over-consolidation ratio | OCR | 1.0 | 1.0 | 1.5 | — |
| Pre-overburden pressure | POP | 0.0 | 0.0 | 0.0 | — |

> In the Interfaces tabsheet, select Manual in the Strength box and enter a value of 0.65 for the parameter Rinter. This parameter relates the strength of the interfaces to the strength of the soil, according to the equations:

$$c_i = R_{inter}\ c_{soil}\ \text{and}\ \tan \varphi_i = R_{inter}\ \tan \varphi_i \leqslant \tan \varphi_{soil}$$

Hence, using the entered $R_{inter}$-value gives a reduced interface friction and interface cohesion (adhesion) compared to the friction angle and the cohesion in the adjacent soil. When the Rigid option is selected in the Strength drop-down, the interface has the same strength properties as the soil ($R_{inter} = 1.0$). Note that a value of $R_{inter} < 1.0$, reduces the strength as well as the stiffness of the interface.

> In the Initial tabsheet, define the OCR-value according to Table 6.2.
> Click OK to close the window.
> In the same way, define the material properties of the "Sand" and "Soft" Clay materials as given by Table 6.2. The Tension cut-off option is activated by default at a value of 0 kN/m. This option is found in the Advanced options on the Parameters tabsheet of the soil window. Here the Tension cut-off value can be changed or the option can be deactivated entirely.
> After closing the Material sets window, click the OK button to close the Modify soil layers window.
> In the Soil mode right click on the upper soil layer. In the appearing right hand mouse button menu, select the Fill option in the Set material menu.
> In the same way, assign the Soft Clay material to the soil layer between $y = -9.5$ m and $y = -11.0$ m.
> Assign the Sand material to the remaining two soil layers.
> Proceed to the Structures mode to define the structural elements.

## 6.7.2  Definition of structural elements

The creation of sheet pile walls, waling, struts and surface loads and ground anchors is described below:

> Create a surface between (30 20 0), (30 32 0), (50 32 0) and (50 20 0).
> Extrude the surface to $z = -1$, $y = -6.5$ and $z = -11$.
> Right-click on the deepest created volume (between $z = 0$ and $z = -11$) and select the Decompose into surfaces option from the appearing menu.
> Delete the top surfaces (2 surfaces). An extra surface is created as the volume is decomposed.
> Hide the excavation volumes (do not delete). The eye button in the Model explorer and the Selection explorer trees can be used to hide parts of the model and simplify the view. A hidden project entity is indicated by a closed eye.
> Click the Create structure button.

✒ Create beams (waling) around the excavation circumference at level $z = -1$ m. Press the <Shift> key and keep it pressed while moving the mouse cursor in the z-direction. Stop moving the mouse as the z-coordinate of the mouse cursor is −1 in the cursor position indicator. Note that as you release the<Shift> key, the z-coordinate of the cursor location does not change. This is an indication that you can draw only on the xy-plane located at $z = -1$.

✒ Click on (30 20 −1), (30 32 −1), (50 32 −1), (50 20 −1), (30 20 −1) to draw the walings. Click on the right mouse button to stop drawing walings.

➤ Create a beam (strut) between (35 20 −1) and (35 32 −1). Press <Esc> to end defining the strut.

▣ Create data sets for the walings and strut according to Table 6.3 and assign the materials accordingly.

⦂⦂⦂ Copy the strut into a total of three struts at $x = 35$ (existing), $x = 40$, and $x = 45$.

Table 6.3  Material properties for the beams

| Parameters | Name | Strut | Waling | Unit |
|---|---|---|---|---|
| Material type | Type | Elastic | Elastic | |
| Young's modulus | $E$ | $2.1 \times 10^8$ | $2.1 \times 10^8$ | kN/m$^2$ |
| Unit weight | $\gamma$ | 78.5 | 78.5 | kN/m$^2$ |
| Cross section area | $A$ | 0.007 367 | 0.008 682 | m$^2$ |
| Moment of Inertia | $I_2$ | $5.073 \times 10^{-5}$ | $3.66 \times 10^{-4}$ | m$^4$ |
| | $I_3$ | $5.073 \times 10^{-5}$ | $1.045 \times 10^{-4}$ | m$^4$ |

**Modelling ground anchors**

In PLAXIS 3D ground anchors can be modelled using the Node-to-node anchor and the Embedded beam options as described in the following:

〰 First the ungrouted part of the anchor is created using the Node-to-node anchor feature. Start creating a node-to-node anchor by selecting the corresponding button in the options displayed as you click on the Create structure button.

➤ Click on the command line and type "30 24 −1 21 24 −7". Press <Enter> and <ESC> to create the ungrouted part of the first ground anchor.

➤ Create a node-to-node anchor between the points (50 24 −1) and (59 24 −7).

➤ The grouted part of the anchor is created using the Embedded beam option. Create embedded beams between (21 24 −7) and (18 24 −9) and between (59 24 −7) and (62 24 −9). Set the Behaviour to Grout body.

➤ Create a data set for the embedded beam and a data set for the node-to-node anchor according to Table 6.4 and Table 6.5 respectively. Assign the data sets to the node-to-node anchors and to the embedded beams.

**Table 6.4  Material properties for the node-to-node anchors**

| Parameter | Name | node-to-node anchor | Unit |
|---|---|---|---|
| Material type | Type | Elastic | |
| Axial stiffness | EA | $6.5 \times 10^5$ | kN |

**Table 6.5  Material properties for the embedded beams ( grout body)**

| Parameter | Name | Grout | Unit |
|---|---|---|---|
| Young's modulus | $E$ | $3 \times 10^7$ | kN/m$^2$ |
| Unit weight | $\gamma$ | 24 | kN/m$^3$ |
| Beam type | — | Predefined | — |
| Predefined beam type | — | Massive circular beam | — |
| Diameter | Diameter | 0.14 | m |
| Axial skin resistance | Type | Linear | — |
| Skin resistance at the top of the embedded beam | $T_{skin,start,max}$ | 200 | kN/m |
| Skin resistance at the bottom of the embedded beam | $T_{skin,end,max}$ | 0.0 | kN/m |
| Base resistance | $F_{max}$ | 0.0 | kN |

The remaining grouted anchors will be created by copying the defined grouted anchor.

➤ Click on the Select button and click on all the elements composing both of the ground anchors keeping the <Ctrl> key pressed.

⠿ Use the Create array function to copy both ground anchors ( 2 embedded beams + 2 node-to-node anchors) into a total of 4 complete ground anchors located at $y = 24$ and $y = 28$ by selection the 1D, in y direction option in the Shape drop-down menu and define the Distance between columns as 4 m.

➤ Mull-select all parts of the ground anchors ( 8 entities in total). While all parts are selected and the <Ctrl> key is pressed, click the right mouse button and select the Group from the appearing menu.

⊞ In the Model explorer tree, expand the Groups subtree by clicking on the ( + ) in front of the groups.

➤ Click the Group 1 and rename it to "Groundanchors".

To define the sheet pile walls and the corresponding interfaces, follow these steps:

➤ Select all four vertical surfaces created as the volume was decomposed. Keeping the <Ctrl> key pressed, click the right mouse button and select the Create plate option from the appearing menu.

▣ Create a data set for the sheet pile walls (plates) according to Table 6.6. Assign the data sets to the four walls.

As all the surfaces are selected, assign both positive and negative interfaces to them using the options in the right mouse button menu. The term positive and negative for interfaces has no physical meaning. It only enables distinguishing between interfaces on each side of a surface.

**Table 6.6  Material properties of the sheet pile walls**

| Parameter | Name | Sheet pile wall | Unit |
|---|---|---|---|
| Type of behavior | Type | Elastic non isotropic | |
| Thickness | $d$ | 0.379 | m |
| Weight | $\gamma$ | 2.55 | kN/m³ |
| Young's modulus | $E_1$ | $1.46 \times 10^7$ | kN/m² |
| | $E_2$ | $7.3 \times 10^5$ | kN/m² |
| Poisson's ratio | $v_{12}$ | 0.0 | — |
| Shear modulus | $G_{12}$ | $7.3 \times 10^5$ | kN/m² |
| | $G_{13}$ | $1.27 \times 10^6$ | kN/m² |
| | $G_{23}$ | $3.82 \times 10^5$ | kN/m² |

Non-isotropic (different stiffnesses in two directions) sheet pile walls are defined. The local axis should point in the correct direction (which defines as the 'stiff' or the 'soft' direction). As the vertical direction is generally the stiffest direction in sheet pile walls, local axis 1 shall point in the z-direction.

⊞ In the Model explorer tree expand the Surfaces subtree, set *Axisfunction* to *Manual* and set Axis1$_z$ to-1. Do this for all the pile wall surfaces. The first local axis is indicated by a red arrow; the second local axis is indicated by a green arrow and the third local axis is indicated by a blue arrow.

◲ Create a surface load defined by the points: (34 19 0), (41 19 0), (41 12 0), (34 12 0). The geometry is now completely defined.

## 6.7.3  Mesh generation

➢ Click on the Mesh tab to proceed to the Mesh mode.

➢ Select the surface representing the excavation, In the *Selection explorer* set the value of *Coarseness* factor to 0.25.

◔ Set the element distribution to *Coarse*, Uncheck the box for *Enhanced mesh refinements*. Generate the mesh.

💬 View the mesh. Hide the soil in the model to view the embedded beams.

➢ Click on the Close tab to close the Output program and go back to the Mesh mode of the Input program.

### 6.7.4  Performance calculations

The calculation consists of 6 phases. The initial phase consists of the generation of the initial stresses using the $K_0$ procedure. The next phase consists of the installation of the sheet piles and the first excavation. Then the waling and struts will be installed. In phase 3, the ground anchors will be activated and prestressed. Further excavation will be performed in the phase after that. The last phase will be the application of the additional load next to the pit.

➢ Click on the Staged construction tab to proceed with definition of the calculation phases.

➢ The initial phase has already been introduced. Keep its calculation type as $K_0$ procedure. Make sure all the soil volumes are active and all the structural elements are inactive.

➢ Add a new phase (phase-1). The default values of the parameters will be used for this calculation phase.

➢ Deactivate the first excavation volume (from $z=0$ to $z=-1$).

➢ In the *Model/explorer*, activate all plates and interfaces by clicking on the checkbox in front of them. The active elements in the project are indicated by a green check mark in the Model explorer.

🛎 Add a new phase (Phase-2). The default values of the parameters will be used for this calculation phase.

➢ In the *Model explorer* activate all the beams.

➢ Add a new phase (Phase-3). The default values of the parameters will be used for this calculation phase.

➢ In the *Mode/explorer* activate the *Groundanchors* group.

🔓 Select one of the node-to-node anchors.

⊞ In the Selection explorer expand the node-to node anchor features.

➢ Click on the *Adjust prestress* checkbox. Enter a prestress force of 200 kN (Figure 6.20).

➢ Do the same for all the other node-to-node anchors.

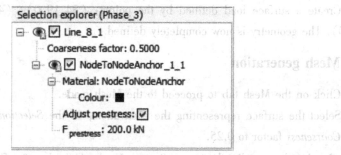

**Figure 6.20  Node-to-node anchor in the *Selection explorer***

✎ Add another phase (Phase-4). The default values of the parameters will be used for this calculation phase.

↳ Select the soil volume to be excavated in this phase (between $z=-1$ and $z=-6.5$).

⊞ In the Selection explorer under *waterconditions* feature, as shown in Figure 6.21, click on the Conditions and select the Dry option from the drop-down menu.

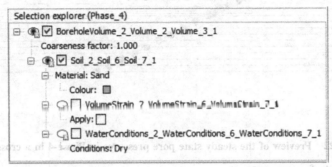

**Figure 6.21   Water conditions in the Selection explorer**

➤ Deactivate the volume to be excavated (between $z=-1$ and $z=-6.5$).

➤ Hide the soil and the plates around the excavation.

↳ Select the soil volume below the excavation (between $z=-6.5$ and $z=-9.5$).

⊞ In *Selection explorer* under *Waterconditions* feature.

➤ Click Conditions and select Head from the drop-down menu. Enter $z_{ref}=-6.5$ m.

↳ Select the soft clay volume below the excavation.

➤ Set the water conditions to *Interpolate*.

✎ Preview this calculation phase.

▦ Click the Vertical cross section button in the *Preview window* and define the cross section by drawing a line across the excavation.

➤ Select the $p_{steady}$ option from the *Stresses menu*.

▥ Display the contour lines for steady pore pressure distribution. Make sure that the Legend option is checked in View menu. The steady state pore pressure distribution is displayed in Figure 6.22. Scroll the wheel button of the mouse to zoom in or out to get a better view.

➤ Change the legend settings to:

Scaling: manual

Maximum value: 0

Number of intervals: 18

➤ Click on the Close button to return to the Input program.

✎ Add another phase (Phase-5). The default values of the parameters will be used for this calculation phase.

➤ Activate the surface load and set $\sigma_z=-20$ kN/m$^2$.

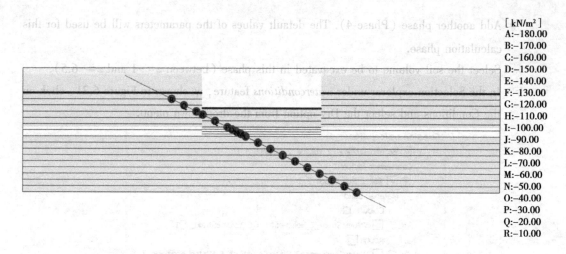

| [ kN/m² ] |
| A:–180.00 |
| B:–170.00 |
| C:–160.00 |
| D:–150.00 |
| E:–140.00 |
| F:–130.00 |
| G:–120.00 |
| H:–110.00 |
| I:–100.00 |
| J:–90.00 |
| K:–80.00 |
| L:–70.00 |
| M:–60.00 |
| N:–50.00 |
| O:–40.00 |
| P:–30.00 |
| Q:–20.00 |
| R:–10.00 |

**Figure 6.22  Preview of the steady state pore pressures in Phase-4 in a cross section**

### Defining points for curves

Before starting the calculation process, some stress points next to the excavation pit and loading are selected to plot a stress strain curve later on.

  ✓ Click the Select points for curves button. The model and Select points window will be displayed in the Output program.

  ➢ Define (37.5 19 −1.5) as Point-of-interest coordinates.

  ➢ Click the Search closest button. The number of the closest node and stress point will be displayed.

  ➢ Click the checkbox in front of the stress point to be selected. The selected stress point will be shown in the list.

  ➢ Select also stress points near the coordinates (37.5 19 −5), (37.5 19 −6) and (37.5 19 −7) and close the Select points window.

  ➢ Click the Update button to close the Output program.

  ⊠ Start the calculation process.

  ⊟ Save the project when the calculation is finished.

  ➢ To plot curves of structural forces, nodes can only be selected after the calculation.

  ➢ Nodes or stress points can be selected by just clicking them. When moving the mouse, the exact coordinates of the position are given in the cursor location indicator bar at the bottom of the window.

## 6.7.5  Viewing the results

After the calculations, the results of the excavation can be viewed by selecting a calculation phase from the Phases tree and pressing the View calculation results button.

  ⊠ Select the final calculation phase (Phase-5) and click the View calculation results button.

  The Output program will open and show the deformed mesh at the end of the last phase.

➤ The stresses, deformations and three dimensional geometry can be viewed by selecting the desired output from the corresponding menus. For example, choose Plastic points from the Stresses menu to investigate the plastic points in the model.

➤ In the Plastic points window, Figure 6.23, select all the options except the Elastic points and the Show only inaccurate points options. Figure 6.24 shows the plastic points generated in the model at the end of the final calculation phase.

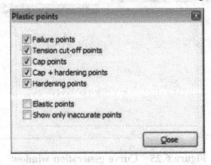

**Figure 6.23　Plastic points window**

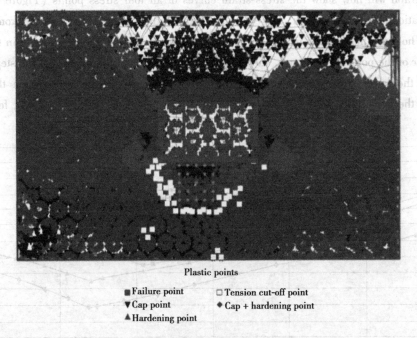

**Figure 6.24　Plastic points at the end of the final phase**

The graph of Figure 6.25 will now show the major principal strain against the major principal stress. Both values are zero at the beginning of the initial conditions. After generation of the initial conditions, the principal strain is still zero; whereas the principal stress is not zero anymore. To plot the curves of all selected stress points in one graph, follow these steps:

- Select Add curve—From current project from right mouse button menu.
- Generate curves for the three remaining stress nodes in the same way.

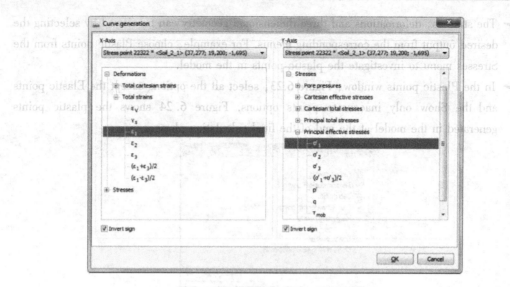

**Figure 6.25   Curve generation window**

The graph will now show the stress-strain curves of all four stress points (Figure 6.26). To see information about the markers, make sure the Value indication option is selected from the View menu and hold the mouse on a marker for a while. Information about the coordinates in the graph, the number of the point in the graph, the number of the phase and the number of the step is given. Especially the lower stress points show a considerable increase in the stress when the load is applied in the last phase. To create a stress path plot for stress node (37.5 19 −1.5), follow these steps:

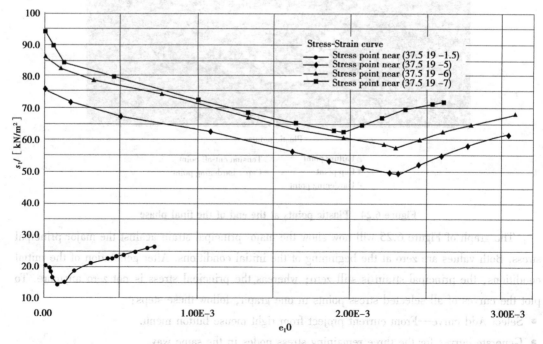

**Figure 6.26   Stress-strain curve**

- Create a new chart.
- In the Curves generation window, select node (37. 5 19 −1.5) from the drop-down menu of the $x$-axis of the graph and $\sigma'_{yy}$ under Cartesian effective stresses.
- Select node (37. 5 19 −1.5) from the drop-down menu of the $y$-axis of the graph. Select $\sigma'_{zz}$ under Cartesian effective stresses.
- Click OK to confirm the input (Figure 6.27).

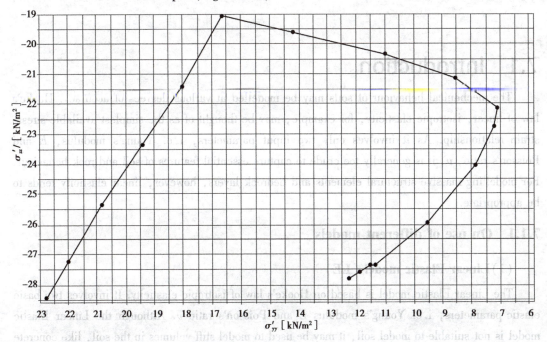

**Figure 6.27  Vertical effective stress($\sigma'_{zz}$) versus horizontal effective stress ($\sigma'_{yy}$) at stress node located near (37.5 19 −1.5)**

# 7  Soil Constitutive Models

## 7.1  Introduction

The mechanical behaviour of soils may be modelled at various degrees of accuracy. Hooke's law of linear, isotropic elasticity, for example, may be thought of as the simplest available stress-strain relationship. As it involves only two input parameters, i. e. Young's modulus, $E$ and Poisson's ratio, $\nu$, it is generally too crude to capture essential features of soil and rock behaviour. For modelling massive structural elements and bedrock layers, however, linear elasticity tends to be appropriate.

### 7.1.1  On use of different models

#### (1) Linear Elastic model (LE)

The Linear Elastic model is based on Hooke's law of isotropic elasticity. It involves two basic elastic parameters, i. e. Young's modulus $E$ and Poisson's ratio $\nu$. Although the Linear Elastic model is not suitable to model soil, it may be used to model stiff volumes in the soil, like concrete walls, or intact rock formations.

#### (2) Mohr-Coulomb model (MC)

The linear elastic perfectly-plastic Mohr-Coulomb model involves five input parameters, i.e. $E$ and $\nu$ for soil elasticity; $\phi$ and $c$ for soil plasticity and $\psi$ as an angle of dilatancy. This Mohr-Coulomb model represents a 'first-order' approximation of soil or rock behaviour. It is recommended to use this model for a first analysis of the problem considered. For each layer, one estimates a constant average stiffness or a stiffness that increases linearly with depth. Due to this constant stiffness, computations tend to be relatively fast and one obtains a first estimate of deformations.

#### (3) Hardening Soil model (HS)

The Hardening Soil model is an advanced model for the simulation of soil behaviour. As for the Mohr-Coulomb model, limiting states of stress are described by means of the friction angle, $\phi$; the cohesion, $c$; and the dilatancy angle, $\psi$. However, soil stiffness is described much more accurately by using three different input stiffnesses: the triaxial loading stiffness, $E_{50}$, the triaxial unloading stiffness, $E_{ur}$, and the oedometer loading stiffness, $E_{oed}$. As average values for various

· 130 ·

soil types, $E_{ur} \approx 3E_{50}$ and $E_{oed} \approx E_{50}$ are suggested as default settings, but both very soft and very stiff soils tend to give other ratios of $E_{oed}/E_{50}$, which can be entered by the user. In contrast to the Mohr-Coulomb model, the Hardening Soil model also accounts for stress-dependency of stiffness moduli. This means that all stiffnesses increase with pressure. Hence, all three input stiffnesses relate to a reference stress, usually taken as 100 kPa ( 1 bar ). Besides the model parameters mentioned above, initial soil conditions, such as pre-consolidation, play an essential role in most soil deformation problems. This can be taken into account in the initial stress generation.

### (4) Hardening Soil model with small-strain stiffness ( HSsmall )

The Hardening Soil model with small-strain stiffness ( HSsmall ) is a modification of the above Hardening Soil model that accounts for the increased stiffness of soils at small strains. At low strain levels, most soils exhibit a higher stiffness than at engineering strain levels, and this stiffness varies non-linearly with strain. This behaviour is described in the HSsmall model by using an additional strain-history parameter and two additional material parameters, i.e. $G_{ref0}$ and $\gamma_{0.7}$. $G_{ref0}$ is the small-strain shear modulus and $\gamma_{0.7}$ is the strain level at which the shear modulus has reduced to about 70% of the small-strain shear modulus. The advanced features of the HSsmall model are most apparent in working load conditions. Here, the model gives more reliable displacements than the HSmodel. When used in dynamic applications, the Hardening Soil model with small-strain stiffness also introduces hysteretic material damping.

### (5) Soft Soil model ( SS )

The Soft Soil model is a Cam-Clay type model especially meant for primary compression of near normally-consolidated clay-type soils. Although the modelling capabilities of this model are generally superseded by the Hardening Soil model, the Soft Soil model is better capable to model the compression behaviour of very soft soils.

### (6) Soft Soil Creep model ( SSC )

The Hardening Soil model is generally suitable for all soils, but it does not account for viscous effects, i. e. creep and stress relaxation. In fact, all soils exhibit some creep and primary compression and is thus followed by a certain amount of secondary compression. The latter is most dominant in soft soils, i. e. normally consolidated clays, silts and peat, and PLAXIS thus implemented a model under the name Soft Soil Creep model. The Soft Soil Creep model has been developed primarily for application to settlement problems of foundations, embankments, etc. For unloading problems, as normally encountered in tunnels and other excavation problems, the Soft Soil Creep model hardly supersedes the simple Mohr-Coulomb model. As for the Hardening Soil model, proper initial soil conditions are also essential when using the Soft Soil Creep model. This also includes data on the pre-consolidation stress, as the model accounts for the effect of over-consolidation. Note that the initial over-consolidation ratio also determines the initial creep rate.

### (7) Modified Cam-Clay model ( MCC )

The Modified Cam-Clay model is a well-known model from international soil modelling

literature; see for example Muir Wood (1990). It is meant primarily for the modelling of near normally-consolidated clay-type soils. This model has been added to PLAXIS to allow for a comparison with other codes.

The Mohr-Coulomb model may be used for a relatively quick and simple first analysis of the problem considered. In many cases, even if good data on dominant soil layers is limited, it is recommended to use the Hardening Soil model or the HS small model in an additional analysis. No doubt, one seldomly has test results from both triaxial and oedometer tests, but good quality data from one type of test can be supplemented by data from correlations and/or in situ testing. Finally, a Soft Soil Creep analysis can be performed to estimate creep, i.e. secondary compression in very soft soils.

### 7.1.2 Limitations

The soil models can be regarded as a qualitative representation of soil behaviour whereas the model parameters are used to quantify the soil characteristics. However, when someone doing numerical analysis, the simulation of reality remains an approximation, which implicitly involves some inevitable numerical and modelling errors. Moreover, the accuracy at which reality is approximated depends highly on the expertise of the user regarding the modelling of the problem, the understanding of the soil models and their limitations, the selection of model parameters, and the ability to judge the reliability of the computational results. Some of the limitations in the currently available models are listed below:

#### (1) Linear Elastic model

Soil behaviour is highly non-linear and irreversible. The linear elastic model is sufficient to capture the essential features of soil. The use of the linear elastic model may, however, be considered to model strong massive structures in the soil or bedrock layers. Stress states in the linear elastic model are not limited in any way, which means that the model shows infinite strength. Be careful using this model for materials that are loaded up to their material strength.

#### (2) Mohr-Coulomb model

The linear elastic perfectly-plastic Mohr-Coulomb model is a first order model that includes only a limited number of features that soil behaviour shows in reality. Although the increase of stiffness with depth can be taken into account, the Mohr-Coulomb model includes neither stress-dependency nor stress-path dependency nor strain dependency of stiffness or anisotropic stiffness. In general, effective stress states at failure are quite well described by using the Mohr-Coulomb failure criterion with effective strength parameters $\phi'$ and $c'$. For undrained materials, the Mohr-Coulomb model may be used with the friction angle $\phi$ set to 0° and the cohesion $c$ set to $c_u(s_u)$, to enable adirect control of undrained shear strength. In that case, note that the model does not automatically include the increase of shear strength with consolidation.

#### (3) Hardening Soil model

Although the Hardening Soil model can be regarded as an advanced soil model, there are a number of features of real soil behaviour the model does not include. It is a hardening model that does not account for softening due to soil dilatancy and de-bondingeffects. In fact, it is an isotropic hardening model so that it models neither hysteretic and cyclic loading nor cyclic mobility. Moreover, the model does not distinguish between large stiffness at small strains and reduced stiffness at engineering strain levels. The user has to select the stiffness parameters in accordance with the dominant strain levels in the application. Last but not least, the use of the Hardening Soil model generally results in longer calculation times, since the material stiffness matrix is formed and decomposed in each calculation step.

### (4) Hardening Soil model with small-strain stiffness

As the Hardening Soil model with small-strain stiffness (HSsmall) incorporates the loading history of the soil and a strain-dependent stiffness, it can, to some extent, be used to model cyclic loading. However, it does not incorporate a gradual softening during cyclic loading, so is not suitable for cyclic loading problems in which softening plays a role. In fact, just as in the Hardening Soil model, softening due to soil dilatancy and debonding effects are not taken into account. Moreover, the HSsmall does not incorporate the accumulation of irreversible volumetric straining nor liquefaction behaviour with cyclic loading. The use of the HSsmall will generally result in calculation times that are even longer than those of the Hardening Soil model.

### (5) Soft Soil model

The same limitations (including the ones for the Soft Soil Creep model) hold in the Soft Soil model. The utilization of the Soft Soil model should be limited to the situations that are dominated by compression. It is not recommended for use in excavation problems, since the model hardly supercedes the Mohr-Coulomb model in unloading problems.

### (6) Soft Soil Creep model

All above limitations also hold true for the Soft Soil Creep model. In addition this model tends to over-predict the range of elastic soil behaviour. This is especially the case for excavation problems, including tunnelling. Care must also be taken with the generation of initial stresses for normally consolidated soils. Although it would seem logical to use OCR = 1.0 for normally consolidated soils, such use would generally lead to an over-prediction of deformations in problems where the stress level is dominated by the initial self-weight stresses. Therefore, for such problems it is recommended to use a slightly increased OCR-value to generate the initial stress state. In fact, in reality most soils tend to show as lightly increased pre-consolidation stress in comparison with the initial effective stress. Before starting an analysis with external loading, it is suggested to perform a single calculation phase with a short time interval and without loading to verify the surface settlement rate based on common practice.

### (7) Modified Cam-Clay model

The same limitations (including those in the Soft Soil Creep model) hold in the Modified Cam-Clay model. Moreover, the Modified Cam-Clay model may allow for unrealistically high shear stresses. This is particularly the case for overconsolidated stress states where the stress path crosses the critical state line. Furthermore, the Modified Cam-Clay model may give softening behaviour for such stress paths. Without special regularization techniques, softening behaviour may lead to mesh dependency and convergence problems of iterative procedures. Moreover, the Modified Cam-Clay model cannot be used in combination with Safety analysis by means of c-phi reduction.

# 7.2 Linear Elastic Perfectly Plastic Model ( Mohr-Coulomb Model)

Soils behave rather non-linearly when subjected to changes of stress or strain. In reality, the stiffness of soil depends at least on the stress level, the stress path and the strain level. Some such features are included in the advanced soil models. The Mohr-Coulomb model, however, is a simple and well-known linear elastic perfectly plastic model, which can be used as a first approximation of soil behavior. The linear elastic part of the Mohr-Coulomb model is based on Hooke's law of isotropic elasticity. The perfectly plastic part is based on the Mohr-Coulomb failure criterion, formulated in a non-associated plasticity framework.

Plasticity involves the development of irreversible strains. In order to evaluate whether or not plasticity occurs in a calculation, a yield function, $f$ is introduced as a function of stress and strain. Plastic yielding is related with the condition $f=0$. This condition can often be presented as a surface in principal stress space. A perfectly-plastic model is a constitutive model with a fixed yield surface, i.e. a yield surface that is fully defined by model parameters and not affected by (plastic) straining. For stress states represented by points within the yield surface, the behavior is purely elastic and all strains are reversible.

## 7.2.1 Formulation of the Mohr-Coulomb model

The Mohr-Coulomb yield condition is an extension of Coulomb's friction law to general states of stress. In fact, this condition ensures that Coulombs friction law be obeyed in any plane within a material element. The full Mohr-Coulomb yield condition consists of six yield functions when formulated in terms of principal stresses (see for instance Smith Griffiths, 1982):

$$f_{1a} = \frac{1}{2}(\sigma_2' - \sigma_3') + \frac{1}{2}(\sigma_2' + \sigma_3')\sin\varphi - c\cos\varphi \leqslant 0 \quad (7.1)$$

$$f_{1b} = \frac{1}{2}(\sigma_3' - \sigma_2') + \frac{1}{2}(\sigma_3' + \sigma_2')\sin\varphi - c\cos\varphi \leqslant 0 \quad (7.2)$$

$$f_{2a} = \frac{1}{2}(\sigma_3' - \sigma_1') + \frac{1}{2}(\sigma_3' + \sigma_1')\sin\varphi - c\cos\varphi \leqslant 0 \quad (7.3)$$

$$f_{2b} = \frac{1}{2}(\sigma_1' - \sigma_3') + \frac{1}{2}(\sigma_1' + \sigma_3')\sin\varphi - c\cos\varphi \leqslant 0 \quad (7.4)$$

$$f_{3a} = \frac{1}{2}(\sigma_1' - \sigma_2') + \frac{1}{2}(\sigma_1' + \sigma_2')\sin\varphi - c\cos\varphi \leqslant 0 \quad (7.5)$$

$$f_{3b} = \frac{1}{2}(\sigma_2' - \sigma_1') + \frac{1}{2}(\sigma_2' + \sigma_1') \sin\varphi - c \cos\varphi \leqslant 0 \qquad (7.6)$$

The two plastic model parameters appearing in the yield functions are the well-known friction angle $\gamma$ and the cohesion $c$. The condition $f=0$ for all yield functions together (where $f_i$ is used to denote each individual yield function) represents a fixed hexagonal cone in principal stress space as shown in Figure 7.1.

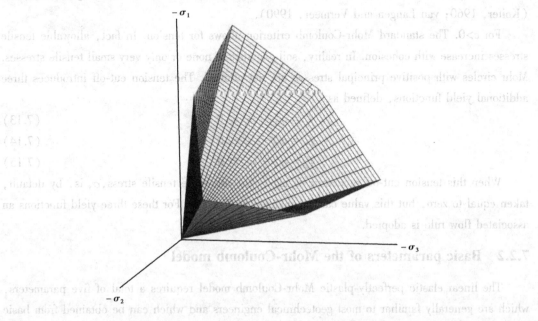

**Figure 7.1  The Mohr-Coulomb yield surface in principal stress space $(c=0)$**

In addition to the yield functions, six plastic potential functions are defined for the Mohr-Coulomb model:

$$g_{1a} = \frac{1}{2}(\sigma_2' - \sigma_3') + \frac{1}{2}(\sigma_2' + \sigma_3') \sin\varphi \qquad (7.7)$$

$$g_{1b} = \frac{1}{2}(\sigma_3' - \sigma_2') + \frac{1}{2}(\sigma_3' + \sigma_2') \sin\varphi \qquad (7.8)$$

$$g_{2a} = \frac{1}{2}(\sigma_3' - \sigma_1') + \frac{1}{2}(\sigma_3' + \sigma_1') \sin\varphi \qquad (7.9)$$

$$g_{2b} = \frac{1}{2}(\sigma_1' - \sigma_3') + \frac{1}{2}(\sigma_1' + \sigma_3') \sin\varphi \qquad (7.10)$$

$$g_{3a} = \frac{1}{2}(\sigma_1' - \sigma_2') + \frac{1}{2}(\sigma_1' + \sigma_2') \sin\varphi \qquad (7.11)$$

$$g_{3b} = \frac{1}{2}(\sigma_2' - \sigma_1') + \frac{1}{2}(\sigma_2' + \sigma_1') \sin\varphi \qquad (7.12)$$

The plastic potential functions contain a third plasticity parameter, the dilatancy angle $\psi$. This parameter is required to model positive plastic volumetric strain increments (dilatancy) as actually observed for dense soils. A discussion of all of the model parameters used in the Mohr-Coulomb model is given in the next section.

When implementing the Mohr-Coulomb model for general stress states, special treatment is required for the intersection of two yield surfaces. Some programs use a smooth transition from one yield surface to another, i.e. the rounding-off of the corners (see for example Smith Griffiths, 1982). For a detailed description of the corner treatment, readers can refer to relevant literature (Koiter, 1960; van Langen and Vermeer, 1990).

For $c>0$. The standard Mohr-Coulomb criterion allows for tension. In fact, allowable tensile stresses increase with cohesion. In reality, soil can sustain none or only very small tensile stresses. Mohr circles with positive principal stresses are not allowed. The tension cut-off introduces three additional yield functions, defined as

$$f_4 = \sigma'_1 - \sigma_t \leqslant 0 \qquad (7.13)$$

$$f_5 = \sigma'_2 - \sigma_t \leqslant 0 \qquad (7.14)$$

$$f_6 = \sigma'_3 - \sigma_t \leqslant 0 \qquad (7.15)$$

When this tension cut-off procedure is used, the allowable tensile stress, $\sigma_t$ is, by default, taken equal to zero, but this value can be changed by the user. For these three yield functions an associated flow rule is adopted.

## 7.2.2  Basic parameters of the Mohr-Coulomb model

The linear elastic perfectly-plastic Mohr-Coulomb model requires a total of five parameters, which are generally familiar to most geotechnical engineers and which can be obtained from basic tests on soil samples. These parameters with their standard units are listed below:

$E$: Young's modulus            $[kN/m]$

$v$: Poisson's ratio            $[-]$

$c$: Cohesion            $[kN/m^2]$

$\varphi$: Friction angle            $[°]$

$\psi$: Dilatancy angle            $[°]$

$\sigma_t$: Tension cut-off and tensile strength $[kN/m^2]$

Instead of using the Young's modulus as a stiffness parameter, alternative stiffness parameters can be entered. These parameters with their standard units are listed below:

$G$: Shear modulus            $[°]$

$E_{oed}$: Oedometer modulus            $[kN/m^2]$

Parameters can either be effective parameters [indicated by a prime sign (')] or undrained parameters (indicated by a subscript $u$), depending on the selected drainage type.

In the case of dynamic applications, alternative and/or additional parameters may be used to define stiffness based on wave velocities. These parameters are listed below:

$V_p$: Compression wave velocity            $[m/s]$

$V_s$: Shear wave velocity            $[m/s]$

## (1)Young's modulus ($E$)

A stiffness modulus has the dimension of stress. The values of the stiffness parameter adopted in a calculation require special attention as many geomaterials show a non-linear behavior from the very beginning of loading. In triaxial testing of soil samples, the initial slope of the stress-strain curve (tangent modulus) is usually indicated as $E_0$ and the secant modulus at 50% strength is denoted as $E_{50}$(see Figure 7.2). For materials with a large linear elastic range, it is realistic to use $E_0$; but for loading of soils, one generally uses $E_{50}$. Considering unloading problems, as in the case of tunneling and excavations, one needs an unload-reload modulus ($E_{ur}$) instead of $E_{50}$.

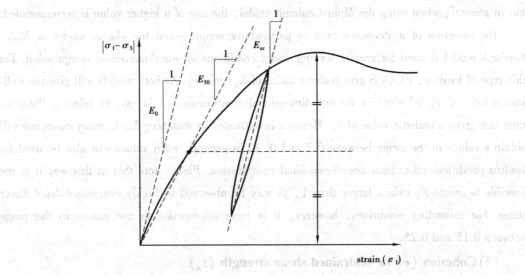

**Figure 7.2 Definition of $E_0$, $E_{50}$, $E_{ur}$ for drained triaxial test results**

For soils, both the unloading modulus, $E_{ur}$ and the first loading modulus, $E_{50}$ tend to increase with the confining pressure. Hence, deep soil layers tend to have greater stiffness than shallow layers. Moreover, the observed stiffness depends on the stress path that is followed. The stiffness is much higher for unloading and reloading than for primary loading. Also, the observed soil stiffness in terms of a Young's modulus may be lower for (drained) compression than for shearing. Hence, when using a constant stiffness modulus to represent soil behavior, one should choose a value that is consistent with the stress level and the stress path development.

In real soils, the stiffness depends significantly on the stress level, which means that the stiffness generally increases with depth. When using the Mohr-Coulomb model, the stiffness is a constant value. In order to account for the increase of the stiffness with depth, the $E_{inc}$-value may be used, which is the increase of the Young's modulus per unit of depth (expressed in the unit of stress per unit depth). At the level given by the ref parameter, and above, the stiffness is equal to the reference Young's modulus, $E_{ref}$, as entered in the Parameters tabsheet. Below, the stiffness is given by

$$E(y) = E_{ref} + (y_{ref} - y) E_{inc} \qquad y < y_{ref} \qquad (7.16)$$

where $y$ represents the vertical direction. The actual value of Young's modulus in the stress points is obtained from the reference value and $E_{inc}$. Note that during calculations, a stiffness increasing with depth does not change as a function of the stress state.

### (2) Poisson's ratio $v$

Standard drained triaxial tests may yield a significant rate of volume decrease at the very beginning of axial loading and, consequently, a low initial value of Poisson's ratio ($v_0$). For some cases, such as particular unloading problems, it may be realistic to use such a low initial value, but in general, when using the Mohr-Coulomb model, the use of a higher value is recommended.

The selection of a Poisson's ratio is particularly simple when the elastic model or Mohr-Coulomb model is used for gravity loading under conditions of one-dimensional compression. For this type of loading, PLAXIS give realistic ratios of $K_0 = \sigma'_h / \sigma'_v$. As both models will give the well-known ratio of $\sigma'_h / \sigma'_v = v/1-v$ for one-dimensional compression, it is easy to select a Poisson's ratio that gives a realistic value of $K_0$. Hence $v$ is evaluated by matching $K_0$. In many cases one will obtain $v$ values in the range between 0.3 and 0.4. In general, such values can also be used for loading conditions other than one-dimensional compression. Please note that in this way it is not possible to create $K_0$ values larger than 1, as may be observed in highly overconsolidated stress states. For unloading conditions, however, it is more appropriate to use values in the range between 0.15 and 0.25.

### (3) Cohesion ($c$) or undrained shear strength ($s_u$)

The cohesive strength has the dimension of stress. In the Mohr-Coulomb model, the cohesion parameter may be used to model the effective cohesion $c$ of the soil (cohesion intercept), in combination with a realistic effective friction angle $\varphi'$ [see Figure 7.3(a)]. This may not only be done for drained soil behavior, but also if the type of material behavior is set to Undrained(A).

Alternatively, the cohesion parameter may be used to model the undrained shear strength $s_u$ of the soil, in combination with $p = u = 0$ when the Drainage type is set to Undrained (B) or Undrained (C). In that case the Mohr-Coulomb failure criterion reduces to the well-known Tresca criterion.

The disadvantage of using effective strength parameters $c'$ and $\varphi'$ in combination with the drainage type being set to Undrained (A) is that the undrained shear strength as obtained from the model may deviate from the undrained shear strength in reality because of differences in the actual stress path being followed. In this respect, advanced soil models generally perform better than the Mohr-Coulomb model, but in all cases it is recommended to compare the resulting stress state in all calculation phases with the present shear strength in reality ($|\sigma_1 - \sigma_3| \leqslant 2s_u$).

On the other hand, the advantage of using effective strength parameters is that the change in shear strength with consolidation is obtained automatically, although it is still recommended to check the resulting stress state after consolidation. The advantage of using the cohesion parameter to model undrained shear strength in combination with $\gamma = 0$ (Undrained (B) or Undrained (C)) is that the user has direct control over the shear strength, independent of the actual stress state and stress path followed. Please note that this option may not be appropriate when using advanced soil models.

## (4) Friction angle ($\varphi$)

The friction angle $\varphi$ is entered in degrees. In general, the friction angle is used to model the effective friction of the soil, in combination with an effective cohesion $c$ [Figure 7.3(a)]. This may not only be done for drained soil behavior, but also if the type of material behavior is set to Undrained (A). Alternatively, the soil strength is modelled by setting the cohesion parameter equal to the undrained shear strength of the soil, in combination with $\varphi = 0$ (Undrained (B) or Undrained (C)) [Figure 7.3(b)]. In that case the Mohr-Coulomb failure criterion reduces to the well-known Tresca criterion.

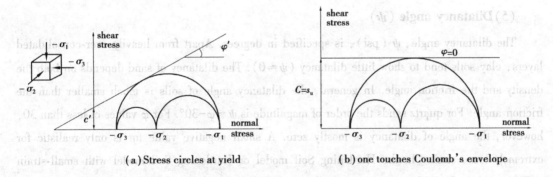

(a) Stress circles at yield          (b) one touches Coulomb's envelope

**Figure 7.3   Failure criterion and the friction angle**

High friction angles, as sometimes obtained for dense sands, will substantially increase plastic computational effort. Moreover, high friction may be subjected to strain-softening behavior, which means that such high friction angles are not sustainable under (large) deformation. Hence, high friction angles should be avoided when performing preliminary computations for a particular "project". The friction angle largely determines the shear strength as shown in Figure 7.3 by means of Mohrs stress circles. A more general representation of the yield criterion is shown in Figure 7.4. The Mohr-Coulomb failure criterion proves to be better for describing soil strength for general stress states than the Drucker-prager approximation. Typical effective friction angles are in the order of 20-30 degrees for clay and silt (the more plastic the clay, the lower the friction), and 30-40 degrees for sand and gravel (the denser the sand, the higher the friction).

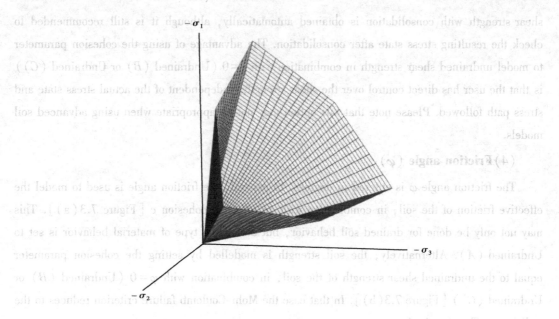

**Figure 7.4   Failure surface in principal stress space for cohesionless soil**

## (5) Dilatancy angle ($\psi$)

The dilatancy angle, $\psi$ (psi), is specified in degrees. Apart from heavily over-consolidated layers, clay soils tend to show little dilatancy ($\psi \approx 0$). The dilatancy of sand depends on both the density and the friction angle. In general, the dilatancy angle of soils is much smaller than the friction angle. For quartz sands the order of magnitude is $\psi \approx \varphi - 30°$. For $\varphi$ values of less than 30, however, the angle of dilatancy is mostly zero. A small negative value for is only realistic for extremely loose sands. In the Hardening Soil model or Hardening Soil model with small-strain stiffness, the end of dilatancy, as generally observed when the soil reaches the critical state, can be modelled using the Dilatancy cut-off. However, this option is not available for the Mohr-Coulomb model. For further information about the link between the friction angle and dilatancy, see Bolton (1986).

A positive dilatancy angle implies that in drained conditions the soil will continue to dilate as long as shear deformation occurs. This is clearly unrealistic, as most soils will reach a critical state at some point and further shear deformation will occur without volume changes. In undrained conditions, a positive dilatancy angle, combined with the restriction on volume changes, leads to a generation of tensile pore stresses. In an undrained effective stress analysis, therefore, the strength of the soil may be overestimated. When the soil strength is modelled as undrained shear strength, $s_u$, and $\varphi = 0$, (Undrained ($B$) or Undrained ($C$)) the dilatancy angle is automatially set to zero. Great care must be taken when using a positive value of dilatancy in combination with drainage type set to Undrained ($A$). In that case the model will show unlimited soil strength due to tensile pore stresses. These tensile pore stresses can be limited by setting the cavitation cut-off.

### (6) Shear modulus ($G$)

The shear modulus, $G$, has the dimension of stress. According to Hooke's law, the relationship between Young's modulus $E$ and the shear modulus is given by (see Eq. 7.17)

$$G = \frac{E}{2(1+v)} \tag{7.17}$$

Entering a particular value for one of the alternatives $G$ or $E_{oed}$ results in a change of the $E$ modulus whilst $v$ remains the same.

### (7) Oedometer modulus ($E_{oed}$)

The oedometer modulus, $E_{oed}$, or constrained modulus has the dimension of stress. According to Hooke's law, the relationship between Young's modulus $E$ and the oedometer modulus is given by (see Eq. 7.18)

$$E_{oed} = \frac{(1-v)E}{(1-2v)(1+v)} \tag{7.18}$$

Entering a particular value for one of the alternatives $G$ or $E_{oed}$ results in a change of the $E$ modulus whilst $v$ remains the same.

# 7.3　Hardening Soil Model (Isotropic Hardening)

In contrast to an elastic perfectly-plastic model, the yield surface of a hardening plasticity model is not fixed in principal stress space, but it can expand due to plastic straining.

Distinction can be made between two main types of hardening, namely shear hardening and compression hardening. Shear hardening is used to model irreversible strains due to primary deviatoric loading. Compression hardening is used to model irreversible plastic strains due to primary compression in oedometer loading and isotropic loading. Both types of hardening are contained in the present model.

The Hardening Soil model is an advanced model for simulating the behaviour of different types of soil, both soft soils and stiff soils (Schanz, 1998). When subjected to primary deviatoric loading, soil shows a decreasing stiffness and simultaneously irreversible plastic strains develop. In the special case of a drained triaxial test, the observed relationship between the axial strain and the deviatoric stress can be well approximated by a hyperbola. Such a relationship was first formulated by Kondner (1963) and later used in the well-known hyperbolic model (Duncan and Chang, 1970). The Hardening Soil model, however, supersedes the hyperbolic model by far: Firstly by using the theory of plasticity rather than the theory of elasticity, secondly by including soil dilatancy and thirdly by introducing a yield cap. Some basic characteristics of the model are listed below:

- Stress dependent stiffness according to a power law    Input parameter m
- Plastic straining due to primary deviatoric loading    Input parameter $E_{50}^{ref}$
- Plastic straining due to primary compression    Input parameter $E_{oed}^{ref}$
- Elastic unloading/reloading    Input parameters $E_{ur}^{ref}$, $v_{ur}$
- Failure according to the Mohr-Coulomb failure criterion    Parameters $c, \varphi$ and $\psi$

A basic feature of the present Hardening Soil model is the stress dependency of soil stiffness. For oedometer conditions of stress and strain, the model implies example the relationship $E_{oed} = E_{oed}^{ref} \left( \dfrac{\sigma}{p^{ref}} \right)^m$. In the special case of soft soils, it is realistic to use $m = 1$. In such situations there is also a simple relationship between the modified compression index $\lambda^*$ as used in models for soft soil and the oedometer loading modulus.

$$E_{oed}^{ref} = \frac{p^{ref}}{\lambda^*} \quad \lambda^* = \frac{\lambda}{1 + e_0}$$

where $p^{ref}$ is a reference pressure. Here we consider a tangent oedometer modulus at a particular reference pressure $p^{ref}$. Hence, the primary loading stiffness relates to the modified compression index $\lambda^*$ or to the standard Cam-clay compression index $\lambda$.

Similarly, the unloading-reloading modulus relates to the modified swelling index $\kappa^*$ or to the standard Cam-clay swelling index $\kappa$. There is the approximate relationship:

$$E_{ur}^{ref} = \frac{2p^{ref}}{\kappa^*} \quad \kappa^* = \frac{\kappa}{1 + e_0}$$

This relationship applies in combination with the input value $m = 1$.

### 7.3.1 Hyperbolic relationship for standard drained triaxial test

A basic idea for the formulation of the Hardening Soil model is the hyperbolic relationship between the vertical strain, $\varepsilon_1$, and the deviatoric stress, $q$ in primary triaxial loading.

Here standard drained triaxial tests tend to yield curves that can be described by

$$- \varepsilon_1 = \frac{1}{E_i} \frac{q}{1 - \dfrac{q}{q_a}} \quad \text{for } q < q_f \tag{7.19}$$

where

$q_a$ is the asymptotic value of the shear strength and $E_i$ the initial stiffness.

$E_i$ is related to $E_{50}$ by

$$E_i = \frac{2 E_{50}}{2 - R_f} \tag{7.20}$$

This relationship is plotted in Figure 7.5. The parameter $E_{50}$ is the confining stress dependent stiffness modulus for primary loading and is given by the equation:

$$E_{50} = E_{50}^{ref} \left( \frac{c \cos \varphi - \sigma_3' \sin \varphi}{c \cos \varphi + p^{ref} \sin \varphi} \right)^m \quad (7.21)$$

where

$E_{50}^{ref}$ is a reference stiffness modulus corresponding to the reference confining pressure $p^{ref}$.

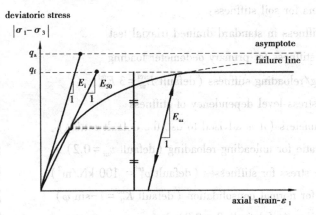

Figure 7.5 Hyperbolic stress-strain relation in primary loading for a standard drained triaxial test

A default setting $p = 100$ stress units is always used. The actual stiffness depends on the minor principal stress, $\sigma_3'$, which is the confining pressure in a triaxial test. Please note that $\sigma_3'$ is negative for compression. The amount of stress dependency is given by the power $m$. In order to simulate a logarithmic compression behavior, as observed for soft clays, the power should be taken equal to 1. 0. Janbu (1963) reported values of $m$ around 0.5 for Norwegian sands and silts, whilst von Soos (1990) reported various different values in the range $0.5 < m < 1.0$.

The ultimate deviatoric stress, $q_f$, and the quantity $q_a$ are defined as

$$q_f = (c \cot \varphi - \sigma_3') \frac{2 \sin \varphi}{1 - \sin \varphi} \quad \text{and} \quad q_a = \frac{q_f}{R_f} \quad (7.22)$$

Again it is remarkable that $\sigma_3'$ is usually negative. The above relationship for $q_f$ is derived from the Mohr-Coulomb failure criterion, which involves the strength parameters $c$ and $\varphi$. As soon as $q = q_f$, the failure criterion is satisfied and perfectly plastic yielding occurs as described by the Mohr-Coulomb model.

The ratio between $q_f$ and $q_a$ is given by the failure ratio $R_f$, which should obviously be smaller than or equal to 1. For unloading and reloading stress paths, another stress-dependent stiffness modulus is used:

$$E_{ur} = E_{ur}^{ref} \left( \frac{c \cos \varphi - \sigma_3' \sin \varphi}{c \cos \varphi + p^{ref} \sin \varphi} \right)^m \quad (7.23)$$

where $E_{ur}^{ref}$ is the reference of Young's modulus for unloading and reloading, corresponding to the reference pressure $p^{ref}$. In many practical cases, it is appropriate to set $E_{ur}^{ref}$ equal to 3 $E_{50}^{ref}$.

## 7.3.2 Parameters of the hardening soil model

Some parameters of the present hardening model coincide with those of the non-hardening Mohr-Coulomb model. These are the failure parameters $c, \varphi$ and $\psi$ (see in section 8.2).

Basic parameters for soil stiffness:

$E_{50}^{ref}$: Secant stiffness in standard drained triaxial test     $[kN/m^2]$

$E_{oed}^{ref}$: Tangent stiffness for primary oedometer loading     $[kN/m^2]$

$E_{ur}^{ref}$: Unloading/reloading stiffness (default $E_{ur}^{ref} = 3 E_{50}^{ref}$)     $[kN/m^2]$

$m$: Power for stress-level dependency of stiffness     $[-]$

Advanced parameters (it is advised to use the default setting):

$v_{ur}$: Poissons ratio for unloading-reloading (default $v_{ur} = 0.2$)     $[-]$

$p^{ref}$: Reference stress for stiffnesses (default $p^{ref} = 100 \ kN/m^2$)     $[kN/m^2]$

$K_0^{nc}$: $K_0$-value for normal consolidation (default $K_0^{nc} = 1 - \sin \varphi$)     $[-]$

$R_f$: Failure ratio $q_f/q_a$ (default $R_f = 0.9$).     $[-]$

$\sigma_{tension}$: Tensile strength (default tension = 0 stress units)     $[kN/m^2]$

$c_{inc}$: As in Mohr-Coulomb model (default $c_{inc} = 0$)     $[kN/m^3]$

Instead of entering the basic parameters for soil stiffness, alternative parameters can be entered. These parameters are listed below:

$C_c$: Compression index     $[-]$

$C_s$: Swelling index or reloading index     $[-]$

$e_{init}$: Initial void ratio     $[-]$

### (1) Stiffness moduli $E_{50}^{ref}$, $E_{oed}^{ref}$, $E_{ur}^{ref}$ and power $m$

The advantage of the Hardening Soil model over the Mohr-Coulomb model is not only the use of a hyperbolic stress-strain curve instead of a bi-linear curve, but also the control of stress level dependency. When using the Mohr-Coulomb model, the user has to select a fixed value of Young's modulus whereas for real soils, this stiffness depends on the stress level. It is therefore necessary to estimate the stress levels within the soil and use these to obtain suitable values of stiffness. With the Hardening Soil model, however, this cumbersome selection of input parameters is not required.

Instead, a stiffness modulus $E_{50}^{ref}$ is defined for a reference minor principle effective stress of $-\sigma_3' = p^{ref}$. This is the secant stiffness at 50% of the maximum deviatoric stress, at a cell pressure equal to the reference stress $p^{ref}$ (Figure 7.6). As a default value, the program uses $p^{ref} = 100 \ kN/m^2$.

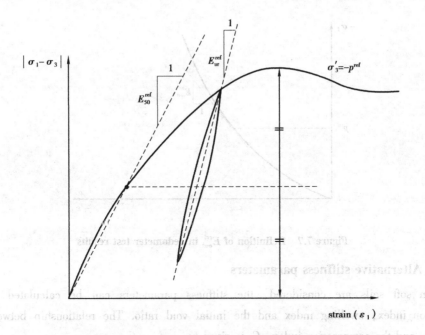

**Figure 7.6　Definition of $E_s$ and $E_w$ for drained triaxial test results**

The elastoplastic Hardening Soil model does not involve a fixed relationship between the (drained) triaxial stiffness $E_{50}$ and the oedometer stiffness $E_{oed}$ for one-dimensional compression. Instead, these stiffnesses can be inputted independently. It is now important to define the oedometer stiffness. Here we use the equation to define oedometer stiffness:

$$E_{oed} = E_{oed}^{ref} \left( \frac{c \cos \varphi - \dfrac{\sigma_3'}{K_0^{nc}} \sin \varphi}{c \cos \varphi + p^{ref} \sin \varphi} \right)^m \tag{7.24}$$

where $E_{oed}$ is a tangent stiffness modulus obtained from an oedometer test, as indicated in Figure 7.7. Hence, $E_{oed}^{ref}$ is a tangent stiffness at a vertical stress of $-\sigma_1' = -\sigma_3'/K_0^{nc} = p^{ref}$. Note that we basically use $\sigma_1'$ rather than $-\sigma_3'$ and that we consider primary loading.

When undrained behaviour is considered in the Hardening Soil model, the Drainage type should preferably be set to Undrained ($A$). Alternatively, Undrained ($B$) can be used in case the effective strength properties are not known or the undrained shear strength is not properly captured by using Undrained ($A$). However, it should be noted that the material loses its stress-dependency of stiffness in that case. Undrained ($C$) is not possible since the model is essentially formulated as an effective stress model.

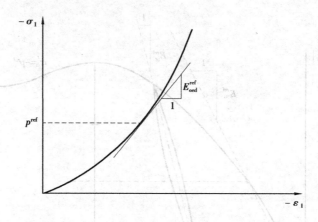

**Figure 7.7   Definition of $E_{oed}^{ref}$ in oedometer test results**

## (2) Alternative stiffness parameters

When soft soils are considered, the stiffness parameters can be calculated from the compression index, swelling index and the initial void ratio. The relationship between these parameters and the compression index, $C_c$ is given by

$$C_c = \frac{2.3(1 + e_{init})p^{ref}}{E_{oed}^{ref}} \tag{7.25}$$

The relationship between the $E_{ur}^{ref}$ and the swelling index, $C_s$ is given by

$$C_s \approx \frac{2.3(1 + e_{init})(1 + v)(1 - 2v)p^{ref}}{(1 - v) E_{ur}^{ref} K_0} \tag{7.26}$$

Regardless the previous value of $E_{50}$, a new value will be automatically assigned according to

$$E_{50}^{ref} = 1.25 E_{oed}^{ref} \tag{7.27}$$

Although for Soft soils, $E_{oed}^{ref}$ could be as high as $2E_{oed}^{ref}$, this high value could lead to a limitation in the modeling; therefore a lower value is used. Changing the value of $C_s$ will change the stiffness parameter $E_{ur}$.

## (3) Dilatancy cut-off

After extensive shearing, dilating materials arrive in a state of critical density where dilatancy has come to an end, as indicated in Figure 7.8. This phenomenon of soil behavior can be included in the Hardening Soil model by means of a dilatancy cut-off. In order to specify this behavior, the initial void ratio, $e_{init}$ and the maximum void ratio, $e_{max}$ of the material must be entered as general parameters. As soon as the volume change results in a state of maximum void, the mobilized dilatancy angle, $\psi_m$ is automatically set back to zero, as indicated in Figure 7.8.

For $e < e_{max}$:

$$\sin \psi_m = \frac{\sin \varphi_m - \sin \varphi_{cv}}{1 - \sin \varphi_m \sin \varphi_{cv}} \quad (7.28a)$$

where

$$\sin \varphi_{cv} = \frac{\sin \varphi - \sin \psi}{1 - \sin \varphi \sin \psi} \quad (7.28b)$$

For $e \geqslant e_{max}$: $\quad\quad\quad\quad\quad \psi_m = 0$

The void ratio is related to the volumetric strain, $\varepsilon_v$, by the relationship

$$- (\varepsilon_v - \varepsilon_v^{init}) = \ln \left( \frac{1 + e}{1 - e_{init}} \right) \quad (7.29)$$

where an increment of $\varepsilon_v$, is positive for dilatancy.

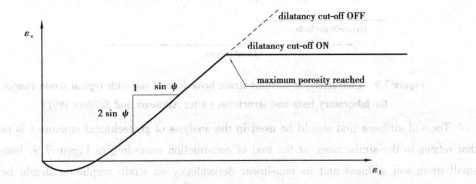

**Figure 7.8　Resulting strain curve for a standard drained triaxial test when including dilatancy cut-off**

The initial void ratio, $e_{init}$, is the in-situ void ratio of the soil body. The maximum void ratio is the void ratio of the material in a state of critical void (critical state). As soon as the maximum void ratio is reached, the dilatancy angle is set to zero. The minimum void ratio, $e_{min}$ of a soil can also be input, but this general soil parameter is not used within the context of the Hardening Soil model.

# 7.4　Hardening Soil Model with Small-strain Stiffness (HS-small)

The original Hardening Soil model assumes elastic material behavior during unloading and reloading. However, the strain range in which soils can be considered truly elastic, i.e. where they recover from applied straining almost completely, is very small. With increasing strain amplitude, soil stiffness decays non-linearly. Plotting soil stiffness against log (strain) yields characteristic S shaped stiffness reduction curves. Figure 7.9 gives an example of such a stiffness reduction curve. It also outlines the characteristic shear strains that can be measured near geotechnical structures and the applicable strain ranges of laboratory tests. It turns out that at the minimum strain which can be reliably measured in classical laboratory tests, i.e. triaxial tests and oedometer tests without special instrumentation, soil stiffness is often decreased to less than half of its initial value.

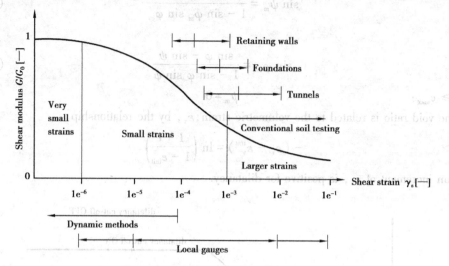

**Figure 7.9  Characteristic stiffness-strain behaviour of soil with typical strain ranges**
**for laboratory tests and structures (after Atkinson and Sallfors 1991)**

The soil stiffness that should be used in the analysis of geotechnical structures is not the one that relates to the strain range at the end of construction according to Figure 7.9. Instead, very small-strain soil stiffness and its non-linear dependency on strain amplitude should be properly taken into account. In addition to all features of the Hardening Soil model, the Hardening oil model with small-strain stiffness offers the possibility to do so. In fact, only two additional parameters are needed to describe the variation of stiffness with strain:

- the initial or very small-strain shear modulus $G_0$.
- the shear strain level $\gamma_{0.7}$ at which the secant shear modulus $G_s$ is reduced to about 70% of $G_0$.

## 7.4.1  Describing Small-strain stiffness with a simple hyperbolic law

In soil dynamics, small-strain stiffness has been a well-known phenomenon for a long time. In static analysis, the findings from soil dynamics have long been considered not to be applicable.

Seeming differences between static and dynamic soil stiffness have been attributed to the nature of loading (e.g. inertia forces and strain rate effects) rather than to the magnitude of applied strain which is generally small in dynamic conditions (earthquakes excluded). As inertia forces and strain rate have only little influence on the initial soil stiffness, dynamic soil stiffness and small-strain stiffness can in fact be considered as synonyms. Probably the most frequently used model in soil dynamics is the Hardin-Drnevich relationship. From test data, sufficient agreement is found that the stress-strain curve for small strains can be adequately described by a simple hyperbolic law. The following analogy to the hyperbolic law for larger strains by Kondner (1963) was proposed by Hardin Drnevich (1972):

$$\frac{G_s}{G_0} = \frac{1}{1 + \left|\dfrac{\gamma}{\gamma_r}\right|} \qquad (7.30)$$

where the threshold shear strain $\gamma_r$ is quantified as

$$\gamma_r = \frac{\tau_{max}}{G_0} \qquad (7.31)$$

with $\tau_{max}$ being the shear stress at failure. Essentially, Eqs.7.30 and 7.31 relate large (failure) strains to small-strain properties which often work well.

More straightforward and less prone to error is the use of a smaller threshold shear strain. Santos & Correia (2001), for example suggest to use the shear strain $\gamma_r = \gamma_{0.7}$, at which the secant shear modulus Gs is reduced to about 70 of its initial value. Eq. 7.30 then is rewritten as

$$\frac{G_s}{G_0} = \frac{1}{1 + a\left|\dfrac{\gamma}{\gamma_{0.7}}\right|} \qquad (7.32)$$

where $a = 0.385$.

In fact, using $a = 0.385$ and $\gamma = \gamma_{0.7}$ gives $G_s/G_0 = 0.722$. Hence, the formulation "about 70%" should be interpreted more accurately as 72.2%.

Figure 7.10 shows the fit of the modified Hardin-Drnevich relationship (Eq. 7.30) to normalised test data.

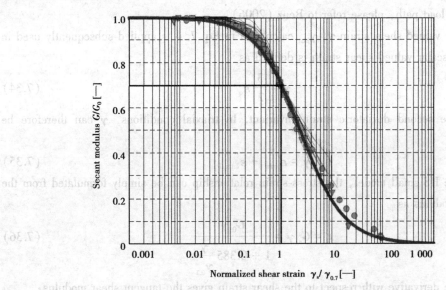

**Figure 7.10   Results from the Hardin-Drnevich relationship compared to test data by Santos & Correia (2001)**

### 7.4.2 Applying the Hardin-Drnevich relationship in the HS model

The decay of soil stiffness from small strains to larger strains can be associated with loss of intermolecular and surface forces within the soil skeleton. Once the direction of loading is reversed, the stiffness regains a maximum recoverable value which is in the order of the initial soil stiffness. Then, while loading in the reversed direction is continued, the stiffness decreases again. A strain history dependent, multi-axial extension of the Hardin-Drnevich relationship is therefore needed in order to apply it in the Hardening Soil model. Such an extension has been proposed by Benz (2006) in the form of the small-strain overlay model. Benz derived a scalar valued shear strain $\gamma_{\text{hist}}$ by the following projection:

$$\gamma_{\text{hist}} = \sqrt{3} \, \frac{\|\overline{\overline{H}}\Delta\underline{e}\|}{\|\Delta\underline{e}\|} \tag{7.33}$$

where $\Delta\underline{e}$ is the actual deviatoric strain increment and $\overline{\overline{H}}$ is a symmetric tensor that represents the deviatoric strain history of the material. Whenever a strain reversal is detected, the tensor $\overline{\overline{H}}$ is partially or fully reset before the actual strain increment $\Delta\underline{e}$ is added. As the criterion for strain reversals serves a criterion similar to that in Simpsons brick model (1992): All three principal deviatoric strain directions are checked for strain reversals separately which resemble three independent brick models. When there is no principal strain rotation, the criterion reduces to two independent brick-models. For further details on the strain tensor $\overline{\overline{H}}$ and its transformation at changes in the load path, please refer to Benz (2006).

The scalar valued shear strain $\gamma = \gamma_{\text{hist}}$ calculated in Eq. 7. 4 is applied subsequently used in Eq. 8.33, the scalar valued shear strain is defined as

$$\gamma = \frac{3}{2} \varepsilon_q \tag{7.34}$$

where $\varepsilon_q$ is the second deviatoric strain invariant. In triaxial conditions, $\gamma$ can therefore be expressed as

$$\gamma = \varepsilon_{\text{axial}} - \varepsilon_{\text{lateral}} \tag{7.35}$$

Within the HS small model, the stress-strain relationship can be simply formulated from the secant shear modulus as

$$\tau = G_s \gamma = \frac{G_0 \gamma}{1 + 0.385 \dfrac{\gamma}{\gamma_{0.7}}} \tag{7.36}$$

Taking the derivative with respect to the shear strain gives the tangent shear modulus:

$$G_t = \frac{G_0}{\left(1 + 0.385 \dfrac{\gamma}{\gamma_{0.7}}\right)^2} \tag{7.37}$$

This stiffness reduction curve reaches far into the plastic material domain. In the Hardening

Soil model and HS small model, stiffness degradation due to plastic straining is simulated with strain hardening. In the HS small model, the small-strain stiffness reduction curve is therefore bound by a certain lower limit, determined by conventional laboratory tests:

- The lower cut-off of the tangent shear modulus $G_t$ is introduced at the unloading reloading stiffness $G_{ur}$ which is defined by the material parameters $E_{ur}$ and $v_{ur}$:

$$G_t \geqslant G_{ur} \text{ where } G_{ur} = \frac{E_{ur}}{2(1 + v_{ur})} \text{ and } G_t = \frac{E_t}{2(1 + v_{ur})} \quad (7.38)$$

- The cut-off shear strain cut-off can be calculated as

$$\gamma_{cut\text{-}off} = \frac{1}{0.385}\left(\sqrt{\frac{G_0}{G_{ur}}} - 1\right)\gamma_{0.7} \quad (7.39)$$

Within the HS small model, the quasi-elastic tangent shear modulus is calculated by integrating the secant stiffness modulus reduction curve over the actual shear strain increment. An example of a stiffness reduction curve used in the HS small model is shown in Figure 7.11.

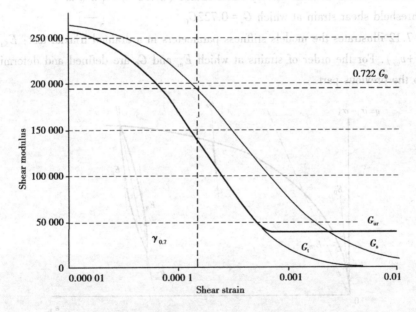

**Figure 7.11   Secant and tangent shear modulus reduction curve**

Moreover, the tangent shear modulus $G_t$ and corresponding Young's modulus $E_t$ (considering a constant Poisson's ratio $v_{ur}$) is stress-dependent, and follows the same power law. For primary loading situations, the model uses same hardening plasticity formulations as the Hardening Soil model, where $E_{ur}$ is replaced by $E_t$ as described above.

### 7.4.3  Model parameters

Compared to the standard Hardening Soil model, the Hardening Soil model with small-strain stiffness requires two additional stiffness parameters as input: $G_0^{ref}$ and $\gamma_{0.7}$. All other parameters, including the alternative stiffness parameters, remain the same as in the standard Hardening Soil model. $G_0^{\text{ref}}$ defines the shear modulus at very small strains e. g. $\varepsilon < 10^{-6}$ at a reference minor principal stress of $-\sigma_3' = p^{\text{ref}}$.

Poisson's ratio $v_{\text{ur}}$ is assumed a constant, so that the shear modulus $G_0^{\text{ref}}$ can also be calculated from the very small strain Young's modulus as $G_0^{\text{ref}} = E_0^{\text{ref}} / [ 2 ( 1 + v_{\text{ur}} ) ]$. The threshold shear strain $\gamma_{0.7}$ is the shear strain at which the secant shear modulus $G_s^{\text{ref}}$ is decayed to $0.722 G_0^{\text{ref}}$. The threshold shear strain $\gamma_{0.7}$ is to be supplied for virgin loading. In summary, the input stiffness parameters of the Hardening Soil model with small-strain stiffness are listed below:

$m$: Power for stress-level dependency of stiffness $\qquad$ [—]

$E_{50}^{\text{ref}}$: Secant stiffness in standard drained triaxial test $\qquad$ [kN/m²]

$E_{\text{oed}}^{\text{ref}}$: Tangent stiffness for primary oedometer loading $\qquad$ [kN/m²]

$E_{\text{ur}}^{\text{ref}}$: unloading/reloading stiffness from drained triaxial test $\qquad$ [kN/m²]

$v_{\text{ur}}$: Poisson's ratio for unloading-reloading $\qquad$ [—]

$G_0^{\text{ref}}$: reference shear modulus at very small strains ($\varepsilon < 10^{-6}$) [kN/m²]

$\gamma_{0.7}$: threshold shear strain at which $G_s = 0.722 G_0$ $\qquad$ [—]

Figure 7.12 illustrates the models stiffness parameters in a drained triaxial test: $E_{50}$, $E_{\text{ur}}$ and $E_0 = 2 G_0 ( 1 + v_{\text{ur}} )$. For the order of strains at which $E_{\text{ur}}$ and $G_0$ are defined and determined, one may refer to the previous part.

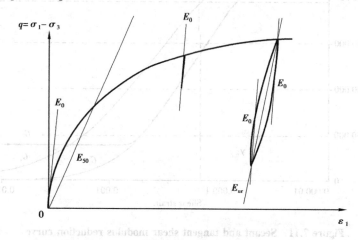

**Figure 7.12 Stiffness parameters $E_{50}$, $E_{\text{ur}}$ and $E_0 = 2 G_0 ( 1 + v_{\text{ur}} )$ of the Hardening Soil model with small-strain stiffness in a triaxial test**

A first estimation of the Hssmall (or HS) parameters for quartz sand based on the relative density (RD) is given in Brinkgreve et al. (2010).

**On the parameters $G_0$ and $\gamma_{0.7}$**

A number of factors influence the small-strain parameters $G_0$ and $\gamma_{0.7}$. Most importantly, they are influenced by the materials actual state of stress and void ratio $e$. In the Hssmall model, the stress dependency of the shear modulus $G_0$ is taken into account with the power law:

$$G_0 = G_0^{\text{ref}} \left( \frac{c \, \cos \varphi - \sigma_3' \, \sin \varphi}{c \, \cos \varphi + p^{\text{ref}} \, \sin \varphi} \right)^m \tag{7.40}$$

which resembles the ones used for other stiffness parameters. The threshold shear strain $\gamma_{0.7}$ is taken independently of the mean stress.

Assuming that within a HS Small (or HSS) computation void ratio changes are rather small, the material parameters are not updated for changes in the void ratio. Knowledge of a material's initial void ratio can nevertheless be very helpful in deriving its small-strain shear stiffness $G_0$. Many correlations are offered in the literature (Benz, 2006). A good estimation for many soils is for example the relation given by Hardin & Black(1969):

$$G_0^{\text{ref}} = 33 \times \frac{(2.97 - e)^2}{1 + e} [\text{MPa}] \quad \text{for } p^{\text{ref}} = 100 [\text{kPa}] \tag{7.41}$$

Alpan (1970) empirically related dynamic soil stiffness to static soil stiffness (see Figure 7.13). The dynamic soil stiffness in Alpan's chart is equivalent to the small-strain stiffness $G_0$ or $E_0$. Considering that the static stiffness $E_{\text{static}}$ defined by Alpan equals approximately the unloading/reloading stiffness $E_{\text{ur}}$ in the HS small model, Alpan's chart can be used to guess a soils small-strain stiffness entirely based on its unloading/reloading stiffness $E_{\text{ur}}$. Although Alpan suggests that the ratio $E_0/E_{\text{ur}}$ can exceed 10 for very soft clays, the maximum ratio $E_0/E_{\text{ur}}$ or $G_0/G_{\text{ur}}$ permitted in the Hssmall model is limited to 20.

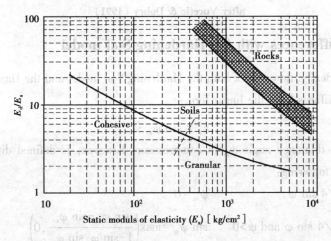

Figure 7.13 Relation between dynamic ($E_d = E_0$) and
static soil stiffness($E_s \approx E_{\text{ur}}$) after Alpan (1970)

In the absence of test data, correlations are also available for the threshold shear strain $\gamma_{0.7}$. Figure 7.14, for example, gives a correlation between the threshold shear strain and the Plasticity Index. Using the original Hardin-Drnevich relationship, the threshold shear strain $\gamma_{0.7}$ might also be related to the models failure parameters. Applying the Mohr-Coulomb failure criterion in function yields

$$\gamma_{0.7} \approx \frac{1}{9\,G_0}[\, 2\,c'(1 + \cos(2\,\varphi')) - \sigma'_1(1 + K_0)\sin(2\,\varphi')\,]\qquad(7.42)$$

where

$K_0$ is the earth pressure coefficient at rest and is the effective vertical stress (pressure negative).

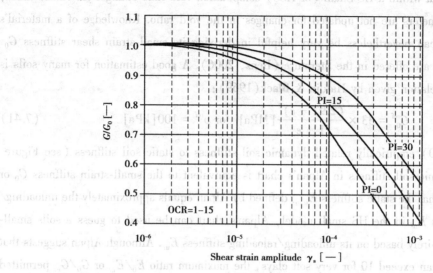

Figure 7.14  Influence of plasticity index (PI) on stiffness reduction
after Vucetic & Dobry (1991)

## 7.4.4  Other differences with the Hardening Soil model

The shear hardening flow rule of both the Hardening Soil model and the Hardening Soil model with small-strain stiffness have the linear form:

$$\dot{\varepsilon}^{\mathrm{p}}_{\mathrm{v}} = \sin\psi_{\mathrm{m}}\,\dot{\gamma}^{\mathrm{p}}\qquad(7.43)$$

The mobilized dilatancy angle $\psi_{\mathrm{m}}$ in compression, however, is defined differently. The HS model assumes the following:

For $\sin\varphi_{\mathrm{m}} < 3/4\,\sin\varphi$  $\qquad\qquad \psi_{\mathrm{m}} = 0$

For $\sin\varphi_{\mathrm{m}} \geqslant 3/4\,\sin\varphi$ and $\psi > 0$  $\qquad \sin\psi_{\mathrm{m}} = \max\left(\dfrac{\sin\varphi_{\mathrm{m}} - \sin\varphi_{\mathrm{cv}}}{1 - \sin\varphi_{\mathrm{m}}\sin\varphi_{\mathrm{cv}}}, 0\right)\qquad(7.44)$

For $\sin\varphi_{\mathrm{m}} \geqslant 3/4\,\sin\varphi$ and $\psi \leqslant 0$  $\qquad \psi_{\mathrm{m}} = \psi$

If $\varphi = 0$  $\qquad\qquad\qquad\qquad\qquad \psi_{\mathrm{m}} = 0$

where

$\varphi_{\mathrm{cv}}$ is the critical state friction angle, being a material constant independent of density;

$\psi_{\mathrm{m}}$ is the mobilised friction angle:

$$\sin\psi_{\mathrm{m}} = \frac{\sigma'_1 - \sigma'_3}{\sigma'_1 + \sigma'_3 - 2c\cot\varphi}\qquad(7.45)$$

For small mobilised friction angles and for negative values of $\psi_m$, as computed by Rowe's formula, $\psi_m$ in the Hardening Soil model is taken zero. Bounding the lower value of $\psi_m$ may sometimes yield too little plastic volumetric strains though. Therefore, the Hardening Soil model with small-strain stiffness adapts an approach by Li and Dafalias (2000) whenever $\psi_m$, as computed by Rowe's formula, is negative. In that case, the mobilised dilatancy in the Hardening Soil model with small-strain stiffness is calculated by the following equation:

$$\sin \psi_m = \frac{1}{10}\left( - M \exp\left[ \frac{1}{15}\ln\left( \frac{\eta}{M}\frac{q}{q_a} \right) \right] + \eta \right) \tag{7.46}$$

where $M$ is the stress ratio at failure and $\eta - \max \left( \frac{q}{p}, \frac{1-K_0}{\frac{1}{3}(1+2 K_0)} \right)$.

Eq. 7.18 is a simplified version of the void ratio dependent formulation by Li and Dafalias (2000).

The mobilized dilatancy as a function of $\psi_m$ for the HS small model is visualised in Figure 7.15.

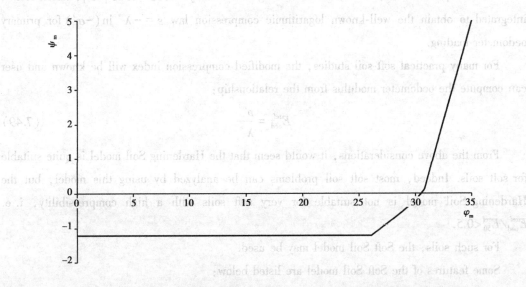

Figure 7.15　Plot of mobilised dilatancy angle m and mobilized friction angle mm
for HSS model (Brinkgreve, 2017)

# 7.5　The Soft Soil Model

As for soft soils, we consider near-normally consolidated clays, clayey silts and peat. A special feature of such materials is their high degree of compressibility. This is best demonstrated by oedometer test data as reported for instance by Janbu (1985) in his Rankine lecture.

Considering tangent stiffness moduli at a reference oedometer pressure of 100 kPa, he reported for normally consolidated clays, $E_{oed} = 1$ to 4 MPa, depending on the particular type of clay considered. The differences between these values and stiffnesses for $N_c$-sands are considerable as here we have values in the range of $10 \sim 50$ MPa, at least for non-cemented laboratory samples. Hence, in oedometer test, normally consolidated clays behave ten times softer than normally consolidated sands. This illustrates the extreme compressibility of soft soils.

A feature of soft soils is the linear stress-dependency of soil stiffness. According to the Hardening Soil model we have

$$E_{oed} = E_{oed}^{ref} \left( \frac{-\sigma_1'}{p^{ref}} \right)^m \tag{7.47}$$

at least for $c = 0$ and $-\sigma_3' = K_0^{nc} \sigma_1'$, a linear relationship is obtained for $m = 1$. Indeed, on using an exponent equal to unity, the above stiffness law reduces to

$$E_{oed} = \frac{-\sigma_1'}{\lambda^*} \text{ where } \lambda^* = \frac{p^{ref}}{E_{oed}^{ref}} \tag{7.48}$$

For this special case of $m = 1$, the Hardening Soil model yields $\dot{\varepsilon} = \lambda^* \sigma_1'/\sigma_1'$, which can be integrated to obtain the well-known logarithmic compression law $\varepsilon = -\lambda^* \ln(-\sigma_1')$ for primary oedometer loading.

For many practical soft-soil studies, the modified compression index will be known and user can compute the oedometer modulus from the relationship:

$$E_{oed}^{ref} = \frac{p^{ref}}{\lambda^*} \tag{7.49}$$

From the above considerations, it would seem that the Hardening Soil model is quite suitable for soft soils. Indeed, most soft soil problems can be analyzed by using this model, but the Hardening Soil model is not suitable for very soft soils with a high compressibility, i. e. $E_{oed}^{ref}/E_{50}^{ref} < 0.5$.

For such soils, the Soft Soil model may be used.

Some features of the Soft Soil model are listed below:

- Stress dependent stiffness (logarithmic compression behavior).
- Distinction between primary loading and unloading-reloading.
- Memory for pre-consolidation stress.
- Failure behavior according to the Mohr-Coulomb criterion.

## 7.5.1  Isotropic states of stress and strain ($\sigma'_1 = \sigma'_2 = \sigma'_3$)

In the Soft Soil model, it is assumed that there is a logarithmic relation between changes in volumetric strain, $\varepsilon_v$, and changes in mean effective stress, $p'$, which can be formulated as

$$\varepsilon_v - \varepsilon_v^0 = -\lambda^* \ln \left( \frac{p' + c \cot \varphi}{p^0 + c \cot \varphi} \right) \qquad \text{(virgin compression)} \qquad (7.50)$$

In order to maintain the validity of Eq.7.50, a minimum value of $p'$ is set equal to a unit stress. The parameter $\lambda^*$ is the modified compression index, which determines the compressibility of the material in primary loading. Note that $\lambda^*$ differs from the index $\lambda$ as used by Burland (1965). The difference is that Eq.7.50 is a function of volumetric strain instead of void ratio. Plotting Eq.7.50 gives a straight line.

During isotropic unloading and reloading, a different path (line) is followed, which can be formulated as

$$\varepsilon_v^e - \varepsilon_v^{e0} = -\kappa^* \ln \left( \frac{p' + c \cot \varphi}{p^0 + c \cot \varphi} \right) \qquad \text{(unloading and reloading)} \qquad (7.51)$$

Again, a minimum value of $p'$ is set equal to a unit stress. The parameter $\kappa^*$ is the modified swelling index, which determines the compressibility of the material in unloading and subsequent reloading. Note that $\kappa^*$ differs from the index $n$ as used by Burland. The ratio $\lambda^*/\kappa^*$ is, however, equal to Burland's ratio $\lambda^*/\kappa$. The soil response during unloading and reloading is assumed to be elastic as denoted by the superscript e in Eq. 7.51. The elastic behaviour is described by Hooke's law. Eq. 7.51 implies linear stress dependency on the tangent bulk modulus as

$$K_{ur} = \frac{E_{ur}}{3(1 - 2v_{ur})} = \frac{p' + c \cot \varphi}{\kappa^*} \qquad (7.52)$$

In which the subscript $ur$ denotes unloading/reloading. Note that effective parameters are considered rather than undrained soil properties, as might be suggested by the subscripts $ur$. Neither the elastic bulk modulus, $K_{ur}$, nor the elastic Young's modulus, $E_{ur}$ is used as an input parameter. Instead, $v_{ur}$ and $\kappa^*$ are used as input constants for the part of the model that computes the elastic strains.

An infinite number of unloading/reloading lines may exist in Figure 7.16, each corresponding to a particular value of the isotropic pre-consolidation stress $p_p$. The pre-consolidation stress represents the largest stress level experienced by the soil. During unloading and reloading, this pre-consolidation stress remains constant. In primary loading, however, the pre-consolidation stress increases with the stress level causing irreversible (plastic) volumetric strains.

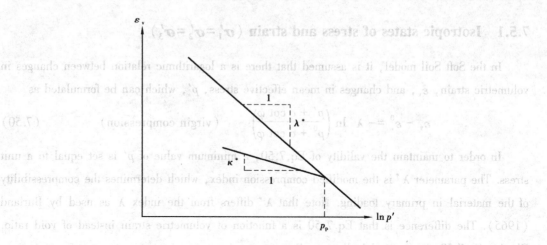

**Figure 7.16  Logarithmic relation between volumetric strain and mean stress**

## 7.5.2  Yielding founction

The yield function of the Soft Soil model is defined as

$$f = \bar{f} - p_p \tag{7.53}$$

where $\bar{f}$ is a function of the stress state $(p', \tilde{q})$ and $p_p$, the pre-consolidation stress is a function of plastic strain such that

$$\bar{f} = \frac{\tilde{q}^2}{M^2(p' + c \cot \varphi)} + p' \tag{7.54}$$

$$p_p = p_p^0 \exp\left(\frac{-\varepsilon_v^p}{\lambda^* - \kappa^*}\right) \tag{7.55}$$

and $\tilde{q}$ is a similar deviatoric stress quantity as defined for the cap yield surface in the Hardening Soil model: $\tilde{q} = \sigma_1' + (\alpha - 1)\sigma_2' - \alpha \sigma_3'$ where $\alpha = (3 + \sin \varphi)/(3 - \sin \varphi)$.

The yield function $(f = 0)$ describes an ellipse in the $p'$-$\tilde{q}$-plane, as illustrated in Figure 7.17. The parameter $M$ in Eq. 7.54 determines the height of the ellipse. The height of the ellipse is responsible for the ratio of horizontal to vertical stresses in primary one-dimensional compression.

As a result, the parameter $M$ determines largely the coefficient of lateral earth pressure $K_0^{nc}$. In view of this, the value of $M$ can be chosen so that a known value of $K_0^{nc}$ is matched in primary one-dimensional compression. Such an interpretation and use of $M$ differs from the original critical state line idea, but it ensures a proper matching of $K_0^{nc}$. The tops of all ellipses are located on a line with slope $M$ in the $p'$-$\tilde{q}$-plane. In (Burland, 1965) and (Burland, 1967) the $M$-line is referred to as the critical state line and presents stress states at post peak failure. The parameter $M$ is then based on the critical state friction angle. In the Soft Soil model, however, failure is not necessarily related to critical state. The Mohr-Coulomb failure criterion is a function of the strength parameters $\varphi$ and $c$, which might not correspond to the $M$-line. The isotropic pre-consolidation stress $p_p$ determines the extent of the ellipse along $p'$ axis. During loading, an infinite number of ellipses may exist (see Figure 7.17), each corresponding to a particular value of $p_p$. In tension ($p'<0$), the ellipse extends to $c\cot\varphi$ (Eq. 7.55). In order to make sure that the right hand side of the ellipse (i.e. the 'cap') will remain in the 'compression' zone ($p'$-0), a minimum value of $c\cot\varphi$ is adopted for $p_p$. For $c=0$, a minimum value of $p_p$ equal to a stress unit is adopted. Hence, there is a threshold ellipse as illustrated in Figure 7.17.

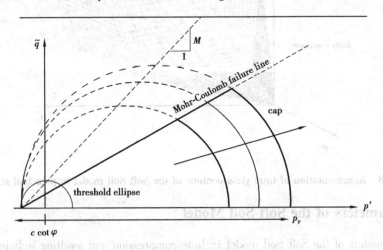

**Figure 7.17  Yield surface of the Soft Soil model in $p'$-$\tilde{q}$-plane**

The value of $p_p$ is determined by volumetric plastic strain following the hardening relation. This equation reflects the principle that the pre-consolidation stress increases exponentially with decreasing volumetric plastic strain (compaction). $p_p^0$ can be regarded as the initial value of the pre-consolidation stress, and according to Eq. 7.56, the initial volumetric plastic strain is assumed to be zero.

In the Soft Soil model, the yield function, Eq. 7.53 describes the irreversible volumetric strain in primary compression, and forms the cap of the yield contour. To model the failure state, a perfectly-plastic Mohr-Coulomb type yield function is used. This yield function represents a straight line in $p'$-$\tilde{q}$-plane as shown in Figure 7.17. The slope of the failure line is smaller than the slope of the $M$-line. The total yield contour, as shown by the bold lines in Figure 7.17, is the boundary of the elastic stress area. The failure line is fixed, but the cap may increase in primary compression. Stress paths within this boundary give only elastic strain increments, whereas stress paths that tend to cross the boundary generally give both elastic and plastic strain increments.

For general states of stress ($p'$, $\tilde{q}$), the plastic behaviour of the Soft Soil model is defined by the combination of the cap yield function and the Mohr-Coloumb yield functions. The total yield contour in principal stress space is indicated in Figure 7.18.

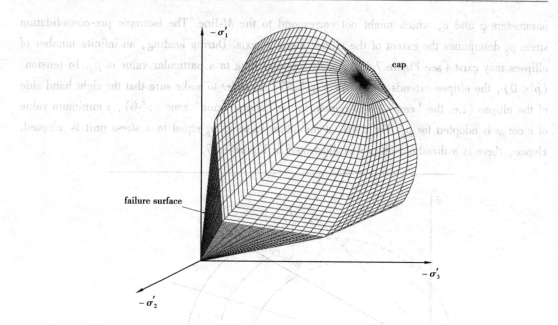

**Figure 7.18   Representation of total yield contour of the Soft Soil model in principal stress space**

## 7.5.3   Parameters of the Soft Soil Model

The parameters of the Soft Soil model include compression and swelling indicies, which are typical for soft soils, as well as the Mohr-Coulomb model failure parameters. In total, the Soft Soil model requires the following parameters to be determined:

Basic parameters:

| | | |
|---|---|---|
| $\lambda^*$ | Modified compression index | [—] |
| $\kappa^*$ | Modified swelling index | [—] |
| $c$ | Effective cohesion | [kN/m²] |
| $\varphi$ | Friction angle | [°] |
| $\psi$ | Dilatancy angle | [°] |
| $\sigma_t$ | Tensile strength | [kN/m²] |

Advanced parameters (use default settings):

| | | |
|---|---|---|
| $v_{ur}$ | Poisson's ratio for unloading/reloading | [—] |
| $K_0^{nc}$ | Coefficient of lateral stress in normal consolidation | [—] |
| $M$ | $K_0^{nc}$-parameter | [—] |

### (1) Modified swelling index and modified compression index

These parameters can be obtained from an isotropic compression test including isotropic unloading. When plotting the logarithm of the mean stress as a function of the volumetric strain for clay-type materials, we can approximate it by two straight lines. The slope of the primary loading line gives the modified compression index and the slope of the unloading (or swelling) line gives the modified swelling index. Note that there is a difference between the modified indices $\kappa^*$ and

$\lambda^*$ and the original Cam-clay parameters $\kappa$ and $\lambda$. The latter parameters are defined in terms of the void ratio $e$ instead of the volumetric strain $\varepsilon_v$.

Apart from the isotropic compression test, the parameters $\kappa^*$ and $\lambda^*$ can be obtained from a one-dimensional compression test. Here a relationship exists with the internationally recognized parameters for one-dimensional compression and swelling, $C_c$ and $C_s$. These relationships are summarized in Table 7.1.

**Table 7.1(a)  Relationship to Cam-clay parameters**

| | |
|---|---|
| $1.\lambda^* = \dfrac{\lambda}{1+e}$ | $2.\kappa^* = \dfrac{\kappa}{1+e}$ |

**Table 7.1(b)  Relationship to internationally normalized parameters**

| | |
|---|---|
| $3.\lambda^* = \dfrac{C_c}{2.3(1+e)}$ | $4.\kappa^* = \dfrac{2C_s}{2.3(1+e)}$ |

Remarks on Table 7.1:

①In relations 1 and 2, the void ratio, $e$ is assumed to be constant. In fact, $e$ will change during a compression test, but this will give a relatively small difference in void ratio. For $e$, one can use the average void ratio that occurs during the test or just the initial value.

②In relation 4 there is no exact relation between $\kappa^*$ and the one-dimensional swelling index $C_s$, because the ratio of horizontal and vertical stresses changes during one-dimensional unloading. For this approximation, it is assumed that the average stress state during unloading is an isotropic stress state, i.e. the horizontal and vertical stresses are equal.

③In practice, swelling is often assumed to be equivalent to recompression behavior, which may not be right. Hence $\kappa^*$ should be based on $C_s$ rather than the recompression index $C_r$.

④The factor 2.3 in relation 3 is obtained from the ratio between the logarithm of base 10 and the natural logarithm.

⑤The ratio $\dfrac{\lambda^*}{\kappa^*} = \left(\dfrac{\lambda}{\kappa}\right)$ ranges, in general, between 2.5 and 7.

## (2) Cohesion

The cohesion has the dimension of stresses. A small effective cohesion may be used, including a cohesion of zero. Entering a cohesion will result in an elastic region that is partly located in the tension zone, as illustrated in Figure 7.17. The left hand side of the ellipse crosses the $p'$-axis at a value of $c \cot \varphi$. In order to maintain the right hand side of the ellipse (i.e. the cap) in the 'pressure' zone of the stress space, the isotropic pre-consolidation stress $p_p$ has a minimum value of $c \cot \varphi$. This means that entering a cohesion larger than zero may result in a state of over-consolidation, depending on the magnitude of the cohesion and the initial stress state. As a result, a stiffer behavior is obtained during the onset of loading. It is not possible to specify undrained shear strength by means of high cohesion and a friction angle of zero. Input of model parameters should always be based on effective values. Please note that the resulting effective stress path may not be accurate, which may lead to an unrealistic undrained shear strength. Hence, when using Undrained (A) as drainage type, the resulting stress state must be checked against a known undrained shear strength profile.

### (3) Friction angle

The effective angle of internal friction represents the increase of shear strength with effective stress level. It is specified that in degrees, zero friction angle is not allowed. On the other hand, care should be taken in the use of high friction angles. It is often recommended to use $\varphi_{cv}$, i.e. the critical state friction angle, rather than a higher value based on small strains. Moreover, using a high friction angle will substantially increase the computational requirements.

### (4) Dilatancy angle

For the type of materials, which can be described by the Soft Soil model, the dilatancy degrees is considered in the standard settings of the Soft Soil model.

### (5) Poisson's ratio

In the Soft Soil model, the Poisson's ratio $v$ is the well-known pure elastic constant rather than the pseudo-elasticity constant as used in the linear elastic perfectly-plastic model. Its value will usually be in the range between 0.1 and 0.2. If the standard setting for the Soft Soil model parameters is selected, then $v_{ur} = 0.15$ is automatically used. For loading of normally consolidated materials, Poisson's ratio plays a minor role, but it becomes important in unloading problems. For example, for unloading in a one-dimensional compression test (oedometer), the relatively small Poisson's ratio will result in a small decrease of the lateral stress compared with the decrease in vertical stress. As a result, the ratio of horizontal and vertical stress increases, which is a well-known phenomenon in overconsolidated materials. Hence Poisson's ratio should not be based on the normal consolidated $K_0^{nc}$-value, but on the ratio of the horizontal stress increment to the vertical stress increment in oedometer unloading and reloading test such that

$$\frac{v_{ur}}{1 - v_{ur}} = \frac{\Delta \sigma_{xx}}{\Delta \sigma_{yy}} \qquad \text{(unloading and reloading)} \qquad (7.56)$$

### (6) $K_0^{nc}$-parameter

The parameter $M$ is automatically determined based on the coefficient of lateral earth pressure in normally consolidated condition, $K_0^{nc}$, as entered by the user. The exact relation between $M$ and $K_0^{nc}$ is as follows (Brinkgreve, 1994):

$$M = 3 \sqrt{\frac{(1 - K_0^{nc})^2}{(1 + 2 K_0^{nc})^2} + \frac{(1 - K_0^{nc})(1 - 2 v_{ur})\left(\frac{\lambda^*}{\kappa^*} - 1\right)}{(1 + 2 K_0^{nc})(1 - 2 v_{ur})\frac{\lambda^*}{\kappa^*} - (1 - K_0^{nc})(1 + v_{ur})}} \qquad (7.57)$$

The value of $M$ is indicated in the input window. As can be seen from Eq. 7.58, $M$ is also influenced by the Poisson's ratio $v_{ur}$ and by the ratio $\dfrac{\lambda^*}{\kappa^*}$. However, the influence of $K_0^{nc}$ is dominant. Eq. 7.58 can be approximated by

$$M \approx 3.0 - 2.8\,K_0^{nc} \qquad (7.58)$$

# 7.6  Modified Cam-clay Model

The Modified Cam-clay model is described in several textbooks on critical state soil mechanics, for example Muir Wood (1990). In this chapter, a short overview is given of the basic equations.

## 7.6.1  Formulation of the Modified Cam-clay model

In the Modified Cam-clay model, a logarithmic relation is assumed between void ratio $e$ and the mean effective stress $p'$ in virgin isotropic compression, which can be formulated as

$$e - e_0 = -\lambda \ln\left(\frac{p'}{p^0}\right) \qquad \text{(virgin isotropic compression)} \qquad (7.59)$$

The parameter $\lambda$ is the Cam-clay isotropic compression index, which determines the compressibility of the material in primary loading. When plotting relation Eq., 7.60 in a $e$-ln $p'$ diagram one obtains a straight line. During unloading and reloading, a different line is followed, which can be formulated as

$$e - e_0 = -\kappa \ln\left(\frac{p'}{p^0}\right) \qquad \text{(isotropic unloading and reloading)} \qquad (7.60)$$

The parameter $\kappa$ is the Cam-clay isotropic swelling index, which determines the compressibility of material in unloading and reloading. In fact, an infinite number of unloading and reloading lines exists in $p'$-$e$-plane, each corresponding to a particular value of the preconsolidation stress $p_p$.

The yield function of the Modified Cam-clay model is defined as

$$f = \frac{q^2}{M^2} + p'(p' - p_p) \qquad (7.61)$$

The yield surface ($f=0$) represents an ellipse in $p'$-$q$-plane as indicated in Figure 7.19. The yield surface is the boundary of the elastic stress states. Stress paths within this boundary only give elastic strain increments, whereas stress paths that tend to cross the boundary generally give both elastic and plastic strain increments.

In $p'$-$q'$-plane, the top of the ellipse intersects a line that we can be written as

$$q = Mp' \qquad (7.62)$$

This line is called the critical state line (CSL) and gives the relation between $p'$ and $q$ in a state of failure (i.e. the critical state). The constant $M$ is the tangent of the critical state line and determines the extent to which the ultimate deviatoric stress, $q$ depends on the mean effective stress, $p'$. Hence, $M$ can be regarded as a friction constant. Moreover, $M$ determines the shape of the yield surface (height of the ellipse) and influences the coefficient of lateral earth pressure $K_0^{nc}$ in a normally consolidated stress state under conditions of one-dimensional compression.

The preconsolidation stress, $p_p$ determines the size of the ellipse. In fact, an infinite number of ellipses exist, each corresponding to a particular value of $p_p$.

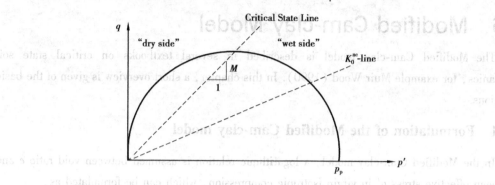

**Figure 7.19   Yield surface of the Modified Cam-clay model in $p'$-$q$-plane**

The left hand side of the yield ellipse (often described as the "dry side" of the critical state line) may be thought of as a failure surface. In this region, plastic yielding is associated with softening, and therefore failure. The values of $q$ can become unrealistically large in this region.

For more detailed information on Cam-clay type models, the readers can be referred to Muir Wood (1990).

## 7.6.2   Parameters of the Modified Cam-clay model

The Modified Cam-clay model is based on five parameters:

| | |
|---|---|
| $v_{ur}$ | Poisson's ratio |
| $\kappa$ | Cam-clay swelling index |
| $\lambda$ | Cam-clay compression index |
| $M$ | Tangent of the critical state line |
| $e_{init}$ | Initial void ratio |

### (1) Poisson's ratio

Poissons ratio $v_{ur}$ is a real elastic parameter and not a pseudo-elasticity constant as used in the Mohr-Coulomb model. Its value will usually be in the range between 0.1 and 0.2.

### (2) Compression index and swelling index

These parameters can be obtained from an isotropic compression test including isotropic unloading. When plotting the natural logarithm of the mean stress as a function of the void ratio for clay-type materials, we can approximate the polt by two straight lines. The slope of the primary loading line gives the compression index and the slope of the unloading line gives the swelling index. These parameters can be obtained from a one-dimensional compression test.

### (3) Tangent of the critical state line

In order to obtain the correct shear strength, the parameter $M$ should be based on the friction angle $\varphi$. The critical state line is comparable with the Drucker-prager failure line, and represents a (circular) cone in principal stress space. Hence, the value of $M$ can be obtained from $\varphi$:

$$M = \frac{6 \sin \varphi}{3 - \sin \varphi} \quad \text{(for initial compression stress states)} \quad (\sigma_1' \leqslant \sigma_2' = \sigma_3') \quad (7.63a)$$

$$M = \frac{6 \sin \varphi}{3 + \sin \varphi} \quad \text{(for triaxial extension stress states)} \quad (\sigma_1' = \sigma_2' \leqslant \sigma_3') \quad (7.63b)$$

$$M \approx \sqrt{3} \sin \varphi \quad \text{(for plain strain stress states)} \quad (\sigma_1' = \sigma_2' \leqslant \sigma_1') \quad (7.63c)$$

In addition to determining the shear strength, the parameter $M$ has an important influence on the value of the coefficient of lateral earth pressure, $K_0^{nc}$ in a state of normal consolidation. In general, when $M$ is chosen such that the model predicts the correct shearing strength, the resulting value of $K_0^{nc}$ is too high.

In addition to the output of standard stress and strain, the Modified Cam-clay model provides output (when being used) on state variables such as the isotropic pre-consolidation stress $p$, and the isotropic over-consolidation ration OCR. These parameters can be visualized by selecting the State parameters option from the Stresses menu. An overview of available state parameters is given below:

| $p_{eq}$ | Equivalent isotropic stress $p_{eq} = p' - \dfrac{q^2}{M^2(p' - c \cot \varphi)}$ | $[kN/m^2]$ |
|---|---|---|
| $p_p$ | Isotropic preconsolidation stress | $[kN/m^2]$ |
| OCR | Isotropic over-consolidation ratio $(OCR = p_p/p_{eq})$ | $[-]$ |

# 8　Dewatering of Excavations

## 8.1　Introduction

According to investigations, most problems encountered in deep excavation have direct or indirect relations with groundwater. Therefore, whether groundwater has been properly dealt with is a crucial point for the success of an excavation. Groundwater-induced problems in an excavation may arise from insufficient investigation of groundwater or geological conditions that lead to inability to fully control the groundwater. It may also arise from misunderstanding of the influence of groundwater, so that a wrong excavation method is adopted. Thus, it is necessary to perform detailed investigations of groundwater and its influences on soils or structures during excavation.

The permeability of clay is lower than $10^{-6}$ cm/s, from which it follows that the flow velocity of groundwater in clay is rather slow. As a result, to lower the groundwater level or decrease the water content of clay below the groundwater level in a short period requires large amounts of energy or chemical methods to force the movement of groundwater in clay. When the groundwater level is lowered or the water content is decreased, the properties of clay will change significantly. The shear strength will increase and the compressibility will decline. In geotechnical engineering, the methods are often categorized as soil improvement methods. However, soil improvement methods are not the subject of this chapter. This chapter is confined to the introduction of dewatering methods, which are employed to lower groundwater levels. Besides, with the low permeability of clay, that is, the flow velocity of groundwater in clay is small, the possibility of groundwater leaking into the excavation zone, which will cause much inconvenience during construction, need not be considered. There is no occurrence of boiling in clay, either. Therefore, when clay is encountered in an excavation, the groundwater level can be ignored in practical engineering applications, except in cases of the strengthening of soil.

The permeability of sand or gravel is usually greater than $10^{-3}$ cm/s, which follows that the flow velocity of groundwater in sand or gravel is rather high and groundwater probably will leak into the excavation zone during excavation and cause much trouble. In the worst cases, it may bring about the loosening of soils, sand boiling, or upheaval failure. To avoid such conditions, it is necessary to design comprehensive dewatering schemes before or during excavation. On the other hand, with the higher flow velocity of sand or gravel, simple methods such as pumping are usually enough to lower the groundwater levels.

Goals of dewatering can be summarized as follows:

①**To keep the excavation bottom dry**. With the higher flow velocity of groundwater in sand or gravel, groundwater may well flow into the excavation zone, which will cause inconvenience for construction. To keep the excavation bottom dry, the groundwater level is generally lowered to 0.5-1 m below the excavation surface. Groundwater flows so slowly in clayey soils that flowing of groundwater into the excavation zone will not occur. There is no need to lower the groundwater level in clay.

②**To prevent leakage of groundwater or soils**. To excavate in sandy or gravelly soil, with a high groundwater level, using either soldier piles or sheet piles, which are not satisfactorily watertight, or diaphragm walls or bored piles with joints that may have defects will risk the possibility of the leaking of groundwater into the excavation zone through the retaining wall. The leakage of groundwater and soils may lead to disastrous results and bring about collapses and failures of excavations in the worst cases, when the leaking is great enough to enlarge the holes in the retaining wall.

③**To avoid sand boiling**. To keep the excavation bottom dry when excavating in sandy or gravelly soils requires lowering the groundwater level within the excavation zone to 0.5-1 m below the excavation bottom at least. While excavation proceeds, the difference between the groundwater levels within and outside the excavation zone grows larger. When the hydraulic gradient around the excavation bottom grows larger or equals the critical hydraulic gradient of soils, sand boiling will occur. Many methods are available to avoid sand boiling. One of them is to lower the groundwater level outside the excavation zone. However, the possibility of ground settlement outside the excavation zone has to be considered.

④**To forestall the upheaval failure**. There exists a permeable layer (such as sand or gravel) underlying the clayey layer. The water pressure from the permeable layer will generate an upheaving force against the clayey layer. When the water pressure acting on the bottom of the clayey layer is larger than the total weights of the clayey layers, upheaval failure will occur. One of the methods to prevent the occurrence of upheaval failure is to lower the piezometric pressure of the permeable layer by pumping.

⑤**To keep the basement from floating**. With the completion of excavation, one starts the construction of basements. In sandy soils, with the light weight of structures during the stage of basement construction (the weight of the basement only), the phenomenon of the floating of the basement is likely to happen if the weight of structures is smaller than the water pressure acting on the foundation base. Once the floating phenomenon has happened, with the differential heaves of the foundation, the floated basement will not necessarily sink back to the original elevation while building construction proceeds, which may lead to damage of the structures. In the worst condition, the basement may need to be demolished or reconstructed. Therefore, dewatering is usually required at the stage of basement construction to keep the upheaving force on foundation bottoms smaller than the weight of structures during construction. Dewatering is to be continued till the upheaving force is smaller than the weight of structures during construction.

# 8.2 Dewatering Methods

Some commonly used methods of dewatering in excavations are in Table 8.1. The following will introduce three methods: the open sump or ditch method, the deep well method, and the well point method.

**Table 8.1  Application conditions for various dewatering methods**

| Methods | Suitable dewatering depth (m) | Coefficient of permeability (m/sec) | Methods | Suitable dewatering depth (m) | Coefficient of permeability (m/sec) |
|---|---|---|---|---|---|
| Open sump method | 0-2 | $>10^{-3}$ | Electro-osmosis method | Depend on wells | $10^{-7}$-$10^{-4}$ |
| Well point method | 3-6 | $10^{-5}$-$10^{-1}$ | Deep well method | $>15$ | $>10^{-4}$ |
| Vacuum well point method | 3-6 | $10^{-6}$-$10^{-3}$ | Deep well+auxiliary vacuum pumps | $>25$ | $>10^{-6}$ |

## 8.2.1  Open sumps or ditches

The open sump method is to collect the groundwater seeping onto the excavation bottom in an open sump, placed in the excavation bottom, by gravity or other natural means and pump the collected water out. If the excavation area is very large or the base is a long narrow shape, several sumps may be placed along the longer side or simply use a long narrow sump, which is called an open ditch. Both the open sump and the ditch are gravity draining methods, which can be the least expensive dewatering method if conditions are favorable (Powers et al., 2007). In practice, the former is more widely used than the latter.

As shown in Figure 8.1, open sumps are usually placed near the retaining wall at the lowest ground or excavation surface. The depth of an open sump is generally 0.6-1 m.

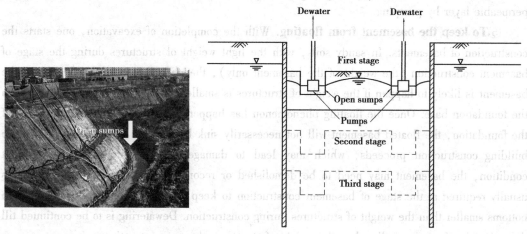

**Figure 8.1  Dewatering method: the open sump method**

The open sump method is the most common and economical method of dewatering. However, its application is confined to permeable layers such as sandy and gravelly soils. Since the bottom of the open sump is lower than the excavation bottom, it will shorten the seepage path along which groundwater from outside seeps into the excavation zone and as a result the exit gradient on the sump bottom will be larger than that on the excavation surface. This fact may bring about sand boiling on the sump bottom, which has to be prevented.

## 8.2.2 Well points

The well point method is also called the vacuum well point method. The method is to place the collecting points connected to the pumping pipe inside a small-diameter well and have them arranged in a line or in a rectangle. The collecting points are usually separated by 0.8-2.0 m. The top of the pumping pipe is connected to the common collecting pipe, which is then pumped out under vacuum, drawing out pore water from the soil and lowering the groundwater level accordingly. Figure 8.2 is a schematic diagram illustrating dewatering the well point method. The method is a type of forced draining method.

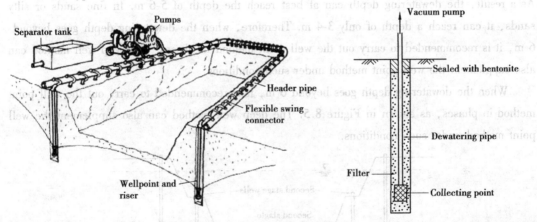

Figure 8.2  Dewatering method: the deep well point method     Figure 8.3  Configuration of a well point

The collecting pipes are usually arranged around the vicinity of an excavation site. Collecting points are the main structures of the well point method. A collecting point is usually 100 cm long, has an external diameter of 5-7 cm, and many small holes are bored along the side to collect groundwater. On the front end is usually installed a spurting device, which is used to help the collecting point penetrate into the ground using high pressure water. As shown in Figure 8.3, the space between a collecting point and the well side are backfilled with filters to protect the collecting point from obstruction. The well can be enveloped by bentonite near the ground surface to increase the degree of vacuum and the efficiency of the well accordingly. Figure 8.4 is the photos of field application.

**Figure 8.4   Well points application and pump for an excavation**

Theoretically, the dewatering depth of the well point method can reach as deep as 10.33 m (1 atmospheric pressure). It is, however, impossible to achieve the condition of total vacuum inside the well. Besides, head loss is unavoidable since there exists friction, which is generated by groundwater flowing through soils, filters, collecting points, pumping pipes, and collecting pipes. As a result, the dewatering depth can at best reach the depth of 5-6 m. In fine sands or silty sands, it can reach a depth of only 3-4 m. Therefore, when the dewatering depth goes beyond 6 m, it is recommended to carry out the well point method in phases. The deep well method can also supplement the well point method under such conditions.

When the dewatering depth goes beyond 6 m, it is recommended to carry out the well point method in phases, as shown in Figure 8.5. The deep well method can also supplement the well point method under such conditions.

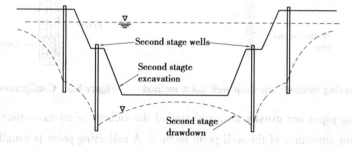

**Figure 8.5   Two-stage well point dewatering**

## 8.2.3   Deep wells

As shown in Figure 8.6, the deep well method is to drill a well near the excavation zone and pump water out of it to make groundwater around the well flow into it under the influence of gravity. As a result, the groundwater level in the vicinity of the well will decline. The deep well method is another gravity draining method.

The diameter of a deep well is about 150-200 mm. If the goal of pumping is confined to lowering groundwater level to keep the excavation bottom dry, the depth of the well could be set around 2.0-5.0 m below the excavation surface and not lower than the bottom of the retaining wall. The wells are to be located around the vicinity of the excavation zone. The pumps used in a deep well can be submerged pumps or centrifugal pumps. Concerning the design, functions, and application limits of various pumps, please consult related documents. According to the type and arrangement of pumps used in a deep well, the pumping depth can reach more than 30 m. These types of wells are typically not spaced closer than 15 m. Figure 8.7 is a photo of deep wells for a deep shaft excavation.

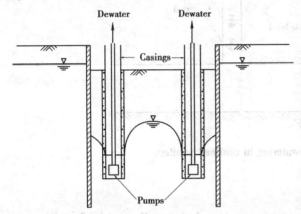

Figure 8.6　Dewatering via the deep well method

Figure 8.7　Application of deep wells

# 8.3　Well Theory

The well theory is aimed at the relationship between the discharge quantity and the drawdown. The relationship is affected by a number of factors: the numbers of wells, their structures, the geological conditions, and pumping time. The relationship between the discharge quantity and drawdown can be either solved by mathematical well formulas or numerical groundwater modeling. The formers are inferred by assuming ideal conditions, whereas the latter is applicable to any geological or pumping condition and is therefore widely adopted. There are a great number of well formulas and it is impossible to introduce all of them here. Only the most widely used well formulas are to be discussed here for reference and application.

## 8.3.1　Full penetration wells

Permeable layers are also designated as aquifers. When both above and below an aquifer, with the piezometric level higher than the top of the aquifer and impermeable layers being found, the aquifer is called a confined aquifer.

When pumping an aquifer with a single well, the piezometric level will decline and a virtual drawdown cone will be formed, centered at the well, as shown in Figure 8.8. From the beginning of pumping, the area of the virtual drawdown cone will grow with the continuation of pumping. After a long period, the expansion rate of the area of the virtual drawdown cone will decline or stop expanding. This section is to introduce the pumping-induced drawdown according to the penetration depth of the well into an aquifer and the expansion conditions of the virtual draw-down cone.

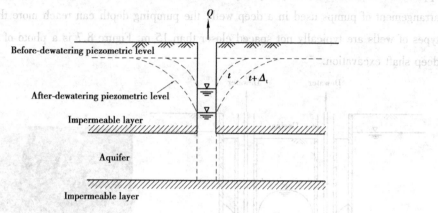

Figure 8.8   Dewatering in confined aquifers

## 1) Full penetration wells

A full penetration well is one that fully penetrates through the aquifer and its flow direction is thus horizontal [Fig 8.9(a)]. An equation of the drawdown curve with regard to time by Theis (1935) can be derived based on the following assumptions:

①The aquifer has a uniform thickness and is a homogeneous and isotropic confined aquifer.

②The well has to be one that fully penetrates through the aquifer.

③The well is 100% efficient. That is to say, no drop exists between the water table in the well and the drawdown curve.

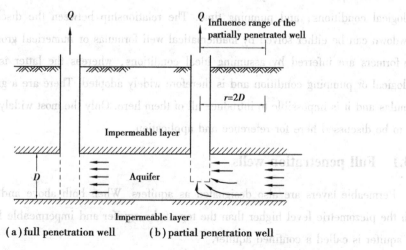

(a) full penetration well        (b) partial penetration well

Figure 8.9   Flow directions around a well in confined aquifers

④Meets the Dupuit-Thiem assumption. That is, the hydraulic gradient of any point on the drawdown curve is the slope at the point.

⑤No recharge water is within the influence range.

⑥The radius of the well is so small that the amount of retained water can be ignored.

Then the equation is written as follows (Figure 8.10):

$$s = \frac{Q}{4\pi kD}W(u) = \frac{Q}{4\pi T}W(u) \tag{8.1}$$

$$W(u) = \int_u^\infty \frac{e^{-u}}{u}du = -0.577\ 2 - \ln u - \sum_{n=1}^\infty (-1)^n \frac{u^n}{n \times n!} \tag{8.2}$$

$$u = \frac{r^2 S}{4Tt} \tag{8.3}$$

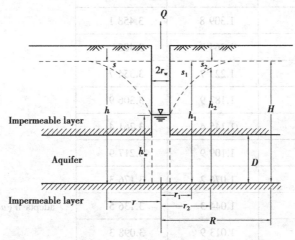

Figure 8.10   Symbol of dewatering in confined aquifers

where

$s$ = drawdown;

$Q$ = discharge quantity;

$k$ = coefficient of permeability;

$D$ = thickness of the aquifer;

$T = kD$ = coefficient of transmissivity;

$W(u)$ = well function;

$u$ = parameter of the well function;

$r$ = distance to the center of the well;

$S$ = coefficient of storage or storativity. The coefficient of storage, which generally ranges between 0.000 5-0.001, is defined as the drained volume of the pore water due to lowering a unit head per unit surface area of an aquifer;

$t$ = time since pumping started.

Eq. 8.1 is Theis's (1935) nonequilibrium equation. Table 8.2 lists the values of well function calculated from Eq. 8.2. Figure 8.11 is the $W(u)-(1/u)$ relation curve calculated from Eq. 8.2, which is also called the standard curve.

### Table 8.2 Well function $W(u)$ (Ferris et al., 1962)

| $N$ | $u=N\times1$ | $u=N\times10^{-1}$ | $u=N\times10^{-1}$ | $u<10^{-2}$ |
|---|---|---|---|---|
| 1.2 | 0.158 4 | 1.659 5 | 3.857 6 | |
| 1.3 | 0.135 5 | 1.588 9 | 3.778 5 | |
| 1.4 | 0.116 2 | 1.524 1 | 3.705 4 | |
| 1.5 | 0.1 | 1.464 5 | 3.637 4 | |
| 1.6 | 0.086 31 | 1.409 2 | 3.573 9 | |
| 1.7 | 0.074 65 | 1.357 8 | 3.514 3 | |
| 1.8 | 0.064 74 | 1.309 8 | 3.458 1 | |
| 1.9 | 0.056 2 | 1.264 9 | 3.405 | |
| 2.0 | 0.048 90 | 1.222 7 | 3.354 7 | |
| 2.1 | 0.042 61 | 1.182 9 | 3.306 9 | |
| 2.2 | 0.037 19 | 1.145 4 | 3.261 4 | |
| 2.3 | 0.032 5 | 1.109 9 | 3.217 9 | |
| 2.4 | 0.028 44 | 1.076 2 | 3.176 3 | |
| 2.5 | 0.024 91 | 1.044 3 | 3.136 5 | adopts $W(u)=-0.577\,2-\ln u$ |
| 2.6 | 0.021 85 | 1.013 9 | 3.098 3 | |
| 2.7 | 0.019 18 | 0.984 9 | 3.061 5 | |
| 2.8 | 0.016 86 | 0.957 6 | 3.026 1 | |
| 2.9 | 0.014 82 | 0.930 9 | 2.999 2 | |
| 3.0 | 0.013 05 | 0.905 7 | 2.959 1 | |
| 3.1 | 0.011 49 | 0.881 5 | 2.927 3 | |
| 3.2 | 0.010 13 | 0.858 3 | 2.896 5 | |
| 3.3 | 0.008 939 | 0.836 1 | 2.866 8 | |
| 3.4 | 0.007 891 | 0.814 7 | 2.837 9 | |
| 3.5 | 0.006 97 | 0.794 2 | 2.809 9 | |
| 3.6 | 0.006 16 | 0.774 5 | 2.782 7 | |
| 3.7 | 0.005 448 | 0.755 4 | 2.756 3 | |
| 3.8 | 0.004 82 | 0.737 1 | 2.730 6 | |

continued

| N | $u=N\times1$ | $u=N\times10^{-1}$ | $u=N\times10^{-1}$ | $u<10^{-2}$ |
|---|---|---|---|---|
| 3.9 | 0.004 267 | 0.719 4 | 2.705 6 | |
| 4.0 | 0.003 779 | 0.702 4 | 2.681 3 | |
| 4.1 | 0.003 349 | 0.685 9 | 2.657 6 | |
| 4.2 | 0.002 969 | 0.67 | 2.634 4 | |
| 4.3 | 0.002 633 | 0.654 6 | 2.611 9 | |
| 4.4 | 0.002 336 | 0.639 7 | 2.589 9 | |
| 4.5 | 0.002 073 | 0.625 3 | 2.568 4 | |
| 4.6 | 0.001 841 | 0.611 4 | 2.547 4 | |
| 4.7 | 0.001 635 | 0.597 9 | 2.526 8 | |
| 4.8 | 0.001 453 | 0.584 8 | 2.506 8 | |
| 4.9 | 0.001 291 | 0.572 1 | 2.487 1 | |
| 5.0 | 0.001 148 | 0.559 8 | 2.467 9 | |
| 5.1 | 0.001 021 | 0.547 8 | 2.449 1 | |
| 5.2 | 0.000 908 6 | 0.536 2 | 2.430 6 | |
| 5.3 | 0.000 808 6 | 0.525 | 2.412 6 | adopts $W(u)=-0.577\ 2-\ln u$ |
| 5.4 | 0.000 719 8 | 0.514 | 2.394 8 | |
| 5.5 | 0.000 640 9 | 0.503 4 | 2.377 5 | |
| 5.6 | 0.000 570 8 | 0.493 | 2.360 4 | |
| 5.7 | 0.000 508 4 | 0.483 | 3.343 7 | |
| 5.8 | 0.000 453 2 | 0.473 2 | 2.327 3 | |
| 5.9 | 0.000 403 9 | 0.463 7 | 2.311 1 | |
| 6.0 | 0.000 360 1 | 0.454 4 | 2.295 3 | |
| 6.1 | 0.000 321 1 | 0.445 4 | 2.279 7 | |
| 6.2 | 0.002 864 | 0.436 6 | 2.264 5 | |
| 6.3 | 0.000 255 5 | 0.428 | 2.249 4 | |
| 6.4 | 0.000 227 9 | 0.419 7 | 2.234 6 | |
| 6.5 | 0.000 203 4 | 0.411 5 | 2.220 1 | |
| 6.6 | 0.000 181 6 | 0.403 6 | 2.205 8 | |

· □Deep Braced Excavations and Earth Retaining Systems ·

continued

| $N$ | $u=N\times1$ | $u=N\times10^{-1}$ | $u=N\times10^{-1}$ | $u<10^{-2}$ |
|---|---|---|---|---|
| 6.7 | 0.000 162 1 | 0.395 9 | 2.191 7 | |
| 6.8 | 0.000 144 8 | 0.388 3 | 2.177 9 | |
| 6.9 | 0.000 129 3 | 0.381 | 2.164 3 | |
| 7.0 | 0.000 115 5 | 0.373 8 | 2.150 8 | |
| 7.1 | 0.000 103 2 | 0.366 8 | 2.137 6 | |
| 7.2 | 0.000 092 19 | 0.359 9 | 2.124 6 | |
| 7.3 | 0.000 082 39 | 0.353 2 | 2.111 8 | |
| 7.4 | 0.000 073 64 | 0.346 7 | 2.099 1 | |
| 7.5 | 0.000 065 83 | 0.340 3 | 2.086 7 | |
| 7.6 | 0.000 058 86 | 0.334 1 | 2.074 4 | |
| 7.7 | 0.000 052 63 | 0.328 | 2.062 3 | |
| 7.8 | 0.000 047 07 | 0.032 1 | 2.050 3 | |
| 7.9 | 0.000 042 1 | 0.316 3 | 2.038 6 | |
| 8.0 | 0.000 037 67 | 0.310 6 | 2.026 9 | |
| 8.1 | 0.000 033 7 | 0.305 | 2.015 5 | adopts $W(u)=-0.577\,2-\ln u$ |
| 8.2 | 0.000 030 15 | 0.299 6 | 2.004 2 | |
| 8.3 | 0.000 026 99 | 0.294 3 | 1.993 | |
| 8.4 | 0.000 024 15 | 0.289 1 | 1.982 | |
| 8.5 | 0.000 021 62 | 0.284 | 1.971 1 | |
| 8.6 | 0.000 019 36 | 0.279 | 1.960 4 | |
| 8.7 | 0.000 017 33 | 0.274 2 | 1.948 8 | |
| 8.8 | 0.000 015 52 | 0.269 4 | 1.939 3 | |
| 8.9 | 0.000 013 9 | 0.264 7 | 1.928 | |
| 9.0 | 0.000 012 45 | 0.260 2 | 1.918 7 | |
| 9.1 | 0.000 011 15 | 0.255 7 | 1.908 7 | |
| 9.2 | 0.000 009 988 | 0.251 3 | 1.898 7 | |
| 9.3 | 0.000 008 948 | 0.247 | 1.888 8 | |
| 9.4 | 0.000 008 018 | 0.242 9 | 1.879 1 | |

continued

| $N$ | $u=N\times1$ | $u=N\times10^{-1}$ | $u=N\times10^{-1}$ | $u<10^{-2}$ |
|---|---|---|---|---|
| 9.5 | 0.000 007 085 | 0.238 7 | 1.869 5 | |
| 9.6 | 0.000 006 439 | 0.234 7 | 1.859 9 | |
| 9.7 | 0.000 005 771 | 0.230 8 | 1.850 5 | adopts $W(u)=-0.577\ 2-\ln u$ |
| 9.8 | 0.000 005 173 | 0.226 9 | 1.841 5 | |
| 9.9 | 0.000 004 637 | 0.223 1 | 1.832 | |

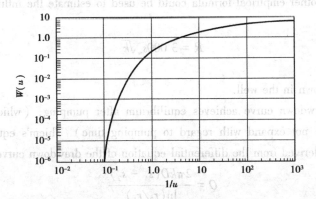

**Figure 8.11   Standard curve of the well function** (Ferris et al., 1962)

Habitually, the coefficients of permeability, transmissivity, and storage are all called hydraulic parameters. The coefficient of permeability is an intrinsic parameter of a soil, the coefficient of transmissivity is the coefficient of permeability multiplied by the thickness of the aquifer. Since the thickness of a confined aquifer is a constant, the coefficient of transmissivity of a confined aquifer is also a constant.

The coefficient of storage is defined as the drained volume of the pore water due to lowering a unit head per unit surface area of an aquifer. Since soil in a confined aquifer is always in the saturated state, the drainage of pore water during dewatering should be caused by the decrease of the thickness of the aquifer. Since the decrease of heads follows the increase of effective stress, the thickness of the aquifer will reduce as a result. Following the above elucidation, the coefficient of storage of a confined aquifer, which generally ranges between 0.000 5-0.001, is similar to the compressibility of soils.

When pumping time ($t$) is rather long or the distance ($r$) is quite short, the parameter u will be very small. Thus, when $u \leqslant 0.05$, the high order of the multinomial at the right side of the equation can be ignored. Equation 9.2 can then be rewritten as

$$s = \frac{Q}{4\pi T}(-0.577\ 2 - \ln u) \qquad (8.4)$$

which can also be written as

$$s = \frac{0.183Q}{T}\log\frac{2.25Tt}{r^2 S} \tag{8.5}$$

Eq. 8.4 or 8.5 is called Jacob's modified nonequilibrium equation (Jacob, 1940), which is only applied when $u \leqslant 0.05$.

Assuming the influence range of pumping is defined as the distance where the draw down just declines to 0, the influence range of pumping-induced drawdown ($R$) can be calculated from Eq. 8.5 as

$$R = \sqrt{\frac{2.25Tt}{S}} \tag{8.6}$$

In addition, another empirical formula could be used to estimate the influence range Sichart (1928)

$$R = 3\,000 s_w \sqrt{k} \tag{8.7}$$

where

$s_w$ is the drawdown in the well.

Assume the drawdown curve achieves equilibrium after pumping, (which follows that the drawdown curve will not expand with regard to pumping time). Thiem's equilibrium (Thiem, 1906) can thus be derived from the differential equation of the drawdown curve:

$$Q = \frac{2\pi k D(s_1 - s_2)}{\ln(r_2/r_1)} \tag{8.8}$$

where

$r_1$ and $r_2$ are the distances of the first observation well and the second observation well to the well center, respectively;

$s_1$ and $s_2$ are drawdowns in the first and the second observation well, respectively.

Let $R$ be the influence range and $R/r$ substitute for $r_2/r_1$, Eq. 8.8 can be derived to compute drawdown at any distance:

$$s = \frac{Q \ln(R/r)}{2\pi k D} \tag{8.9}$$

Let the drawdown in the well be ($H-h_w$), the radius of the well be $r_w$, we can have

$$Q = \frac{2\pi k D(H - h_w)}{\ln(R/r_w)} \tag{8.10}$$

Eq. 8.11 is derived from the differential equation of the drawdown curve, assuming that drawdown after pumping is in equilibrium and that the drawdown curve does not expand with pumping time. Substitute the influence range ($R$) of Eq. 8.6 into Eq. 8.5, Jacob's modified nonequilibrium equation, we then have Eq. 8.8. Thus, Thiem's equilibrium equation is a special case of Jacob's modified nonequilibrium equation. That is to say, when the drawdown curve does not expand any more, the pumping time coming to t and the influence range extending to $R$, Jacob's modified nonequilibrium equation will be identical to Thiem's equilibrium equation.

After pumping for a period of time, turn off the pump and the drawdown curve will gradually return to the original water table. The relationship between the recovered draw down curve and time can obtain the coefficient of transmissivity or permeability, as is designated as the recovery method. According to Theis (1935), the relationship between the residual drawdown and time can be expressed as follows:

$$s' = \frac{Q}{4\pi T} \ln \frac{t}{t'} = \frac{2.3Q}{4\pi T} \lg \frac{t}{t'} \tag{8.11}$$

where

$s'$ = residual drawdown, that is, the distance between the water level in the pumping well and the original groundwater level;

$Q$ = recovered quantity, equivalent to discharge quantity;

$T$ = coefficient of transmissivity;

$t$ = time since pump started;

$t'$ = time since pump is stopped.

The recovery method can be used to examine the results of pumping tests. What's more, during a long period of pumping and dewatering, the activities of pumping may be suspended temporarily for certain reasons and the drawdown curve will recover gradually. The recovery method can be used under such conditions. The results can not only offer extra data but be compared with those of pumping tests.

## 2) Partial penetration well

In practice, the partial penetration well is more widely used than the full penetration well [Figure 8.9(b)]. Within the range of $r \leqslant 2D$, the flow lines in the vicinity of the well are not necessarily horizontal. Instead, some are vertical. For most types of soils, the vertical coefficients of permeability are about one-tenth of the horizontal one. Thus, the well formulas concerning partial penetration wells need modifying.

Considering the nonequilibrium theories for partial penetration wells are quite complicated, we leave them aside and only introduce an equilibrium theory by Kozeny (1953).

As shown in Figure 8.12, when $r > 2D$, the effects of a partial penetration well can be ignored since its drawdown curve can be fully solved when seen as a full penetration well. When $r \leqslant 2D$, however, the amount of drawdown for a partial penetration well is larger than that for a full penetration well, and the quantity of water to be pumped to achieve the designed drawdown can be expressed as follows:

$$Q = \frac{2\pi T(H - h_w)}{\ln(R/r_w)}\mu \tag{8.12}$$

$$\mu = \frac{D_1}{D}\left(1 + 7\sqrt{\frac{r_w}{2D_1}} \cos \frac{\pi D_1}{2D}\right) \tag{8.13}$$

Where

$\mu$ = coefficient of modification;

---

$D_1$ = penetration depth of the well;

$D$ = thickness of the aquifer.

Other parameters mean the same as in the equation of the full penetration well.

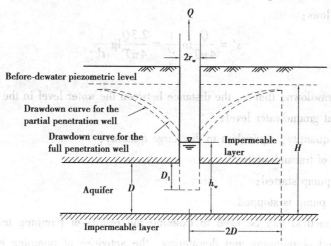

**Figure 8.12  Drawdown curve for partial penetration wells in confined aquifers**

## 8.3.2  Free aquifers

A free aquifer, also called an unconfined aquifer, refers to an aquifer which is exposed to the atmosphere and is underlain by an impermeable layer. The analyses of free aquifers are also divided into full penetration wells and partial penetration wells.

### 1) Full penetration well

The solution for nonequilibrium equations of pumping-induced draw down in free aquifers is quite complicated. The reason is mainly that the coefficient of transmissivity is not a constant and varies with pumping time and distance. In addition, the vertical directions of flow near the well are so crucial that they cannot be ignored in the derivation.

On the other hand, if the drawdown is much smaller than the thickness of aquifers, the thickness of aquifers can thus be assumed to be a constant during dewatering. Since the coefficient of transmissivity is the product of the coefficient of permeability and the thickness of the aquifer, the coefficient of transmissivity is also a constant. Theis's nonequilibrium equation or Jacob's modified nonequilibrium equation can also be applied to pumping in free aquifers accordingly. The required hydraulic parameters are also the coefficients of transmissivity ($T$) and storage ($S$).

As in Section 8.3.1, the coefficient of storage is given as the drained volume of the pore water due to lowering a unit head per unit surface area of an aquifer. The drainage per unit surface area of a free aquifer is caused by the decline of groundwater level. Therefore, in a free aquifer, the coefficient of storage is the same as the ratio of free water in the soil to the soil volume (Powers, 1992). Its value is either smaller than or the same as the porosity of soil. For example, if the porosity of soil is 30% and two-thirds of pore water is drained as a result of the lowering of groundwater level by pumping, the coefficient of storage will then be 20% or 0.2. For most free aquifers, the coefficient of storage would be around 0.2-0.3.

As shown in Figure 8. 13, under the Dupuit-Thiem assumption (Thiem, 1906), the equilibrium equation of drawdown of a full penetration well in a free aquifer is

$$Q = \frac{\pi k (h_2^2 - h_1^2)}{\ln(r_2/r_1)} \tag{8.14}$$

Where $h_1$ and $h_2$ are the heights of the water levels at the distances of $r_1$ and $r_2$, respectively. Substitute the influence range ($R$) and the radius of the well ($r_w$) for $r_2$ and $r_1$, respectively, in the above equation, then we have

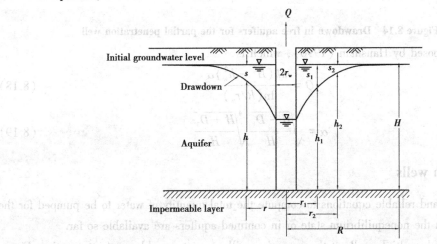

**Figure. 8.13   Drawdown in free aquifers for the full penetration well**

$$Q = \frac{\pi k (H^2 - h_w^2)}{\ln(R/r_w)} \tag{8.15}$$

Rewrite equation to obtain the groundwater level or drawdown at any distance ($r$) as follows:

$$H^2 - h^2 = \frac{Q \ln(R/r)}{\pi k} \tag{8.16}$$

or

$$h^2 - h_w^2 = \frac{Q \ln(r/r_w)}{\pi k} \tag{8.17}$$

where

$h$ is the groundwater level at the distance of $r$.

## 2) Partial penetration well

The same as the full penetration well, the nonequilibrium equation of drawdown of a partial penetration well is rather complicated and is not to be discussed here. This section only introduces

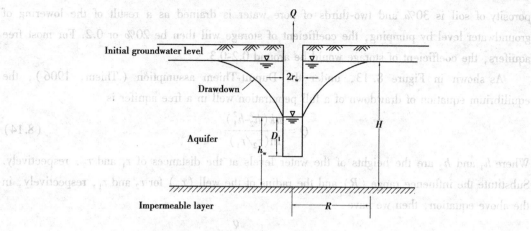

Figure 8.14　Drawdown in free aquifers for the partial penetration well

an equation proposed by Hausman ( 1990 ), which is

$$Q = \frac{\pi k (H^2 - h_w^2) \alpha}{\ln(R/r_w)} \tag{8.18}$$

$$\alpha = \sqrt{\frac{H - D_1}{H}} \sqrt[4]{\frac{H + D_1}{H}} \tag{8.19}$$

### 8.3.3　Group wells

No simple and reliable equations to compute the total quantity of water to be pumped for the multiple wells in the nonequilibrium state or in confined aquifers are available so far.

Pumping with multiple wells at the same time will encounter problems of intervention. Figure 8.15(a) diagrams the multiple wells in a free aquifer. Assuming only well #1 is used in pumping, in the equilibrium state, the required quantity of water to be pumped for well #1 can be computed according to Eq. 8.20:

$$Q = \frac{\pi k (H^2 - h_1^2)}{\ln(R/r_1)} \tag{8.20}$$

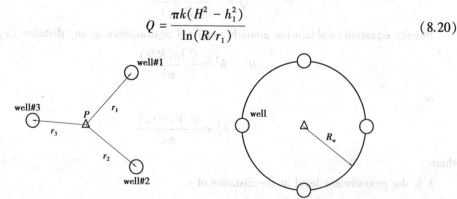

( a ) notations　　　　　　　( b ) circular arrangement of wells

Figure 8.15　Group wells

Similarly, if well #2 is used exclusively, the required quantity of water to be pumped can also be computed using the same equation. Assuming there exist n wells, pumping can also be computed using the same equation. Assuming there exist n wells, pumping would be (Forchheimer, 1930)

$$Q_{tot} = \frac{\pi k (H^2 - h^2)}{\ln R - (1/n) \ln r_1 r_2 \dots r_n} \tag{8.21}$$

where

$Q_{tot}$ = total quantity of water to be pumped;

$n$ = number of wells;

$h$ = groundwater level at point $P$;

$r_1$, $r_2$, ..., $r_n$ = distances from well #1, #2, ⋯ to point $P$.

When the group is arranged in a circle [see Figure 8.15(b)], the above equation can be simplified. The total quantity of water to be pumped at the center $P$ is

$$Q = \frac{\pi k (H^2 - h^2)}{\ln (R/R_w)} \tag{8.22}$$

However, no simple and reliable equations to compute the total quantity of water to be pumped for the multiple wells in the nonequilibrium state or in confined aquifers are available so far.

# 8.4 Pumping Test

Two types of pumping tests, that is, the step drawdown and constant rate tests, are usually carried out before dewatering. The major objective of a step drawdown test is to determine the capacity of a well while the main goal of a constant rate test is to obtain the hydraulic parameters, such as the coefficients of permeability, transmissivity, and storage. Although there are full and partial penetration well theories, as introduced in Section 8.3, the test well for a pumping test should fully penetrate into the aquifer in order for its results to be analyzed by using the well theories of the full penetration well, considering the equations for a partial penetration well, are too complicated and some of them are empirical formulas.

## 8.4.1 Step drawdown tests

In a step drawdown test, the well is pumped at several successively higher pumping rates and the drawdown for each rate is recorded. The entire test is usually carried out within one day. Usually, 5 to 8 pumping rates are applied, each lasting 1-2 h.

Figure 8.16 displays the relationships between pumping rates and drawdowns in a hypothetical case. As shown in the figure, line45 is of steeper slope than line123. Obviously, when the pumping rate is equal to 90 m³/h or 100 m³/h, the drawdown in the well increases substantially and the well is in the condition of overpumping. The pumping rate at the intersection of line123 and line45 is thus the capacity of the well.

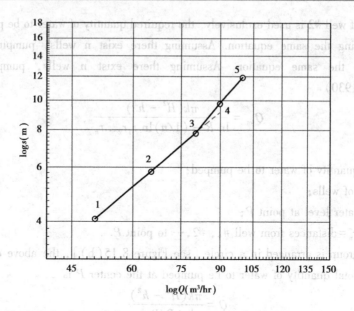

**Figure 8.16   Relation between drawdowns and pumping rates for a step drawdown test**

## 8.4.2   Constant rate tests

Based on the result of a step drawdown test, we can select an appropriate pumping rate slightly smaller than the capacity of the well. Pump the water for a long period of time, and record the drawdown for each observation well. This section will introduce the analytical methods for constant rate pumping tests in confined and free aquifers, respectively.

### 1) Confined aquifers

Given the results of pumping tests of a confined aquifer, the coefficients of transmissivity and storage of the aquifer can be obtained by using the nonequilibrium and the equilibrium equations. Assuming the drawdown curve is under the nonequilibrium condition, according to Theis's nonequilibrium equation, we have

$$\lg s - \lg W(u) = \lg\left(\frac{Q}{4\pi T}\right) = \text{constant} \qquad (8.23)$$

$$\lg(t/r^2) - \lg\left(\frac{1}{u}\right) = \lg\left(\frac{S}{4T}\right) = \text{constant} \qquad (8.24)$$

The relational curve of $\lg s - \lg Q$ is called the standard curve, as shown in Figure 8.11. The relational curve of $\lg(t/r^2) - \lg(1/u)$ is called the data curve. From Eqs. 8.23 and 8.24, we can see that the standard curve and the data curve are quite similar, given the same scale. If we superpose the two curves onto the same diagram of coordinates, the curves of $\lg s$ and $\lg W(u)$ are identical except for a shift of constant $\lg(Q/4\pi T)$. Similarly, the curves of $\lg(t/r^2)$ and $\lg(1/u)$ are also identical except for a shift of constant $\lg(S/4T)$. Taking advantage of the characteristic between the standard curve and the data curve, the hydraulic parameters can be derived as follows:

①Depict the standard curve $W(u)-(1/u)$ on the coordinate diagram on the logarithm scale.

②Depict the data curve $s-t/r^2$ in the same scale as that of the standard curve on another coordinate diagram on the logarithm scale. If there exists only one observation well, the values of $s$ and $t$ obtained from the well can be directly adopted (i.e. $r$ is a constant; $t$ is a variable). If there is more than one observation well, we can also adopt the values of $s-t/r^2$ at a specific time (i.e. $t$ is a constant; $r$ is a variable).

③Shift the coordinate sheets to make the data and standard curves meet. Be sure to keep the coordinates parallel to each other.

④Pick a point, called the matching point, and read the values of $W(u)$, $1/u$, $s$ and $t/r^2$ at the point.

⑤Substitute the values of $W(u)$, $1/u$, $s$ and $t/r^2$ into Eqs 8.1 and 8.3, obtain $T$ and $S$

Since the computing of Theis's nonequilibrium equation is too complicated, it is rarely adopted, and Jacob's nonequilibrium equation is most widely used. When $t$ is so great or $r$ is small enough that $u \leqslant 0.05$, Jacob's nonequilibrium equation can be used to obtain hydraulic parameters. Similarly, we can depict the relations between $s$ and $\lg t/r^2$, whose values are obtained from a certain observation well (i.e. $r$ is a constant; $t$ is a variable). If there is more than one well, the $s-t/r^2$ relation at a specific time (i.e. $t$ is a constant; $r$ is a variable) can also be adopted. According to the $s-\lg t/r^2$ curve and Eq. 8.5, we have

$$T = 0.183 \frac{Q}{(\Delta s)_J} \tag{8.25}$$

where

$(\Delta s)_J$ = slope of $s-\lg t/r^2$ curve, the value of which equals the decreased amount of $s$ when $t/r^2$ increases by a factor of ten.

The coefficient of transmissivity can thus be derived using the above equation. If we extend the line segment of $s-\lg t/r^2$, corresponding value of $t/r^2$ is $(t/r^2)_{s=0}$ where the extended line intersects with $s=0$ at $(t/r^2)_{s=0}$. According to Eq. 8.5, we can compute the coefficient of storage by the following equation:

$$S = 2.25T(t/r^2)_{s=0} \tag{8.26}$$

where

$(t/r^2)_{s=0}$ is the corresponding value of $t/r^2$ when $s=0$.

If the drawdown curve of a pumping test reaches equilibrium or changes little, Thiem's equilibrium equation can then be adopted for the calculation of hydraulic parameters. According to Eq. 8.8, the coefficient permeability can be obtained as follows:

$$k = \frac{Q \ln(r_2/r_1)}{2\pi D(s_2 - s_1)} = \frac{0.366Q}{D(s_2 - s_1)} \lg \frac{r_2}{r_1} \qquad (8.27)$$

and the coefficient of transmissivity is

$$T = kD = \frac{Q \ln(r_2/r_1)}{2\pi(s_2 - s_1)} \qquad (8.28)$$

If there are two or more observation wells, first establish the relations between the drawdown ($s$) and the distance on the logarithm scale ($\log r$) for each well. Then the coefficient of transmissivity can be derived from Eq. 8.27 as

$$T = \frac{0.366Q}{(\Delta s)_T} \qquad (8.29)$$

where

$(\Delta s)_T$ is the slope of slope of $s - \lg r$ curve and its value equals the decreased amount of $s$ when $r$ increases ten times.

Similarly, if we extend the line segment of the $s - \lg r$ curve, the extended line will intersect with $s = 0$ at $R$. The same as earlier discussions, $R$ represents the influence range of the drawdown. According to the influence range in Eq. 8.6, the coefficient of storage can be derived as

$$S = \frac{2.25Tt}{R^2} \qquad (8.30)$$

The coefficient of transmissivity can also be derived from the recovery method. Similarly, depict the $s' - \lg(t/t')$ relation according to the observation of each well. Using Eq. 8.11, we can derive

$$T = \frac{0.183Q}{(\Delta s')_r} \qquad (8.31)$$

where

$(\Delta s')_r$ is the slope of slope of $s' - \lg(t/t')$ curve and its value equals the decreased amount of $s$ when $r$ increases ten times.

## 2) Free aquifers

If the pumping-induced drawdown is much smaller than the thickness of the free aquifer, both Theis's nonequilibrium equation and Jacob's modified nonequilibrium equation for confined aquifers are applicable and the required hydraulic parameters are also the coefficients of transmissivity and storage.

If there are two observation wells, let the distances between the pumping well and the two observation wells be $r_1$, and $r_2$, respectively and the water levels in the two observation wells are separately $h_1$ and $h_2$. According to Eq. 8.14, the coefficient of permeability will be

$$k = \frac{2.3Q}{\pi(h_2^2 - h_1^2)} \lg \frac{r_2}{r_1} \qquad (8.32)$$

# 8.5 Dewatering Plan for an Excavation

## 8.5.1 Selection of dewatering methods

The applicable soil types and dewatering depths of the above three methods can be found in Table 8.1. In practical engineering, one or more methods can be adopted in a single phase or many phases.

## 8.5.2 Determination of hydraulic parameters

For equilibrium equations ( e. g. Thiem's equilibrium equation ), the coefficient of permeability ( $k$ ) is the only required hydraulic parameter. The coefficient of permeability can be obtained from laboratory constant head tests, falling head tests, empirical formulas, or pumping tests.

Nonequilibrium equations ( e. g. Theis's and Jacob's nonequilibrium equations ) require the coefficients of transmissivity ( $T$ ) and coefficient of storage ( $S$ ). The coefficient of transmissivity of a confined aquifer equals the product of the coefficient of permeability and the thickness of the aquifer and is thereby a constant. If the pumping induced an unconfined aquifer far less than the thickness of an aquifer, the thickness of the aquifer can be assumed to be constant and one can obtain the coefficient of transmissivity accordingly.

The coefficient of storage of a confined aquifer, which ranges from 0.000 5 to 0.001, is about the same as the compressibility of the soil. The coefficient of storage of an unconfined aquifer is a little smaller than the porosity of a soil and ranges from 0.2 to 0.3.

## 8.5.3 Determination of the capacity of wells

The efficiency of a well cannot reach 100% since there exist friction losses. As a result, we have to estimate the pumping capacity of each well.

The quantity of groundwater that flows into a deep well ( $Q$ ) can be expressed as follows:

$$Q = 2\pi r_w h_w k i_e \tag{8.33}$$

where

$i_e$ = the entry hydraulic gradient of groundwater flowing into a well;

$h_w$ = the groundwater level at which groundwater flows into the well ( see Figure 8.17)

According to Sichart and Kyrieleis (1930), the entry hydraulic gradient cannot be larger than

$$i_{e,\max} = \frac{1}{15\sqrt{k}} \tag{8.34}$$

where

$k$ is the coefficient of permeability.

If one substitutes $i_{e,\max}$ for $i_e$ in Eq. 8.33, the pumping capacity of a deep well ( $Q_w$ ) will be

$$Q_w = 2\pi r_w \frac{\sqrt{k}}{15} \tag{8.35}$$

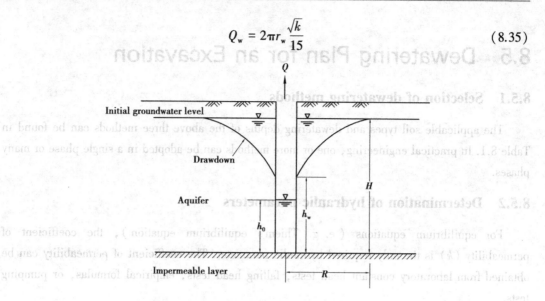

**Figure 8.17  Working merhaniss of pumping**

Because the value of $h_w$ is difficult to estimate, assume it to be about the groundwater level in the deep well at the preliminary estimation. That is, $h_w = h_0$. Then examine it using the results of the pumping test. When pumping is carried out, the capacity of each deep well in the multiple wells may be smaller than the above calculated $Q_w$ for there exists intervention among wells. The influence among the wells in the multiple wells can be estimated by pumping test or numerical simulation of groundwater.

The well point method is also affected by friction loss and group effect, which relate to the type of soil or the coefficient of permeability. Table 8.3 lists some suggested values of the capacities of the well point method (JSA, 1988).

**Table 8.3  Relationship between coefficient of permeability and $Q_w$ for well points (JSA, 1988)**

| $k$ (cm/s) | $Q_w(\times 10^{-3} \text{ m}^3/\text{min})$ |
|---|---|
| $1.0 \times 10^{-3}$ | 1—5 |
| $5.0 \times 10^{-3}$ | 5—10 |
| $1.0 \times 10^{-2}$ | 10—20 |
| $5.0 \times 10^{-2}$ | 40— |

## 8.5.4  Estimation of the number of wells

There have to be enough pumping wells at an excavation site to lower the groundwater level to a depth below the excavation surface. Generally speaking, the groundwater level has to be at least

0.5-1 m below the excavation surface. The design of deep wells is elucidated as follows:

## (1) Compute the total quantity of water to be pumped in the excavation area

Assume the excavation site is an imaginary well, with a radius of $R_w$ (see Figure 8.18) and compute the required total quantity ( $Q_{tot}$ ) of water to be pumped in the imaginary well using either equilibrium or nonequilibrium equations. The computation is discussed in Sections 8.3.1 and 8.3.2.

The radius of the imaginary well ( $R_w$ ) can be computed by means of the equivalent area or perimeter, for example,

$$R_w = \sqrt{\frac{a \times b}{\pi}} \tag{8.36}$$

or

$$R_w = \frac{a + b}{\pi} \tag{8.37}$$

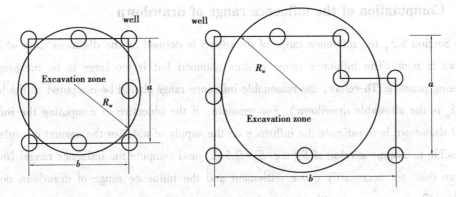

**Figure 8.18　Radius of an imaginary well**

## (2) Compute the pumping capacity of each well

Determine the radius of each deep well and assume its water level to be $h_w$, which should be a little lower than the water level, usually 0.5-1 m below the excavation surface, at the excavation center. Therefore, the capacity of each well ( $Q_w$ ) can be estimated following the method introduced in Section 8.4.3.

## (3) Compute the number of deep wells

The number is

$$n = \frac{Q_{tot}}{Q_w} \tag{8.38}$$

## (4) Examine the assumed water level in each well

The total quantity of water and the number of wells determined, the assumed water level ( $h_w$ ) has to be examined. Under equilibrium and in a free aquifer, the water level in a well can be obtained following the methods introduced in Section 8.3.3. If the derived water level differs from the assumed $h_w$, repeat step 2 to 4 till the computed and assumed values match each other.

### (5) **Examine the drawdown at the excavation center**

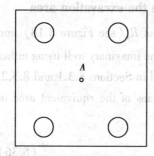

**Figure 8.19    An excavation site with pumping wells**

Following the same method in step 4, compute the drawdown at the excavation center and comers and check if the computed values are smaller than the designed values.

As shown in Figure 8.19, there are four full penetration wells in an excavation site. To obtain the drawdown at the excavation center (point $A$) with the four wells pumping simultaneously, use Eq. 8.1 or 8.4 to compute the drawdown of each well at point $A$ and add up the values to find the total drawdown at point $A$ if pumping time is not long and the drawdown curve is under nonequilibrium condition.

## 8.5.5    Computation of the influence range of drawdown

In Section 8.3, the influence range of drawdown is defined as the distances from which the drawdown is none. The influence range is thus obtained but is too large to be meaningful in engineering practice. Therefore, the reasonable influence range should be computed on the basis of $s = S_a$ ($S_a$ is the allowable drawdown). For example, if the objective of computing the influence range of drawdown is to estimate the influence on the supply of water or the amount of settlement, it is feasible to assume a value of $S_a$ (e.g. $S_a = 0.5$ m) and compute the influence range. Note that drawdown does not necessarily cause settlement and the influence range of drawdown does not equal the influence range of settlement, either.

Figure 8.20(a) shows a diaphragm wall penetrating through the impermeable layer. Since pumping is confined to the excavation area, it will not have influence outside the excavation zone and the influence range of drawdown is minimal accordingly. Figure 8.20(b) shows the pumping is carried out outside of an excavation zone and the influence range of drawdown extends by a distance outside the excavation zone as a result. Figure 8.20(c) shows the pumping is executed under an impermeable layer. While water is pumped in a confined aquifer, the decrease of piezometric level in the confined aquifer may cause settlement, though the groundwater level above the impermeable layer does not necessarily come down. Figure 8.20(d) is a diaphragm wall located in a permeable layer and the well depth is less than the depth of the wall. As a result, pumping will not cause drawdown of groundwater level outside the excavation zone. Figure 8.20 (e) is a case in which the well depth is larger than the depth of the wall and the influence range of drawdown may extend outside the excavation zone by a distance.

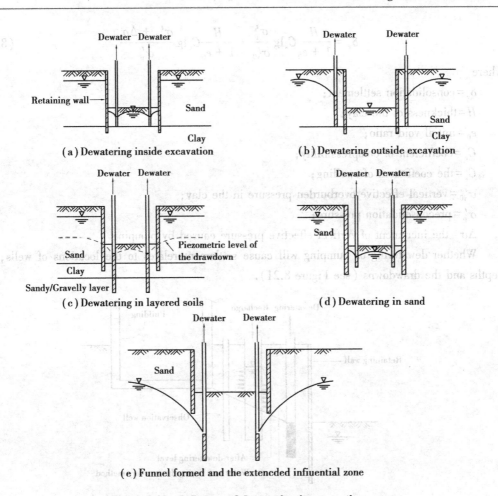

(a) Dewatering inside excavation

(b) Dewatering outside excavation

(c) Dewatering in layered soils

(d) Dewatering in sand

(e) Funnel formed and the extencded infiuential zone

**Figure 8.20   Influence of dewatering in excavations**

# 8.6   Dewatering and Ground Settlement

Dewatering will decrease the porewater pressure and increase the effective stress of soils accordingly. In sandy or gravelly soils, the increase of effective stress will produce elastic settlement. In clayey soils, not only elastic settlement but consolidation settlement will be induced. Generally speaking, the amount of elastic settlement is far less than that of consolidation settlement and usually can be neglected. As far as dewatering and pumping are concerned, the consolidation settlement should be considered.

The amount of consolidation settlement induced by pumping can be computed by using Terzaghi's one-dimensional consolidation theory as follows:

For normally consolidated clay:

$$\delta_v = \frac{H}{1+e_0} C_c \lg \frac{\sigma'_{v0} + \Delta\sigma}{\sigma'_{v0}} \tag{8.39}$$

For overconsolidated clay:

$$\delta_v = \frac{H}{1 + e_0} C_s \lg \frac{\sigma'_p}{\sigma'_{v0}} + \frac{H}{1 + e_0} C_c \lg \frac{\sigma'_{v0} + \Delta\sigma}{\sigma'_p} \qquad (8.40)$$

where

$\delta_v$ = consolidation settlement;

$H$ = thickness of the clay;

$e_0$ = initial void ratio;

$C_c$ = coefficient of compressibility;

$C_s$ = the coefficient of swelling;

$\sigma'_{v0}$ = vertical effective overburden pressure in the clay;

$\sigma'_p$ = preconsolidation pressure;

$\Delta\sigma$ = the increment of vertical effective pressure caused by pumping.

Whether dewatering or pumping will cause settlement relates to the locations of wells, the depths and the drawdowns (see Figure 8.21).

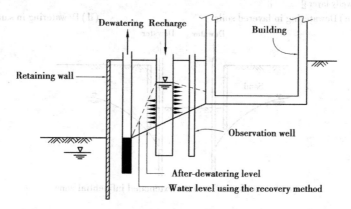

Figure 8.21　Recovery method

Theoretically, the influence range of dewatering-induced settlement is the distance from which settlement declines to 0. However, the definition has little meaning in engineering practice. As explicated, drawdown is the major cause of settlement. If we apply Theis's nonequilibrium theory to compute the influence range of drawdown, the range will work out as extremely far. The result of Jacob's modified nonequilibrium equation is similar. Does it follow that the influence range of settlement is actually extremely far? Certainly not. Thus, the reasonable settlement influence range should be determined on condition that $\delta = \delta_a$ ($\delta_a$ is the allowable settlement) or when the angular distortion is small.

( 1 ) **Example 8.1**

The allocation of the pumping wells and observation wells are shown in Figure 8.22. The thickness of the permeable layer is 10 m; the distances between the pumping well center and the observation wells K-1, K-2, and K-3 are 10, 20, and 30 m, respectively. The radius of the well $r_w = 0.5$ m; the punmping rate $Q = 2\ 000$ cm$^2$/s. Table 8.4 lists the drawdowns observed in the pumping well and the three observation wells. Table 8.5 lists the recovered water levels after

· 192 ·

pumping is stopped. Compute the coefficients of transmissivity and storage using the methods of Theis, Jacob, and Thiem, and the recovery method, respectively.

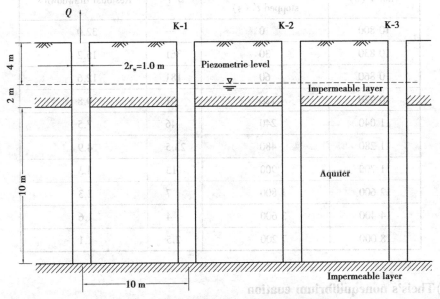

**Figure 8.22  Pumping test and the soil condition**

**Table 8.4  Results of the pumping test**

| Time $t$ (s) | Drawdown $s$ (cm) | | | |
| --- | --- | --- | --- | --- |
| | Pumping well | Observation well K-1 | Observation well K-2 | Observation well K-3 |
| 0 | 0 | 0 | 0 | 0 |
| 30 | 10.0 | 0.3 | 0 | 0 |
| 60 | 12.5 | 0.6 | 0.1 | 0 |
| 120 | 55.5 | 1.1 | 0.3 | 0.1 |
| 240 | 18.1 | 2.1 | 0.7 | 0.3 |
| 480 | 20.2 | 3.5 | 1.2 | 0.6 |
| 900 | 22.4 | 5.0 | 2.0 | 1.0 |
| 1 800 | 24.8 | 6.6 | 3.4 | 1.8 |
| 3 600 | 27.5 | 8.4 | 4.9 | 3.1 |
| 5 400 | 29.2 | 9.4 | 6.0 | 4.1 |
| 7 200 | 30.9 | 10.2 | 6.7 | 4.7 |
| 10 800 | 32.0 | 11.3 | 7.7 | 5.6 |

**Table 8.5  Recovered water levels in the pumping well**

| Time $t$ (s) | Time after pumping is stopped $t'$ (s) | $t/t'$ | Residual drawdown $s'$ |
|:---:|:---:|:---:|:---:|
| 10 800 | 0 | — | 32.0 |
| 10 830 | 30 | 361 | 15.2 |
| 10 860 | 60 | 181 | 12.6 |
| 10 920 | 120 | 91 | 9.8 |
| 11 040 | 240 | 46 | 7.5 |
| 11 280 | 480 | 23.5 | 4.9 |
| 11 700 | 900 | 13 | 3.3 |
| 12 600 | 1 800 | 7 | 2.3 |
| 14 400 | 3 600 | 4 | 1.6 |
| 18 000 | 7 200 | 2.5 | 1.1 |

## (2) Theis's nonequilibrium euation

Let K-2 be the object of analysis and $t$ the variable, compute the $s$-$t/r^2$ relation as shown in Table 8.6. Depict the standard and data curves and superimpose one onto the other. As shown in Figure 8.23, we have the coordinates of the match point:

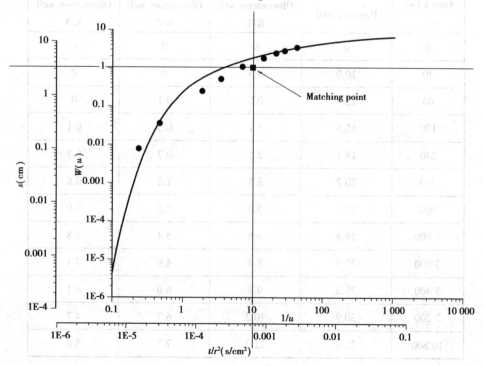

**Figure 8.23  Matching point in theis's method**

$$W(u) = 1, \ 1/u = 10, \ s = 2.4 \text{ cm}, \ t/r^2 = 0.000 \ 65 \text{ s/ cm}^2,$$

$$T = \frac{QW(u)}{4\pi s} = \frac{2 \ 000 \times 1}{4\pi \times 2.4} = 66.3 \text{ cm}^2/\text{s}$$

$$k = \frac{T}{D} = \frac{66.3}{1 \ 000} = 0.066 \ 3 \text{ cm/s}$$

$$S = 4Tu\left(\frac{t}{r^2}\right) = 4 \times 66.3 \times 0.1 \times 0.000 \ 65 = 0.017 \ 2$$

The readers could try deducing the procedure using $r$ as a variable at a specific time $t$, and the result is not unique but similar.

**Table 8.6  Computation based on time as variable**

| Time $t$ (s) | $t/r^2$ (s/cm$^2$) | Drawdown $s$ |
|:---:|:---:|:---:|
| 30 | $7.5 \times 10^{-6}$ | 0 |
| 60 | $1.5 \times 10^{-5}$ | 0.1 |
| 120 | $3.0 \times 10^{-5}$ | 0.3 |
| 240 | $6.0 \times 10^{-5}$ | 0.7 |
| 480 | $1.2 \times 10^{-4}$ | 1.2 |
| 900 | $2.3 \times 10^{-4}$ | 2.0 |
| 1 800 | $4.5 \times 10^{-4}$ | 3.4 |
| 3 600 | $9.0 \times 10^{-4}$ | 4.9 |
| 5 400 | $1.35 \times 10^{-3}$ | 6.0 |
| 7 200 | $1.8 \times 10^{-3}$ | 6.7 |
| 10 800 | $2.7 \times 10^{-3}$ | 7.7 |

### (3) Jacob's modified nonequilibrium equation

Jacob's modified nonequilibrium equation also has two ways to analyze: use either $t$ or $r$ as the variable. Using $t$ as the variable, the computation is as follows: according to the data of K-2, plot $s$-log $t/r^2$ relation is as shown in Figure 8.24, where we have $(\Delta s)_J = 5.61$ cm and $(t/r^2)_{s=0} = 0.000 \ 12$ s/cm$^2$.

then

$$T = 0.183 \frac{Q}{(\Delta s)_J} = 0.183 \frac{2 \ 000}{5.61} = 65.2 \text{ cm}^2/\text{s}$$

$$k = \frac{T}{D} = \frac{65.2}{1\ 000} = 0.065\ 2 \text{ cm/s}$$

$$S = 2.25T\left(\frac{t}{r^2}\right)_{s=0} = 2.25 \times 65.2 \times 0.000\ 12 = 0.017\ 6$$

The readers could try deducing the procedure using $r$ as a variable at a specific time $t$.

### (4) Thiem's equilibrium equation

Assume all wells reach equilibrium when $t = 10\ 800$ s, from Table 8.2, depict the $s$-log $r$ curve as shown in Figure 8.25 where we have $(\Delta s)_T$, the $s$ decreases to marginal when $r$ incresses by ten times, equal to 12.1 cm. That is to say, when $\log r_1/r_2 = 1.0$, $s_1 - s_2 = * 12.1$ cm, then

$$k = \frac{0.366Q}{D(s_1 - s_2)}\lg\frac{r_2}{r_1} = \frac{0.366 \times 2\ 000}{1\ 000 \times 12.1} = 0.060\ 5 \text{ cm/s}$$

The coefficient of transmissivity can be derived from $T = kD$, or find $R$ on the Figure 8.25 $R = 87$ m.

$$T = \frac{0.366Q}{(\Delta s)_T} = \frac{0.366 \times 2\ 000}{12.1} = 60.2 \text{ cm}^2/\text{s}$$

$$S = \frac{2.25T}{R^2} = \frac{2.25 \times 60.5 \times 10\ 800}{8\ 700^2} = 0.019\ 4$$

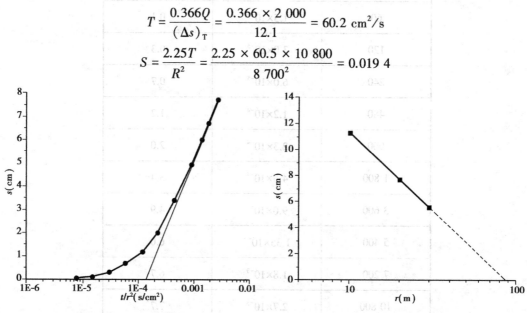

Figure 8.24　Jacob's method　　　　　　Figure 8.25　Thiem's method

### (5) The recovery method

According to the data in Table 9.3, plot $s'$-$\log (t/t^2)$ curve as shown in Figure 8.26 where we obtain $(\Delta s')_r = 8.5$ cm.

$$T = \frac{0.183Q}{(\Delta s')_r} = \frac{0.183 \times 2\ 000}{8.5} = 43.1 \text{ cm}^2/\text{s}$$

$$k = \frac{T}{D} = \frac{43.1}{1\ 000} = 0.043\ 1 \text{ cm/s}$$

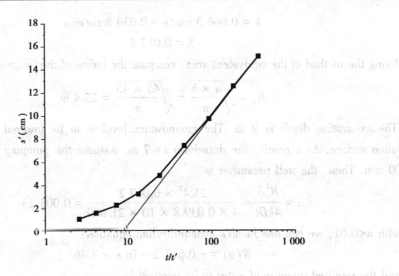

Figure 8.26 The recovery method

## (6) Example 8.2

According to the analysis results of Theis's method in Example 8.1, estimate the required total quantity of water to be pumped in Figure 8.27.

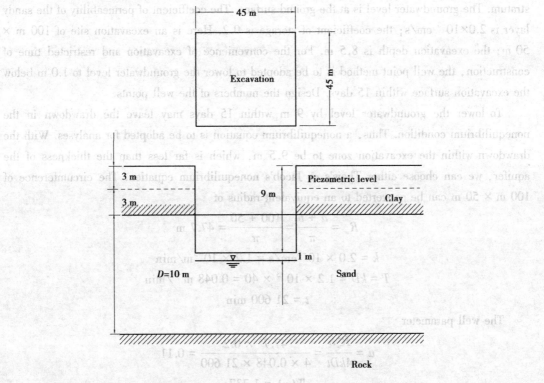

Figure 8.27 Plan and profile of an excavation

From Example 8.1, we know

$$k = 0.066\ 3\ \text{cm/s} = 0.039\ 8\ \text{m/min}$$
$$S = 0.017\ 2$$

Using the method of the equivalent area, compute the radius of the imaginary well as follows:

$$R_w = \sqrt{\frac{a \times b}{\pi}} = \sqrt{\frac{45 \times 45}{\pi}} = 25.4\ \text{m}$$

The excavation depth is 9 m. The groundwater level is to be lowered to 1 m below the excavation surface. As a result, the drawdown $s = 7$ m. Assume the pumping time $t = 15$ days = 21 600 min. Then, the well parameter is

$$u = \frac{R_w^2 S}{4kDt} = \frac{25.4^2 \times 0.017\ 2}{4 \times 0.039\ 8 \times 10 \times 21\ 600} = 0.003\ 23$$

with $u < 0.01$, we can use Jacob's nonequilibrium equation:

$$W(u) = -0.577\ 2 - \ln u = 7.46$$

and the required quantity of water to be pumped is

$$Q_{tot} = \frac{4kDts}{W(u)} = \frac{4 \times \pi \times 0.039\ 8 \times 10 \times 7}{7.46} = 4.69\ \text{m}^3/\text{min}$$

**(7) Example 8.3**

The thickness of a sandy soil layer is about 40 m, below which is an impermeable rocky stratum. The groundwater level is at the ground surface. The coefficient of permeability of the sandy layer is $2.0 \times 10^{-3}$ cm/s; the coefficient of storage is 0.2. Here is an excavation site of 100 m × 50 m; the excavation depth is 8.5 m. For the convenience of excavation and restricted time of construction, the well point method is to be adopted to lower the groundwater level to 1.0 m below the excavation surface within 15 days. Design the numbers of the well points.

To lower the groundwater level by 9 m within 15 days may leave the drawdown in the nonequilibrium condition. Thus, a nonequilibrium equation is to be adopted for analyses. With the drawdown within the excavation zone to be 9.5 m, which is far less than the thickness of the aquifer, we can choose either Theis's or Jacob's nonequilibrium equation. The circumference of 100 m × 50 m can be converted to an equivalent radius of

$$R_w = \frac{a + b}{\pi} = \frac{100 + 50}{\pi} = 47.7\ \text{m}$$
$$k = 2.0 \times 10^{-3}\text{cm/s} = 1.2 \times 10^{-3}\text{m/min}$$
$$T = kD = 1.2 \times 10^{-3} \times 40 = 0.048\ \text{m}^{-3}/\text{min}$$
$$t = 21\ 600\ \text{min}$$

The well parameter

$$u = \frac{R_w^2 S}{4kDt} = \frac{47.7^2 \times 0.2}{4 \times 0.048 \times 21\ 600} = 0.11$$
$$W(u) = 1.737$$

With the pumping height limited to about 5 m for well points and the required amount of dewatering about 9 m, dewatering has to be carried out in two phases:

Assume the discharge quantity of each well point $Q_w = 0.01$ m³/min. The dewatering height of the first phase $s = 5$ m (water level lowered from GL0.0 to GL−5.0 m). The required quanitiy of water to be pumped will be

$$Q_1 = \frac{4kTs}{W(u)} = \frac{4 \times \pi \times 0.048 \times 5}{1.737} = 1.74 \text{ m}^3/\text{min}$$

and the required number of well points will be

$$n_1 = \frac{Q_1}{Q_w} = \frac{1.74}{0.01} = 174$$

Assume the collecting pipes are arranged along the two longer sides of the excavation site, which is 200 m long in total. The distance between two well points will be

$$a = \frac{200}{174} = 1.15 \text{ m} \quad \text{adopt 1.1 m}$$

The dewatering height of the second phases = 4.5 m (the groundwater level lowered from GL-5to GL-9 m). A possible drawdown curve is also as shown in Figure 8.12. The quantity of water to be pumped will be

$$Q_2 = \frac{4kTs}{W(u)} = \frac{4 \times \pi \times 0.048 \times 4.5}{1.737} = 1.56 \text{ m}^3/\text{min}$$

$$n_2 = \frac{Q_1}{Q_w} = \frac{1.56}{0.01} = 156$$

Assume the collecting pipes are also arranged along the two longer sides of the excavation site. The distance between two well points will be

$$a = \frac{200}{156} = 1.28 \text{ m} \quad \text{adopt 1.2 m}$$

# 9 Soil Improvement by Grouting

## 9.1 Introduction

Unreasonable design and construction plan may cause exceeded deformation and excavation failure, both leading to economic loss and even threat to life. As discussed in the provious Chapters, the excavation stability and excavation induced deformation is affected by various factors, such as soils characteristics, retaining wall stiffness, wall embedment depth, and construction methods, which should be considered comprehensively in design and construction.

Strengthening retaining system is a common practice to improve the excavation performance. On the another hand, improving the soils also works and may be more cost-effective.

The chemical grouting method, the deep mixing method, and the jet grouting method are the three main methods of soil improvement in excavations. Although they use different principles, all these methods are similar in their purpose, that is, to improve the soil shear strength or decrease the permeability of soil.

## 9.2 Grouting Equipment

A grouting system is composed of batch and pumping systems, packers, pipes and monitoring system. There are several types of equipment required for introducing grout material into the ground. Much of this depends on the grouting method applied and the desired results for the particular application.

### 9.2.1 Batch and Pumping Systems

Virtually all grouting applications rely on pumps to place the grout and to provide the required pressures for various grouting methods. The grouting pressures may vary widely from a few thousand to tens of thousands of kPa. For cement grouts, the mixture of cement, water, and any other additives must be blended, continuously agitated, and pumped into the ground before the material sets. In these cases, the water is the catalyst and fluidizer, and must be part of the batch. Ideally, the pump system should have a volume capacity to batch all of the grout needed for a single injection process.

### 9.2.2　Packers

In order to maintain grouting pressures and control where the grout is injected into the ground, tight "seals" must be utilized. These seals may be mechanically tightened where the grout hole meets the insertion pipe, or against the pipe wall or hole at a desired depth (downhole packers). Balloon packers are generally hydraulically or pneumatically inflated membranes, which provide a seal above and/or below a grout injection point to control the injection location within a grout hole or grout pipe location. Use of multiple packers may be desirable to isolate the injection point to specific subsurface horizon(s).

### 9.2.3　Pipes

There are a variety of grout pipe configurations available, depending on the type of grouting application. Single point, "push-in" or lance-type driven pipes may be used for certain applications in a wide range of soil conditions. Single point pipes are also commonly inserted in drilled (or jetted) holes, especially for significant depths and hard or difficult-to-penetrate soils, and packers forces the grout past the rubber sleeve, through the weak grout, and into the surrounding ground. The use of sleeved pipes has an additional advantage in that specific horizons may be regrouted by repositioning the injection point.

### 9.2.4　Monitoring system

Real-time computer monitoring of pressures, volumes, and injection locations is now commonplace and has greatly improved efficiency and quality, as well as provided a good record for later review. In addition, control of mixes is critical, and periodic manual tests often are still performed to evaluate apparent viscosity, specific gravity, bleed, cohesion, and other parameters important to quality guarantee and quality control.

# 9.3　Grouting Methods

The chemical grouting method, the deep mixing method, and the jet grouting method are the three main methods for soil improvement in excavations. Although they use different principles, all these methods are similar in their purpose, that is, the improvement of the soil shear strength or the decrease of the permeability of soil.

### 9.3.1　Chemical grouting method

The chemical grouting method is to inject the grouting material into soil strata through existing pores and voids by low pressure. With the condensation of the grouting material, the soils will be strengthened and the permeability and settlement be decreased. The grouting pressure for the chemical grouting method is usually lower than 20 $kg/cm^2$ and the method is, therefore, also called the low pressure grouting method. Figure 9.1 diagrams the implementation of the chemical grouting method (the single packer method in the example).

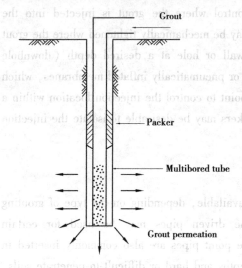

**Figure 9.1   Schematic diagram
of chemical grouting**

Various materials are used for grouting, depending on the purpose of grouting and the properties of rock and soil. Grouting materials are usually categorized into four major types on the basis of the flowing properties and major contents of the materials:

①Suspension type: including cement, clay (or bentonite), or clay (or bentonite) mixed with cement.

② Waterglass solution type: waterglass is a solution of sodium silicate. Adding some chemicals in waterglass brings about different grouting effects. Table 9.1 lists the general categories of the grouting materials for waterglass solution.

③ Suspension and waterglass solution type: mixtures of waterglass solution and other suspension materials, such as waterglass mixed with cement or waterglass mixed with fly ash and cement. The characteristics are a blend of the suspension type and the solution type.

④Polymer solution type: polymer materials include the urea type, the acrylic type, the urethane type, and the lignin type. Generally, polymer materials have good effects. Nevertheless, most of them are highly poisonous and are forbidden for use.

**Table 9.1   Classification of waterglass solutions as grouting materials**

| Alkaline type | Non-alkaline or neutral grouting materials |
| --- | --- |
| Waterglass+acidic reagent (of which there are many to be chosen) | Waterglass+acidic reagent |
| Waterglass+metallic salt reagent (such as calcium chloride) | |
| Waterglass+alkaline reagent (sodium aluminate) | |

If chemical grouting is used in pure sandy soils with a pressure lower than 10 kg/cm$^2$, the grout will permeate into the voids of the sand and achieve a good effect of soil improvement without destroying the soil structure. The method is then sometimes called permeation grouting. The solution type of grouting material is often adopted for this method.

For clayey soils, the permeation method is not applicable. A higher pressure (about 20 kg/cm$^2$) is usually needed to inject the grout into soils, which will produce many cracks. The phenomenon is called hydraulic fracturing. The grouting-induced hydraulic fracturing will form an arborescent structure, which acts like reinforced soil and is certainly useful to improve the strength of soils. The method, sometimes also called fracturing grouting, usually adopts the suspension type of grouting materials.

Table 9.2 lists the grouting materials for various soils, and serves as a reference for a preliminary plan of chemical grouting.

**Table 9.2　Grouting materials for different soils**

| Soil type | Grouting material |
|---|---|
| Clayey soils, including silt clay and loam | Cement suspension type<br>Waterglass suspension type<br>Non-alkaline waterglass suspension type |
| Sandy soils, including sand, silty sand | Permeable solution type |
| Sand-gravel | Suspension type (large voids)<br>Permeable solution type (small voids) |
| Seam | Cement type<br>Suspension type |

## 9.3.2　Jet grouting method

The jet grouting method implants a grouting pipe, with a jet, into the soil to a certain depth with the help of a boring machine and emits high pressure grout or water (about 20 MPa), pounding and cutting soils at the same time. When the pulsing current, at high pressure and speed, exceeds the strength of soils, the soil particles will be separated from the soil body. Some fine soil particles will flow with water or grout out of the ground while other soil particles will mix with the grout under the influence of pounding, centrifugal force and gravity. The mixed soil particles will be rearranged into a grout-soil mixture. When the grout is congealed, it forms a solid body. Figure 9.2 is a diagram of the sequence of jet grouting.

Some commonly used methods of jet grouting are the single tube method, the double tube method, and the triple tube method, as shown in Figure 9.3. The single tube method is to implant the boring rod with a special jet fixed on the side of the rod bottom with the help of the boring machine and to emit pressurized grout (about 20 MPa) into soil to cut and destroy soil bodies. With the rotation and lift of the rod, the grout and the cut soils will be churned and mixed. After a certain period, the whole of the grout and soil will congeal and form a column. The method is often called the CCP (chemical churning pile) method.

As shown in Figure 9.3(b), the configuration of the double tube is a coaxial double jet. The inner jet emits grouting materials at 20 MPa while the outer one pumps compressed air at 0.7 MPa. Under the double effect of jetting grout and the surrounding air pressure, the capability to destroy and cut soil is highly enhanced. With the rotational jetting and lift of the double jet, grouting materials and cut soils are churned and mixed and form a column-like solid body together. The diameter of the column formed with the double tube method is larger than that formed with the single tube method. The method is often called JSG (Jumbo Special Grouting) method, or JSP (Jumbo Special Pile) method.

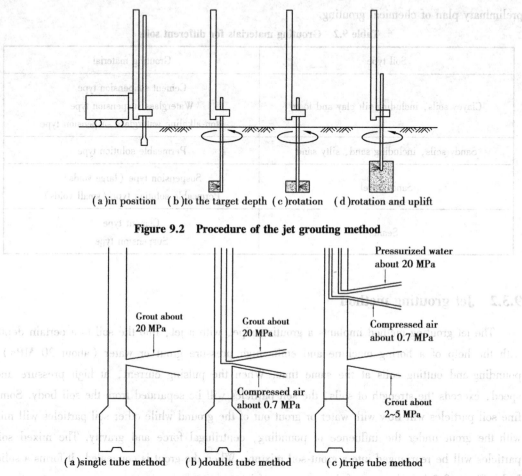

(a)in position  (b)to the target depth  (c)rotation  (d)rotation and uplift

**Figure 9.2   Procedure of the jet grouting method**

(a)single tube method  (b)double tube method  (c)tripe tube method

**Figure 9.3   Types of the jet grouting method**

The triple tube is a jet which emits water, air, and grouting materials simultaneously. As shown in Figure 9.3(c), on the side of the bottom of the grouting pipe is attached a coaxial double jet, the inner one of which emits water at about 20 MPa while the outer one emits compressed air at about 0.7 MPa. With the double contribution of high pressurized water and the surrounding air, the power of cutting and destroying soil bodies is greatly enhanced. Grouting materialsof 2-5 MPa are emitted from another jet right below the double jet. With the rotation and lift of the jet, the grouting materials and the cut soils will be churned and mixed and form a column-like solid body. The diameter of the column formed with the triple tube method is larger than that formed with the double tube method. The method is often called the CJG (Column Jet Grout) method.

　　The diameter of jet grouting piles depends on soil properties and jet pressure. The average piles diameter can be estimated according to Table 9.3.

**Table 9.3　Diameter of jet grouting piles**

| Soil type | Standard penetration test value (SPT-N) | Method | | |
|---|---|---|---|---|
| | | Single tube | Double tube | Triple tube |
| Clayey soil | 0< N <5 | 0.5-0.8 | 0.8-1.2 | 1.2-1.8 |
| | 6< N <10 | 0.4-0.7 | 0.7-1.1 | 1.0-1.6 |
| | 11< N <20 | 0.3-0.5 | 0.6-0.9 | 0.7-1.2 |
| Sandy soil | 0< N <10 | 0.6-1.0 | 1.0-1.4 | 1.5-1.6 |
| | 11< N <20 | 0.5-0.9 | 0.9-1.3 | 1.2-1.8 |
| | 21< N <30 | 0.4-0.8 | 0.8-1.2 | 0.9-1.5 |

　　The commonly used grouting materials for jet grouting are lime (calcium oxide or calcium hydroxide) and cement. Sometimes fine aggregate or fly ash is added. Besides, depending on the functions of the machine and the soil properties, fluid materials (such as cement mortar or cement grout) or dry materials (such as cement or lime powders) are employed.

## 9.3.3　Deep mixing method

　　The deep mixing method (DMM) installs mixing vanes connected to a hollow rod with the help of wash boring, which destroys soil bodies with the vanes and jets out grouting materials with certain pressure. The destroyed soil body and the grouting materials will be mixed up completely and form a solid column. Figure 9.4 diagrams the sequence of the implementation of the DMM.

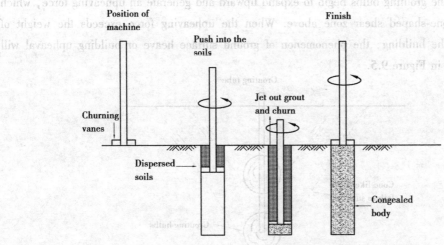

**Figure 9.4　Procedure of the deep mixing method**

Similar to the jet grouting method, the DMM adopts lime and cement for grouting materials. Besides, whether fluid materials such as cement mortar or cement grout, or dry materials such as cement or lime powders are to be adopted depends on the functions of the grouting machine and the soil properties.

## 9.3.4　Compaction grouting method

The compaction grouting method has been applied in property protection for more than 50 years. Compaction Grouting is a technique used mainly for treating granular material (loose sands), where a soil mass is displaced and densified by a low-slump mortar (usually a blend of water, sand, and cement) injected to form continuous grout bulbs.

The method costs low and has been successful in many cases. It is the only grouting method designed specifically not to penetrate soil voids or blend with the native soil. It is also a good option for improving granular foundation materials beneath existing structures, for it is possible to inject from the sides or at inclined angles to reach beneath them. Therefore, compaction grouting can make soils in the vicinity of injection points compact and improve them as a result. It can also uplift a building to restore it from over-settlement. The application of building rectification will be introduced in Section 10.5.1.

The compaction grouting method injects pressurized cement mortar with high consistency, low slump, and low plasticity into soils. The cement mortar does not easily flow into the voids of soils and will therefore form a grouting bulb, which will in turn make an interface with the soils around. Squeezing the cement mortar into soils continuously, the grouting bulb will expand and press the soils around to compact them. Besides, the grouting bulb expands in radiation though faster in lateral direction (because the lateral stress of soils is smaller than the vertical stress). When the soils are compacted to a certain degree, nevertheless, the expansion in the lateral direction will stop. Instead, the grouting bulbs begin to expand upward and generate an upheaving force, which will make a cone-shaped shear zone above. When the upheaving force exceeds the weight of overburden or the building, the phenomenon of ground surface heave or building upheaval will occur as shown in Figure 9.5.

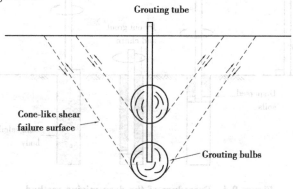

**Figure 9.5　Compaction grouting method**

The diameter of grout bulbs $r$ can be estimated from Eq. 9.1:

$$r = \sqrt[3]{\frac{3kh_1r_0t}{\beta n}} \tag{9.1}$$

where

$k$ = permeability of soil;

$\beta$ = viscosity ratio of grout and water;

$h_1$ = grouting pressure head;

$r_0$ = grouting pipe radius;

$t$ = grouting time;

$n$ = porosity of soil.

The materials used in the compaction grouting method have to be designed to achieve high consistency, low slump, and low plasticity. Otherwise, they may flow into the voids of soils. On the other hand, they should have a certain fluidity to be well transported. Sand mortar and cement mortar are two widely used grouting materials. Fly ash and bentonite are widely adopted additives.

Drawing from the above elucidation, compaction grouting can make soils in the vicinity of injection points compact and improve them as a result. It can also uplift a building to restore it from over-settlement.

# 9.4 Ground Improvement Design

The first step of ground improvement is to determine the location of improvement. Ground improvement outside an excavation zone will decrease the active earth pressure acting on the retaining wall, and within the excavation zone will increase the passive resistance of soils against the retaining wall. The ideal measure is to improve the soils both inside and outside the excavation zone. Nevertheless, the cost may be too high.

According to Ou and Wu's (1990) parametric study using the finite element method, the effects within the excavation zone are better than those outside, given the same conditions. Judged from mechanisms, once excavation is started, the retaining wall will move toward the excavation zone and the active earth pressure is thus produced no matter whether the soils outside the excavation have been improved or not. Therefore, the earth pressure acting on the retaining wall will not be decreased too much. On the other hand, the ground improvement inside the excavation zone will always directly restrain the movement of the retaining wall. Obviously, the effects of ground improvement inside the excavation zone are better than improvement outside.

The location determined, it follows to determine the arrangement of improvement. To resist the forward movement of the retaining wall, some commonly used arrangements include the block type, the column type, and the wall type, as shown in Figure 9.6. The arrangements of these types are elucidated as follows:

①Block type. Within a specific area, improve the soils fully. Replace the soil bodies within the area completely or have them completely combined with chemicals into treated soils.

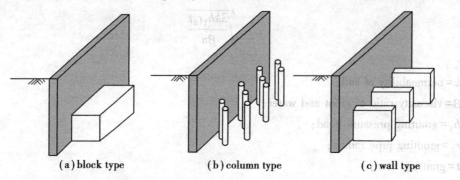

(a) block type         (b) column type         (c) wall type

**Figure 9.6　Typical arrangement of soil improvement in excavations**

②Column type. The pattern of the improved soils is similar to that of piles. The columns of improved soils do not connect with each other.

③Wall type. Connect the columns of improved soils into a wall shape, which joins the retaining wall and forms a counterfort-like wall. The wall can only increase the soil strength in front of the retaining wall. It is not able to raise the moment-resistance stiffness of the wall.

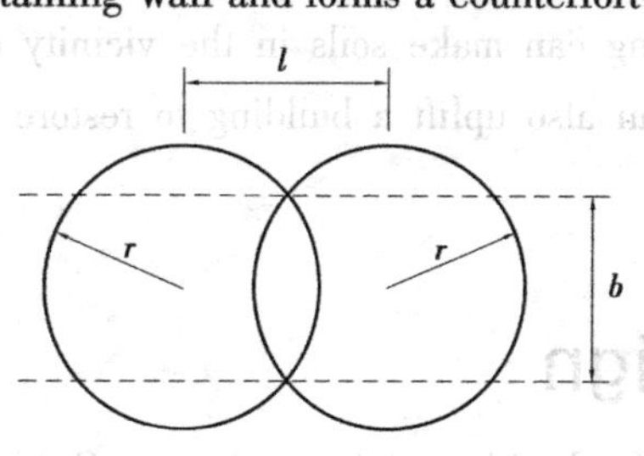

When block type and wall type are adopted, the location of grouting holes should also be arranged according to effective range of grouting to form a continuous grouting body.

**Figure 9.7　Design of a single row grouting**

①Single row: as shown in Figure 9.7, let $l$ be grouting hole distance, and $r$ be effective radius of grouting, then the effective width of grouting body $b$ is

$$b = 2\sqrt{r^2 - \left[l - r + \frac{r - (l - r)}{2}\right]^2} = 2\sqrt{r^2 - \frac{l^2}{4}} \quad (9.2)$$

② Multiple rows: if a single row cannot meet the required effect width, multiple rows arrangement should be employed. To maximize utilization of grouting, both superfluous [Figure 9.8(a)] and inadequate overlap [Figure 9.8 (b)] are supposed be avoided. The optimal

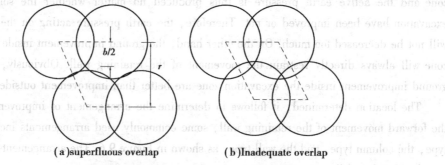

(a) superfluous overlap         (b) Inadequate overlap

**Figure 9.8　Design of double rows grouting**

arrangement is shown in Figure 9.9, and the optimal row distance $r_m$ and effective width $b_m$ are

$$r_m = r + \frac{b}{2} = r + \sqrt{r^2 - \frac{l^2}{4}} \tag{9.3}$$

For double rows:

$$b_m = 2r + b = 2\left(r + \sqrt{r^2 - \frac{l^2}{4}}\right) \tag{9.4}$$

For triple rows:

$$b_m = 2r + 2b = 2\left(r + 2\sqrt{r^2 - \frac{l^2}{4}}\right) \tag{9.5}$$

Drawing from the above elucidation, for odd number rows:

$$b_m = (N - 1)\left(r + \frac{N+1}{N-1}\sqrt{r^2 - \frac{l^2}{4}}\right) \tag{9.6}$$

For even number rows:

$$b_m = N\left(r + \sqrt{r^2 - \frac{l^2}{4}}\right) \tag{9.7}$$

where $N$ is the number of rows.

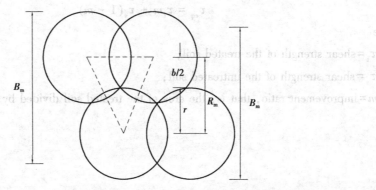

**Figure 9.9　Optimal overlap among grouting**

Figure 9.10 diagrams the plan and profile of the column type of ground improvement. Under such a condition, the passive resistance can be computed on the basis of the properties of the treated soils (composite soils). Both whole [Figure 9.10(a)] and partial [Figure 9.10(b)] improvement within the excavation zone is used, though the passive resistance of soils against the retaining wall under the former condition will be smaller than the latter.

To ensure ground improvement capable of property protection, improvement should be analyzed in terms of the strength of the treated soil, its diameter, span, depth, location, and range. The soils within the area can be viewed as a composite material in analysis. When the composite soil body bears load, according to the principle of force equilibrium, the strength of the composite soil can be written as

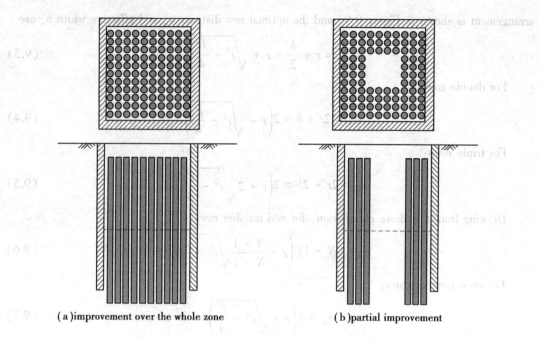

(a)improvement over the whole zone          (b)partial improvement

**Figure 9.10   Soil improvement within excavation zone**

$$\tau_{eq} = \tau_t m + \tau_s(1 - m) \tag{9.8}$$

where

$\tau_t$ = shear strength of the treated soil;

$\tau_s$ = shear strength of the untreated soil;

$m$ = improvement ratio, that is, the area of the treated soil divided by the total area.

# 10   Adjacent Building Protection

## 10.1   Introduction

In the past, with shallower excavations, the influence of excavation on the surrounding environment was not great. Recently, with the increase of excavation depth and scale in urban areas, the magnitude and extent of ground settlement increases, which frequently damages the adjacent buildings. Building damage assessment has been discussed in Chapter 5, and this chapter will introduce methods to the protection of adjacent buildings by means of design and during construction.

## 10.2   Building Protection by Utilizing the Characteristics of Excavation-Induced Deformation

The excavation-induced deformation is affected by various factors, and all design parameters can be optimized to protect adjacent buildings.

### 10.2.1   Reducing the unsupported length of the retaining wall

Figures 10.1 (a) and (b) are identical in the final excavation depth, the number of excavation stages, the number of strut levels, and the location of struts. The only difference between them is the distance between the location of strut and the excavation surface.

In Figure 10.1(a), each level of struts is installed 0.5 m above the excavation surface. When the first stage of excavation is completed, the unsupported length of the wall is 2 m. When the first level of struts is installed at the depth of GL-1.5 m and excavation proceeds to GL-5.5 m, the unsupported length of the wall is 4 m. When the second level of struts is installed and excavation proceeds to the third stage, the unsupported length of the wall is also 4 m. With the completion of the third level of struts, the unsupported length is 4.5 m.

In Figure 10.1(b), each level of struts is installed 1.5 m above the excavation surface. As a result, during the first stage of excavation, the unsupported length of the wall is 3 m. Evaluated similarly, the unsupported lengths of the wall at the second and the third stages are both 5 m. The unsupported length after completion of the final stage of excavation is 4.5 m.

As elucidated above, except for the last stage of excavation where the unsupported lengths in the two cases are identical, the unsupported lengths illustrated in Figure 10.1a are smaller than those in Figure 10.1b at other stages of excavation, with the same earth pressure on the back of the retaining wall. Actually, the total deformation of the retaining wall or ground settlement is the accumulated deformation at every stage of excavation. Thus, the deformation in Figure 10.1a should be smaller than that in Figure 10.1b. That is to say, the location of struts should be as close to the excavation surface as possible. Generally, considering the convenience of strut installation, the distance between struts and the excavation surface is about 0.5 m.

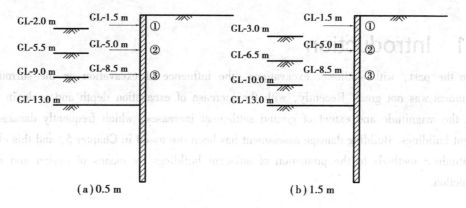

Figure 10.1    Influence of the distance between the strut and
the excavation surface on the unsupported length

## 10.2.2    Decreasing the creep influence

The definition of creep is given as the condition where deformation increases with time, provided the stress remains constant. Creep usually occurs in clayey soils. The softer the soils are, the more obvious the characteristics of creep could be. Creep relates to time and stress. It increases with the increase of the stress level and time and covers a long period of time. Since soils near the retaining wall and on the excavation surface may be on the verge of the ultimate condition, the stress level must be high. As a result, soils with prominent characteristics of creep are susceptible to over settlement.

To prevent creep from occurring, especially when excavating in soft clayey soils, struts have to be installed as soon as each stage of excavation is completed. Strut installation usually takes a few days in braced excavations. If the top-down construction is adopted, it takes a few weeks to construct floor slabs. During the process of strut installation or floor slab construction, creep may continue worsening. The expedient way to handle this is to lay a layer of pure cement, usually at least 10 cm thick, on the excavation surface as soon as the excavation stage is completed (which is also necessary in practice to facilitate the construction machines operating on the excavation surface). It is not easy to examine the effect of such a measure in the prevention of creep though it is considered to be useful to a certain extent from both theoretical and empirical points of views.

### 10.2.3  Taking advantage of corner effect

As elucidated in Section 5.2.2.2, with the lateral stiffness and the resulting arch effect around corners, deformation of diaphragm walls and ground settlement around comers are both smaller than those in the central section. The comers for soldier piles, sheet piles, and bored piles, which are not continuous in the horizontal direction and have no lateral stiffness, accordingly, do not have much difference in deformation or ground settlement from the central section. Thus, if the building is located at a comer or on the shorter side (see Figure 10.2), diaphragm walls can be utilized to take advantage of comer effects for the protection of buildings.

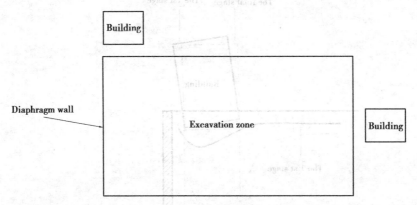

Figure 10.2  Buildings located at corners or along the shorter side of an excavation

### 10.2.4  Building protection by increasing stiffness of the retaining-strut system

Theoretically, the increase of the stiffness of a retaining wall can reduce the wall deformation and the ground settlement. The methods of increasing the system stiffness of the retaining-strut system include increasing the stress or thickness of the retaining wall, decreasing the horizontal span of struts, increasing the stiffness of each strut, and decreasing the vertical span of struts, etc.

When the axial stiffness of the strut is not large enough, increasing the stiffness of struts (such as by cutting the horizontal span or increasing the stiffness of each strut) can effectively reduce the deformation of a retaining wall. However, when the axial stiffness of struts is already rather large, more increase of stiffness will not reduce the deformation accordingly.

Decreasing the vertical distance of struts is an effective way of reducing the deformation of a retaining wall. The reason is that cutting the vertical distance implies the increase of number of strut levels. That is to say, the rigidity of the retaining system is increased. Deformation will decrease as a result. From another perspective, since the deformation of a retaining wall is the accumulated result of the excavation stages, the cutting of the vertical distance will decrease the unsupported length of the retaining wall at each excavation stage. Therefore, deformation will also decrease.

### 10.2.5 Utilizing the characteristics of ground settlement

If ground settlement can be accurately predicted or observed, the characteristics of ground settlement can also be utilized for property protection.

As shown in Figure 10.3, a building is close to the excavation and tilts significantly in the first excavation stage. With the continuation of excavation, the angle of tilt of the building, however, began declining. When the excavation is finished, the building becomes approximately vertical, although the total settlement has increased.

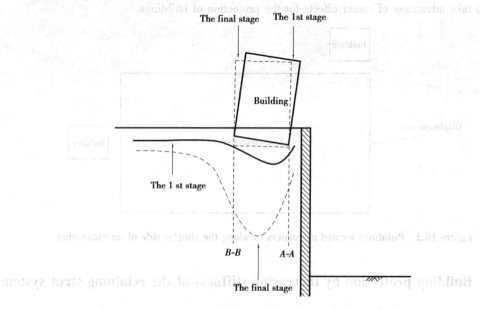

**Figure 10.3  Slanting conditions of the building**

The phenomenon of the declining angle of tilt of the building can be explained by the shape of the settlement profile. In the first stage, the settlement at $A$-$A$ is larger than that at $B$-$B$, causing the tilt. With the excavation proceeding and the increase of excavation depth, settlement at $B$-$B$ exceeds that at $A$-$A$, so the building starts being drawn back and the angle of tilt declining.

From the discussion, it follows that an accurate prediction of settlement during the process of excavation will help property protection. The characteristics of ground surface settlement can be predicted by the methods introduced in Chapters 5 and 6 or be derived from measurements.

# 10.3  Building Protection by Utilizing Auxiliary Methods

The goal of auxiliary methods in excavations is to decrease the wall deformation or ground settlement. In principle, any method that helps decrease the deformation or settlement can be adopted. To utilize auxiliary methods, one has to go through a careful evaluation of the effects.

Otherwise, in addition to increasing construction cost, they may even have no time for property protection. What's more, the incautious implementation of auxiliary methods may be inimical to adjacent properties. The evaluation of auxiliary methods can be done by numerical simulation, mechanism exploration, or case studies. Readers interested in the subject can refer to the related literature.

## 10.3.1　Ground improvement

The main soil improvement methods used in excavations have been introduced in Section 9.3, all of which can be executed to protect adjacent buildings.

## 10.3.2　Counterfort walls

Figure 9.6(c) diagrams the wall type of ground improvement, which usually contacts the retaining wall without forming a whole structure with it. Thus, when the retaining wall is bent and deformed, there will be produced a relative displacement between the wall-type soil body and the retaining wall. That is to say, the wall-type soil body does not increase the moment-resistance of the retaining wall though it increases the strength of soils in front of the retaining wall. If the wall in Figure 9.6(c) is constructed as a counterfort (also called buttress), that is, constructed with the diaphragm wall method (either reinforced or not) and forms a whole structure with the diaphragm wall (retaining wall), like a T-beam in reinforced concrete structures, the counterfort will greatly enhance the capability of moment-resistance.

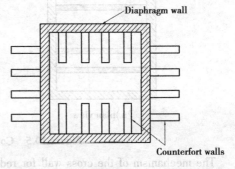

The location of the counterfort can be arranged either at the inner or at the outer side of the retaining wall as shown in Figure 10.4. When arranged at the inner side, it would be mostly subjected to tensile force. Theoretically, the counterfort should be reinforced under such a condition. The actual cases using counterforts, nevertheless, show that unreinforced counterforts are useful as well in increasing the moment-resistance

**Figure 10.4　Locations of counterfort walls**

stiffness. The counterfort constructed at the inner side of the retaining wall should be dismantled with the increase of the excavation depth.

If no property rights are involved, the counterfort can be constructed on the outer side of the wall. The reason is that the counterfort will be subjected to compression force when it is bent or deformed. There is no need to place reinforcements in the counterfort. Besides, it also saves the trouble of dismantling it during the excavation process. If the counterfort need not be dismantled as the excavation is proceeding, it can offer stiffness throughout the whole course of excavation and the construction of the basement.

Theoretically, the counterfort can be cast into a whole structure with the retaining wall.

However, there exists the problem of collapse near the concave corners of trenches. Jet grouting in this area is one of the measures to avoid the collapse of trenches.

Ou and Wang (1997) suggest that the deformation behavior of the counterfort diaphragm wall has much to do with whether it penetrates into the hard soil stratum (sandy or gravelly soils) or not. In other words, it relates to whether the wall bottom is effectively restrained or not. Besides that, the counterfort can only restrain the wall deformation at the part where it is placed. If the deformation of the whole diaphragm wall, that is, along the excavation border, is to be reduced, several counterforts have to be placed evenly along the excavation border. In practical design, the span of the counterforts can be determined referring to successful case histories or may be obtained from the three-dimensional finite element analysis.

### 10.3.3 Cross walls

The arrangement of cross walls is schematically shown in Figure 10.5. One constructs a wall connecting the two opposing retaining walls before excavation is started. It can be constructed using ground improvement techniques (such as jet grouting or deep mixing). To obtain a better construction quality or compressive strength, the cross wall can also be constructed by unreinforced diaphragm walls (the unconfined compression strengths of treated soils are usually between 10 and 20 $kg/cm^2$ whereas that of concrete diaphragm wall can achieve 280 $kg/cm^2$ with a minimum of 100 $kg/cm^2$).

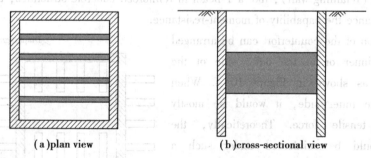

| (a)plan view | (b)cross-sectional view |

**Figure 10.5  Configuration of cross walls**

The mechanism of the cross wall for reducing wall deformation is quite different from that of ground improvement. The designing principle of ground improvement is to enhance the strength of soils in front of the wall, considering them too soft to have enough passive resistance. The cross wall should be viewed as a strut, which exists before excavation and owns a great deal of compression strength (especially for unreinforced diaphragm walls). Theoretically, the locations where cross walls have been placed are less susceptible to deformation because they are restrained from moving. The effects of the cross wall on reducing lateral deformation of a retaining wall are quite obvious.

To ensure that cross walls can effectively reduce deformation of the retaining wall, the design of cross walls has to consider compression strength, depth, and span of two cross walls. Since the behaviors of the diaphragm wall with cross walls are three dimensional, neither the traditional two-dimensional plane strain analysis nor the beam on elastic foundation method can simulate behaviors of cross walls. Thus, to design cross walls, one has to resort to the three-dimensional finite element method or successful case histories.

Ou and Lin (1999) conducted a series of parametric studies on the basis of 14.5 m deep excavations and 31 m deep diaphragm walls to explore the behaviors of cross walls. The results indicate that the deformation of a retaining wall is the smallest at the location where cross walls are constructed, similar to counterfort walls. To save the cost, moreover, cross walls are best constructed between the excavation surface and 5 or 6 m below it.

## 10.3.4  Micro piles

Micro piles are also called soil nails. Since they were first applied in Europe to strengthen or underpin existing buildings, they have been used for more than 40 years. They have also been applied to property protection in many countries. Some successful case histories have been documented in the literature (Woo, 1992) and many more unsuccessful cases were not recorded.

Because the mechanisms of micro piles for property protection are indirect and systematic studies are also lacking, though successfully applied in some excavations, most of them are designed on an empirical basis and without theoretical support.

In practice, the diameter of a micro pile varies from 10 to 30 cm. The reinforcements can be steel bars, steel rails, H steels, or even steel cages. The construction process of micro piles is as follows. First, bore to the designed depth with casings or by other drilling measures and then place reinforcements into the bores. Inject cement mortar into bores under a certain pressure. Pull out the casing little by little and add more mortar. The micro piles are usually arranged in a single row or multiple rows. The distance between piles is three to five times the pile diameter, depending on the soil strength. Whether the arrangement is the single row or the multiple row type, they should be intermingled by 5-30 degrees as shown in Figure 10.6.

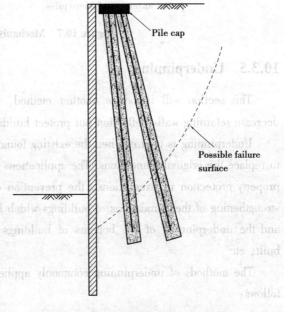

**Figure 10.6  Mechanism of micro piles**

There are two design principles for use of micro piles. The first is they have to pass the potential failure surface so that the shear strength of micro piles and the pull-out resistance can restrain the failure of soils and reduce the possibility of ground settlement accordingly, as shown in

Figure 10.6. The potential failure surface, however, is usually rather large and the micro piles passing it are limited in number. The shear strength and pull-out resistance of the micro piles are not too large. Thus, whether the design is useful remains to be evaluated.

The other principle is to design many small-diameter micro piles enveloping the retaining wall to reinforce the soils through the process of steel placing and grouting. Especially in sandy soils, grouts may permeate into soils extensively and reinforce a larger area of soils. They may even form a quasi-gravity retaining wall. Figure 10.7(a) diagrams the micro piles in two rows, which perform as gravity retaining wall-like structure, as shown in Figure 10.7(b). The method may be capable of stabilizing the retaining wall and reducing the active earth pressure on the retaining wall or increasing the resistance to the failure surface. Thus, ground settlement would be reduced.

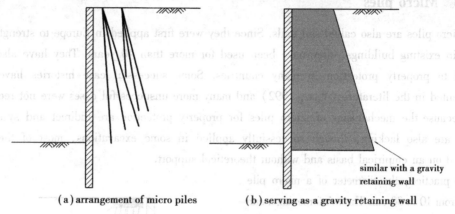

(a) arrangement of micro piles      (b) serving as a gravity retaining wall

similar with a gravity retaining wall

**Figure 10.7   Mechanism of micro piles**

## 10.3.5   Underpinning

This section will introduce another method, underpinning, which may not be helpful to decrease retaining walls deflection but protect buildings from exceeded settlement.

Underpinning is to strengthen the existing foundations of a building, to improve the soils, or, to replace the original foundations. The applications of underpinning are quite extensive, including property protection in excavations, the prevention of natural settlement of heavy buildings, the strengthening of the foundations of buildings which have been unsuitably designed or constructed, and the underpinning of the bottoms of buildings through which a new tunnel has just been built, etc.

The methods of underpinning commonly applied in property protection in excavation are as follows:

### (1) **Improve the soils beneath the foundations of existing buildings**

As shown in Figure 10.8, to prevent over-settlement of the adjacent buildings near excavations, the soils beneath the foundation are treated before excavation. Once excavation is started, the soils behind the retaining wall come to the active condition and the maximum deformation usually occurs near the excavation surface. The potential failure surface outside the excavation zone normally develops from below the excavation surface. Thus, to underpin existing buildings, the depth of the soil improvement should extend from below the foundations to outside the failure surface.

When carrying out soil improvement, the over-pressurized grouting often heaves the building. It also disturbs the soil structure below the building foundations and reduces the soil strength. If the method is not implemented with prudence, it may not only

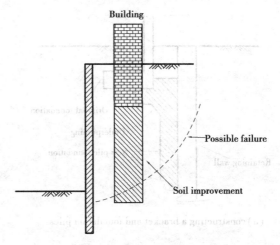

**Figure 10.8  Underpinning an existing building by soil improvement**

turn out to be a failure in strengthening the foundation soils but may also worsen slanting conditions and damage buildings. Thus, a detailed plan referring to successful case histories is required before implementation.

### (2) Add an extra foundation to existing buildings

For fear of the insufficient bearing capacity of the foundations of existing buildings, an extra foundation could be constructed near the original foundation before excavation. Such an underpinning measure is usually implemented before excavation. The implementation, without weakening the original foundations, does not need to be accompanied by the measure of load transfer, which will be introduced in the next.

Figures 10.9 and 10.10 are some possible underpinning methods. Figure 10.9(a) shows a method of constructing new piles beside the retaining wall to support the foundations of existing buildings. Figure 10.9(b) shows a method of constructing a protruding part to the side of the foundations of existing buildings to join the newly constructed piles. Figures 10.10 uses the underpinning methods adopted in the construction of the Singapore mass transit system (Huang, 1992). In Figure 10.10, the government building is a brick building with wood piles of 4.5 m deep and 50-100 mm diameter. The mass transit system passes near the government building. The construction follows the open cut method with an excavation depth of 27 m. The retaining system consists of steel sheet piles and soldier piles. The outside column of the government building is only 3 m from the soldier piles. To protect the building during the construction of the mass transit system, an underpinning measure was used, which was as follows: four micro piles were constructed next to the outside columns. The micro piles were constructed through the pile caps up to 26-28 m below the ground surface, which was 5 m below the 45° failure surface. The resulting void in the ground was filled fully with cement mortar.

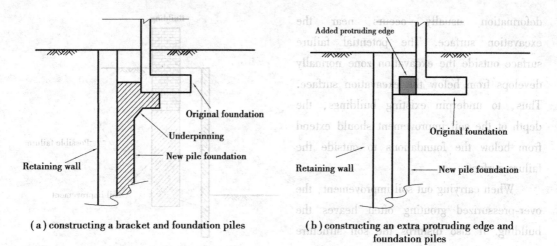

(a) constructing a bracket and foundation piles

(b) constructing an extra protruding edge and foundation piles

**Figure 10.9  Underpinning an existing building**

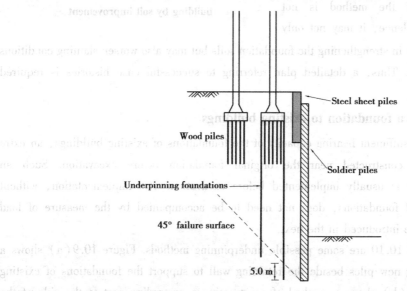

**Figure 10.10  Underpinning foundations of a building near an excavation**

### (3) Construct new foundations under existing buildings

Figure 10.11 diagrams the underpinning method in which an additional foundation is constructed beneath the building's original one. With the new foundation constructed, the original one may weaken or even become useless. The method is as follows:

①Excavate an operation space beside the footing and below the foundation.

②Construct new foundations (piles).

③Install temporary supports.

④Load transfer operation-transfer the weight of the building to the new foundation or the strengthened one. This step is a crucial point in determining whether the underpinning will succeed. A precise preload control and measurement of the behavior of tile structure are to be carried out. The procedure of the load transfer operation is as shown in Figure 10.12.

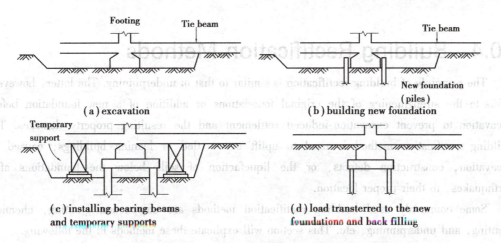

Figure 10.11   Underpinning an existing building by constructing new foundation

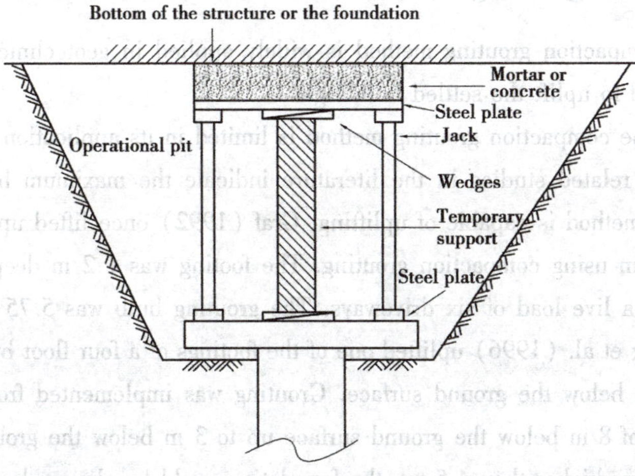

( a ) excavation

( b ) building new foundation

( c ) installing bearing beams and temporary supports

( d ) load transferred to the new foundations and back filling

Figure 10.12   Schematic diagram of load transfer operation

⑤Dismantle the temporary supports. Proceed to grouting and backfilling.

As shown in Figure 10.12, the procedure of load transfer can be described as follows:

①Set jacks and a steel plate on the new foundation.

②Lay sand mortar or concrete between the building bottom and another steel plate, which is set on tile top of the jack.

③Preload tile steel plate or the foundation using jacks. Preloading usually takes the building as the reaction frame. The direct adding of the weight of the building onto the new foundation usually produces settlement. The aim of preloading is to have the new foundation acted on by a preload in advance to accelerate settlement.

④Place temporary supports between the two steel plates.

⑤Hammer wedges into the voids of the temporary supports.

⑥Dismantle the jacks and fill the operation space with concrete. The underpinning is thus completed.

# 10.4 Building Rectification Methods

The principle of building rectification is similar to that of underpinning. The latter, however, refers to the strengthening of the original foundations or addition of a new foundation before excavation to prevent excavation-induced settlement and the resulting property damages. The building rectification method is used to uplift over-settled or leaning buildings, caused by excavation, construction defects, or the liquefaction of soils below the foundations after earthquakes, to their proper location.

Some commonly used building rectification methods are compaction grouting, chemical grouting, and underpinning, etc. This section will explicate these methods in the following.

## 10.4.1 Compaction grouting

Although the compaction grouting method is widely applied in geotechnical engineering, in most cases, it is used to uplift the settled buildings.

Theoretically, the compaction grouting method is limited in its application to buildings under a certain weight. No related studies in the literature indicate the maximum building weight the compaction grouting method is capable of uplifting. Graf (1992) once lifted up a 4.5 m × 4.5 m square footing by 5 cm using compaction grouting. The footing was 1.2 m deep and bore a dead load of 500 tons and a live load of six driveways. The grouting bulb was 5.75 m deep below the ground surface. Wong et al. (1996) uplifted one of the footings of a four floor building by 24 mm, which was about 1 m below the ground surface. Grouting was implemented from the bottom up, starting at the depth of 8 m below the ground surface up to 3 m below the ground surface. When the grouting depth grew higher than 4.5 m, the foundation could be observed uplifted apparently.

A detailed plan of the grouting pressure, grouting rate, grouting depth, arrangement of the grouting points, and the implementation method is required before the building rectification operation. Besides, the building should be equipped with complete monitoring instruments for immediate adjustment of the operation.

To uplift a building, the ways of grouting are two: either bore through the floor of the building and grout vertically or grout slantingly from the building side, as shown in Figure 10.13. The former has better effects since it is more direct. Nevertheless, it will affect the use of the building. The latter is less effective but has the benefit of not disturbing the use of the building.

Basically, the arrangement of the grouting points should be aimed at the place where settlement is the largest. The major grouting point should be arranged at a larger depth. The places where settlements are smaller can be arranged as the minor grouting points where the depths are slightly shallower than the major one, as shown in Figure 10.14 (Nonveiller, 1989). The sequence of grouting should follow the degrees of settlement, beginning with the place where settlement is the largest and then proceeding to the minor points next to it. The operation is to be repeated till the desired amount of lifting is achieved. If the lifting amount is too large, the operation can be carried out in stages.

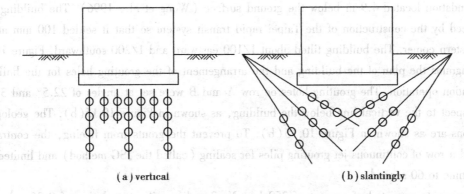

(a) vertical                                (b) slantingly

**Figure 10.13   Grouting positions of the compaction grouting method**

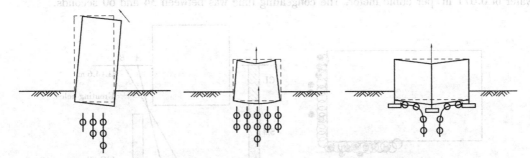

**Figure 10.14   Arrangement of the injection points with the compaction grouting method for the rectification of buildings**

## 10.4.2   Chemical grouting

The development of grout in chemical grouting is not easily controlled and any small incaution might lead to the unpredictable flowing of grouts, which would cause damage to structures and pipes and reduce the effects of grouting. A detailed plan of the grouting pressure, the arrangement of injecting points, the amount of injection, and measures for preventing fugacious flowing of grouts should be made before implementing chemical grouting to rectify a building. Generally speaking, the grouting pressure, the arrangement of injection points, and the amount of injection are all determined on the basis of empirical experience and adjusted according to the monitoring of the heave of the building. No quantitative methods are available so far. Besides, referring to past case histories is recommended. For the prevention of grout flows going where they are not wanted, there are two methods: installing sealing piles (such as sheetpiles) within the grouting range or reducing setting time of grouts.

The cost of chemical grouting to correct a slanting building is relatively low. Nevertheless, the operation of carrying out a chemical grouting for building rectification is highly technical and ingenious, any incaution easily leading to damage to structures and pipes.

We will take a case history from the construction of the Taipei rapid transit system for illustration. The slanting building had 12 floors of superstructure and one floor of basement with a mat foundation located 4.9 m below the ground surface (Wong et al., 1996). The building was influenced by the construction of the Taipei rapid transit system so that it settled 100 mm at the southeastern comer. The building tilted about 1/100 eastward and 1/200 southward. Figure 10.15 (a) diagrams the plan of the building and the arrangement of the grouting holes for the building rectification operation. The grouting holes on row A and B were set at angles of 22.5° and 31.5° with respect to the vertical line below the building, as shown in Figure 10.15(b). The geological conditions are as shown in Figure 10.15(b). To prevent the grouts from fleeing, the contractor installed a row of continuous jet grouting piles for sealing (called the JSG method) and limited the setting time to 60 seconds.

The grouts consisted of cement of 250 kg, No.3 sodium silicate solution of 0.25 $m^3$, and water of 0.671 $m^3$ per cubic meter. The congealing time was between 34 and 60 seconds.

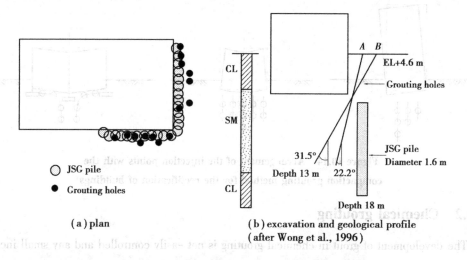

○ JSG pile
● Grouting holes

(a) plan

CL
SM
CL

A  B
EL+4.6 m
—— Grouting holes

31.5°      22.2°
Depth 13 m

JSG pile
Diameter 1.6 m

Depth 18 m

(b) excavation and geological profile
(after Wong et al., 1996)

**Figure 10.15  Building rectification by chemical method**

To set the grouting pressure uniformly on the base of the foundation, three to four grouting machines were used simultaneously, each set to pump between 0.02 and 0.04 $m^3$/min of grouting. First row A was grouted and then row B after row A was totally grouted. The grouting depth of each row was between GL-13-GL-9 m, starting from the bottom and lifting the drilling rod gradually up. With the clayey soil below the depth of GL-13.8 m [see Figure 10.15(b)] and the sandy soil above, the grouting operation took the sandy soil below the depth of GL-13 m as the bearing stratum where a larger amount of pressurized grouts was injected to constitute a solid ground in preparation for uplifting the building. With the lifting of the drilling rod, the grouting rate and pressure were lowered gradually. According to the above principles and experience, the control values of the grouting pressure and flow were determined. After 11 days' building rectification, the building was uplifted.

## 10.4.3    Underpinning

The principle of the underpinning for rectifying buildings is similar to that of the underpinning method introduced in Section 10.3.5. The only difference is that the latter aims at prevention before excavation, transferring the load of the building to a new foundation which is not to be influenced by excavation. Therefore, jacks are not necessarily required to uplift the building. The former is to solve an existing problem, correcting a building problem due to excavation. Since the building has already settled, jacks are required to uplift the building. The procedure of the underpinning method is as shown in Figure 10.16. First excavate a sufficient working space, where underpinning piles are to be constructed, around the mat foundation or the individual footing under the column, then set a jack between the pile cap and the mat slab to uplift the building, and last proceed to the load transfer operation.

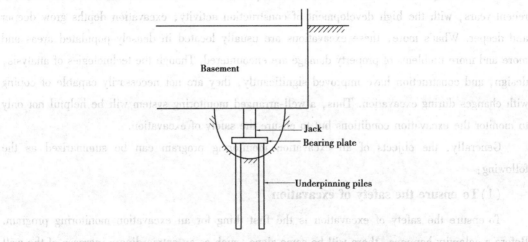

**Figure 10.16    Rectification of buildings by the underpinning method**

The underpinning rectification method can be applied to mat foundations and individual footings. Its strength is that the target or range of rectification is specific, without the problem of grout fleeing, which frequently occurs in grouting methods, or worrying about wrongly uplifting a column. The shortcoming is that it costs much more and requires advanced techniques. If not designed well, the building might settle instead due to the settlement of the bearing pile induced by underpinning. A mat foundation that has been underpinned might become a partial pile foundation and change its designed stress distribution.

# 11 Instrumentation and Monitoring

## 11.1 Introduction

Though in the past theories of excavation analysis and design were not advanced, with conservative design and shallower excavation depths, few construction calamities occurred. In recent years, with the high development of construction activity, excavation depths grow deeper and deeper. What's more, these excavations are usually located in densely populated areas and more and more problems of property damage are encountered. Though the technologies of analysis, design, and construction have improved significantly, they are not necessarily capable of coping with changes during excavation. Thus, a well-arranged monitoring system will be helpful not only to monitor the excavation conditions but to ensure the safety of excavation.

Generally, the objects of an excavation monitoring program can be summarized as the following:

### (1) To ensure the safety of excavation

To ensure the safety of excavation is the first thing for an excavation monitoring program. Before a calamity happens, there will be some signs, such as an extraordinary increase of the wall or soil deformation or stress. A monitoring system can issue an immediate warning to help engineers adopt effective measures to forestall a calamity when these signs appear.

### (2) To ensure the safety of the surroundings

Excavations are usually located in busy commercial areas or densely populated residential areas. It is almost impossible to avoid influencing the surrounding buildings, underground pipes, public utilities, and pedestrians. The designer has to consider all these factors and establish a monitoring system to ensure the safety of people and property during excavation.

### (3) To confirm the design conditions

Since the existing analysis theories are not satisfactorily mature and the geological investigations can not fully represent the in situ conditions and the complicated construction environment, the analysis results do not necessarily meet the actual conditions. Back analyses based on monitoring results can help modify and correct the original design, reduce the cost, shorten the excavation period, and change the basis of design. Further, they can serve as a reference for similar designs in the future and help enhance the excavation techniques.

### (4) To follow long-term behavior

An important construction project finished, the monitoring system can be retained for long-term follow up, studying whether the long-term behavior of the case conforms to the original hypotheses.

### (5) To supply factual materials for legal judgment

Information obtained from monitoring systems, along with the construction records, can serve as evidential data when a calamity or property damage occurs. Understanding the true causes of the events can avoid unnecessary disputes and help the restoration and compensation work.

## 11.2   Element of a Monitoring System

Monitoring field performance is basically the measurement of the physical quantities of some objects, such as deflection, stress and strain. The common monitoring items in excavations are (a) movement of the structure or soil, (b) stress or strain of the structure or soil, and (c) water pressure and level. The monitoring objects of movement include the lateral deformation of the retaining structure and soils, the tilt of the building, the settlement of the ground surface and the building, the heave of the excavation bottom and the uplift of the central post, etc. The measurements of stress and strain include those of the strut load, the stress of the retaining structure, and the earth pressure on the wall. The measurement of water pressure refers to the water pressures within the excavation zone, outside the excavation zone, and on the retaining wall.

In principle, the types of monitoring instruments in an excavation are not special. As long as they meet the budget and the criterion of accuracy, any instruments or methods are feasible. Most of the instruments can be divided into electronic and non-electronic types. The electronic type is more sensitive and easy to read. They can also be divided into an automatic or semi-automatic monitoring system. Nevertheless, the precision is easily influenced by the installation process and the surroundings. The durability of electronic instruments needs examination because an excavation usually needs to be monitored over a long period. Thus, when installing an electronic instrument in the field for long-term monitoring, durability needs special attention.

Generally speaking, the instruments for plane surveying, such as tapes, levels, and theodolites can be used for the measurement of movement. The measurement of the tilt angle can also resort to those for plane surveying (same as above) or electronic instruments. Electronic instrument sensors include strain gauges and force balance accelerometers, etc. The most common instruments to measure stress or strain are the strain gauge or the electronic transducer that takes the strain gauge as the measurement unit. This chapter will omit general instruments for plane surveying. Please refer to related books on plane surveying if interested in the subject.

# 11.3  Measurement of Movement

## 11.3.1  Lateral deformation of retaining walls and soils

The lateral deformation of a retaining wall (such as soldier piles, diaphragm wall) is one of the important monitoring items of excavation. The lateral deformation of a retaining wall relates closely to the ground settlement (or the settlement of the buildings). The magnitude and shape of lateral deformation of the retaining wall can be used for the judgment of the safety of the retaining wall or the buildings in the vicinity.

To explore the characteristics of an excavation or for some special objectives, it is sometimes required to measure the lateral deformation of soils. The strain at a specific point in the soil can be computed if extensometers, used to measure vertical movement at a point in the soil, are installed near an inclinometer casing. The results can be used to understand the tendency to movement of soil. This data is important for excavation studies though difficult to apply to the judgment of excavation safety.

The inclinometer casing and the inclinometer are commonly used devices for the measurement of the lateral deformation of retaining walls or soils. An inclinometer casing has four tracks, which constitute the two perpendicular axes, along each of which two sets of wheels of inclinometer are embedded. The two ends of an inclinometer casing are the male and female joints (Figure 11.1), respectively, by which several inclinometers can be connected to measure greater depth. The spiral of an inclinometer casing should be as small as possible. Otherwise, the tracks may not be on a vertical plane. This is a factor influencing the precision of measurement. ABS (acrylonitrile/butadiene/styrene) pipes, PVC pipes and aluminum pipes are commonly used materials for inclinometer casings. Aluminum and PVC pipes are usually formed through injection molding. As a result, the measuring tracks are external to the piles [Figure 11.1(a)]. In ABS ones, however, four tracks are hollowed in the pipes using lathes [Figure 11.1(b)].

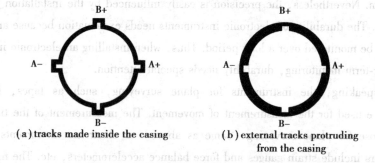

(a) tracks made inside the casing    (b) external tracks protruding
                                          from the casing

**Figure 11.1  Inclinometer casings**

Fig 11.2 illustrates the photo and basic configuration of an inclinometer. As shown, the inclinometer is a four foot (two pairs of wheels) instrument containing a tilt measuring sensor (also called an electronic pendulum). The top of the inclinometer is connected to a cable, which is, in turn, connected to a readout on the ground surface. According to the type of measuring unit of the sensor, the inclinometer can be divided into resistance type, vibrating types, and force balance accelerometer type. The force balance accelerometer type places a pendulum amid the magnetic field of coils of a location detector. When a tilt is produced, the coil magnetic field of the location detector can then detect the displacement of the pendulum with respect to the vertical line. The signal is then converted into a voltage by the detector. The voltage going through the coil will generate a force to have the pendulum recover its original place. Have the measured voltage converted and we can then get the tilt angle.

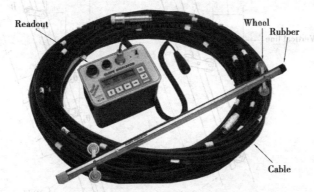

**Figure 11.2　An inclinometer**

When installing an inclinometer casing, one pair of tracks of the inclinometer casing should parallel the direction of deformation of the retaining wall, in other words, should be perpendicular to the retaining wall. The pair of tracks perpendicular to the retaining wall is usually called the *AA* axis. The other pair paralleling the retaining wall is called the *BB* axis, as shown in Figure 11.3.

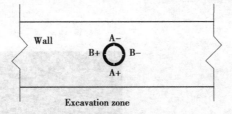

**Figure 11.3　Installation of inclinometer casings with diaphragm walls**

As shown in Figure 11.4, assuming the distance between the two points (or two wheels) is $L$ and the tilt angle measured by an inclinometer is $\theta$, the relative horizontal distance between the two points (ortwo wheels) is $L \sin \theta$. If the tilt angles taken from three serial measurements by the inclinometer are $\theta_1$, $\theta_2$, $\theta_3$, the relative horizontal distance between points A and B is $\sum L \sin \theta = L \sin \theta_1 + L \sin \theta_2 + L \sin \theta_3$.

As discussed above, the value taken from an inclinometer is the relative horizontal displacement between two points. To obtain the real displacement curve, an adjustment of the displacement of the top end of the inclinometer casing must be made or the bottom end of the inclinometer has to be placed at a real fixed point, such as a point in rocks or cobble-gravelly soils. Alternatively, displacement at the top of the casing can be measured using a tape, level, or theodolite against a datum line drawn before excavation. Then, as shown in Figure 11.5, suppose the displacement of the casing top measured using an inclinometer is $d_2$. The amount of $d_2$ is the displacement of the casing top relative to the bottom. If we move the curve laterally by $(d_1-d_2)$, we can obtain the real displacement curve.

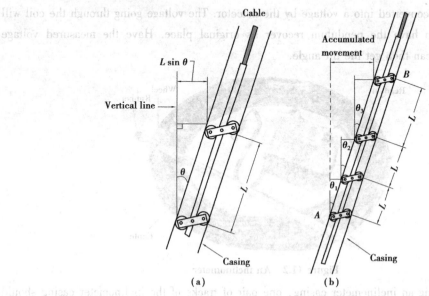

**Figure 11.4　Principle of the measuring of lateral movement by an inclinometer**

**Figure 11.5　Modification of movement of the top end of casings**

**Figure 11.6　Inclinometer casings in a diaphragm wall**

In case of diaphragm walls, the inclinometer casing can be fixed on a main reinforcement of the steel cage of the diaphragm wall, which is then placed in the trench and cast with concrete using

Tremie pipes. This finishes the installation of the inclinometer casing, as shown in Figure 11.6. The inclinometer thus installed will not be embedded deeper than the bottom of the wall. Another way to install the inclinometer casing in the diaphragm wall, as shown in Figure 11.7, is to fix a PVC pipe with 5-10 mm thick walls onto a main reinforcement of the steel cage; the PVC pipe is then lowered into the trench with the cage. When casting the trench, the placement of the PVC pipe is completed. The PVC pipe is then used as a drill guide, allowing a drill to reach the hard soil. The inclinometer casing is then inserted in the PVC pipe or the bore. The space between the inclinometer casing and the bore is filled by suitable filling materials whose properties are similar to the material there. The filling materials in diaphragm wall or hard soil can be cement grouts whereas those in soils should be a mixture of bentonite and cement.

For sheetpiles or soldier piles, which the inclinometer casing is not easily fixed directly on, the inclinometer casing can also be placed in the soils within 2 m of the outer side of the wall to measure the lateral displacement because the movement of the soils near the retaining wall is quite close to that of the wall (Ou et al., 1998), as shown in Figure 11.8.

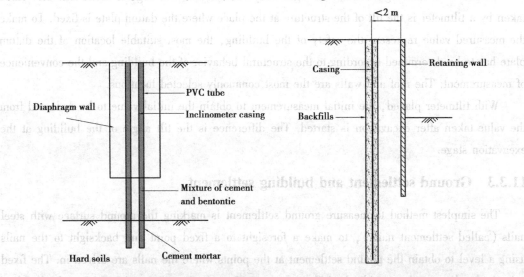

Figure 11.7　Schematic diagram of installation of
the inclinometer casing in a diaphragm wall

Figure 11.8　An inclinometer casing
outside of the diaphragm wall

## 11.3.2　Tilt of buildings

Buildings will slant because of ground settlement and thereby be damaged. The tilting of buildings should be monitored during excavation. The tilt of a building can be estimated by the relative settlement between two reference points, using measuring devices for plane surveying such as the level or theodolite, divided by the horizontal distance between these two reference points. If the two reference points are located on the two adjacent columns or foundation footings, the measured value is the angular distortion. Besides, the tilt of a building can also be monitored by using a tiltmeter, the value produced by which is called the tilt angle.

A commonly used device to measure the tilt of a building is tiltmeter (see Figure 11.9), which is similar to the inclinometer in measurement principle. The same as the inclinometer, the tiltmeter contains a sensor, which can be categorized into resistance type, vibrating type, force balance accelerometer type, etc.

**Figure 11.9　Photos of tiltmeters**

The tilt of a building contains the rigid body rotation and the angular distortion. The value taken by a tiltmeter is the tilt of the structure at the place where the datum plate is fixed. To make the measured value represent the safety of the building, the most suitable location of the datum plate has to be determined according to the structural behavior of the building and the convenience of measurement. The roof and walls are the most commonly selected locations.

With tiltmeter placed, the initial measurement to obtain the initial value to be deducted from the value taken after excavation is started. The difference is the tilt angle of the building at the excavation stage.

### 11.3.3　Ground settlement and building settlement

The simplest method to measure ground settlement is marking the ground surface with steel nails (called settlement nails), to make a foresight to a fixed point and backsight to the nails using a level to obtain the ground settlement at the points where the nails are driven in. The fixed point is the datum point outside the influence range of excavation. A certain position of a building with the pile foundation can be assumed to be a fixed point. If there are not any buildings with pile foundations nearby, a marking object in the distance or outside the influence range of settlement has to be adopted. If necessary, a permanent benchmark has to be set to serve as a fixed point.

The surrounding ground of an excavation may be soil, asphalt pavement, or concrete pavement. The asphalt pavement and concrete pavement have relatively high rigidities. That is, even though the soil below them has settled, the pavements do not necessarily show signs of settlement. Thus, the settlement nails have to be driven through the pavements so that the settlement of the nails can represent the real settlement of the soil. Figure 11.10 illustrates a possible way of setting the settlement nails.

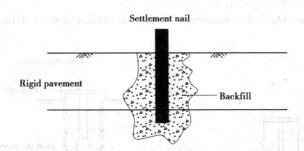

**Figure 11.10  Schematic diagram of the installation of a settlement nail**

The measurement of the settlement of buildings is the same as that of ground settlement except that the settlement nails are to be set on the buildings themselves, on the wall or on the columns, for example.

## 11.3.4  Heave of excavation bottoms

Excavation will cause the excavation bottom to heave. Too much heaving is usually dangerous for the strutting system and causes settlement of the soils in the vicinity. Under worse conditions, it may lead to the failure of the excavation. Thus, the magnitude of the heave of the soils at the excavation bottom should be monitored constantly.

The heave of an excavation bottom is usually measured with heave gauges, which consist of a cross shaped iron plate, an aluminum rod, and a galvanized iron casing, as shown in Figure 11.11. The cross shaped iron plate is placed below the excavation bottom. It is shaped in a cross to enlarge the contact area with the surrounding soils, so that when the soils below the excavation bottom heave, they will cause the heave gauges to move upward accordingly. The top of the cross shaped iron plate is connected to an aluminum rod extending to the ground surface to measure the heave of the gauge. The aluminum rod is protected by a galvanized iron pipe to seclude the aluminum rod from the soils.

The installation of the heave gauge is as follows:

①Connect the cross iron plate with the aluminum rod.

②Drive the galvanized iron casing into the soil 50 cm below the excavation surface using a wash boring machine and hammer it deeper by another 50 cm. It should be 100 cm below the excavation surface in total.

③Keep boring to a depth of about 150 cm below the excavation surface. Pour 100 cm thick bentonite into the bore.

④Put the heave gauge that is already connected to the aluminum rod into the casing and then push into the soil.

⑤Take some protective measures with the top of the heave gauge (i.e. the top of the aluminum rod) to finish the installation of the heave gauge.

The measurement of heave is to measure the elevation differences of the aluminum rod before and after excavation (of a certain stage) using a level against a fixed point or permanent benchmark. Thus, we obtain the movement of the excavation bottom during a certain stage of excavation.

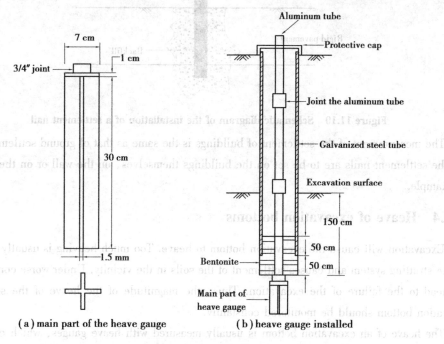

(a) main part of the heave gauge          (b) heave gauge installed

**Figure 11.11  Schematic diagram of the installation of a heave gauge**

# 11.4   Measurement of Stress and Force

## 11.4.1   Sturt load

The loading of the struts has to be monitored constantly during excavation lest it may exceed the allowable value and endanger the safety of the excavation. Some commonly used devices to measure the strut load are the strain gauge, the load cell, and the hydraulic jack gauge.

### (1) **Strain gauge**

The strain gauge is a widely used device to measure the strut load. The types of strain gauges are many, though the commonly used in excavations are the resistance strain gauge and the vibrating wire strain gauge. The two types of strain gauge take advantage of the change in resistance and vibrating frequency of the wire, respectively to measure the strain. The resistance type is more sensitive while the vibrating wire type is less influenced by pollution, humidity and temperature. Therefore, placed outdoors and exposed to sun, rain, and dust, the vibrating type is more suitable and favored.

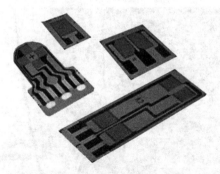

**Figure 11.12　Photo of strain gauges**

## (2) Load cell

The load cell can be divided into the mechanical type, the hydraulic type, and the electronic type. The electronic type is used most commonly. Figure 11.13 is a photo of a load cell and its configuration. As shown in the figure, the main part of the load cell is a circular box made of steel or aluminum alloy where several strain gauges are fixed inside or attached to the exterior. The accuracy of the electric load cell is the same as the strain gauge but it costs more. The load cell is usually applied when the strain gauge is not easily installed, on a wood or steel pipe of struts for example.

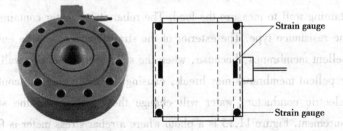

Strain gauge

Strain gauge

**Figure 11.13　Load cell and its schematic configuration**

## (3) Hydraulic jack gauge

After struts have been installed in an excavation, hydraulic jacks are often necessary to preload the struts so that they can be tightly connected with the wales and the lateral deformation of the retaining wall can be decreased. After preloading, the pressure gauge of the hydraulic jack can also be used for the measurement of the strut load. Figure 11.14 is a photo showing how a jack is set between struts. Nevertheless, misalignment of struts, offcenter strut loading, non-parallel bearing plates, and transverse relative movement of bearing plates all cause friction between the valve and cell cylinder, and render the measurement of the strut load inaccurate. Temperature changes and pressure gauge inaccuracy may cause additional error. Using pressure gauges of hydraulic jacks to measure the strut load is not reliable as a result.

**Figure 11.14    Photo of jacks between struts and walls**

## 11.4.2  Stress of the retaining wall

The stress of the retaining wall sometimes has to be measured constantly during the process of excavation lest the stress may exceed the allowable value and endanger the excavation safety.

The stresses of soldier or sheetpiles can be measured by fixing commonly used strain gauges on them. As for column piles or diaphragm walls, the rebar stress meter has to be installed onto the steel inside the retaining wall to measure the load. The rebar stress meter contains a strain gauge, which is usually the resistance type. The exterior of the strain gauge bas to be enveloped in many layers of water repellent membranes, because, when the steel or the retaining wall is acted on by a force and bends, repellent membranes may break, causing the strain gauge to contact water. Since water is also an electric conductor, water will change the resistance of the strain gauge and influence the measurement. Figure 11.15 is a photo where a rebar stress meter is fixed on the steel cage. The figure shows cutting a main reinforcement at the designed depth and collecting a rebar stress meter onto the main reinforcement by welding or a coupler.

**Figure 11.15    Rebar stress meter on a steel cage**

### 11.4.3　Earth pressure on the retaining wall

Before excavation, the earth pressure on the retaining wall is the at-rest earth pressure. After excavation, with the retaining wall moving toward the excavation zone, the earth pressure acting on the back of the wall (outside the excavation zone) decreases. The limiting value of the earth pressure on the back of the wall is called the active earth pressure. On the other hand, the earth pressure on the front of the wall (inside the excavation zone) increases and the limiting value is called the passive earth pressure. When the earth pressure at every depth in front of the wall has reached the passive earth pressure, the retaining wall is called in the ultimate state. The safety factor of the retaining wall in the ultimate state has to be larger than or equal to the values required by codes. Measuring the total earth pressure and effective earth pressure on the retaining wall is helpful to understand the bearing behavior of the wall and the surrounding soils. What's more, it can contribute to the exploration of the deformation and stability characteristics of the excavation. Generally speaking, with the shallow depths of soldier piles and sheetpiles, the earth pressures on them will not be large. Also, as it is difficult to install measuring devices on them, the earth pressures on soldier piles and sheetpiles are not measured in engineering practice. This section will only introduce the earth pressure cell used on diaphragm walls.

The commonly used earth pressure cells are the direct earth pressure cell and the indirect earth pressure cell. The direct earth pressure cell uses strain gauges (of the resistance or vibrating type) in a cell to measure the displacement of the bearing plate, which is then converted into pressure. Thus is derived the pressure on the bearing plate. The indirect earth pressure cell is a cell filled with oil or mercury, which will transmit the earth pressure on the bearing plate. The earth pressure is then measured by a sensor on the exterior of the instrument. Figure 11.16 diagrams the basic configurations of the two types of earth pressure cells.

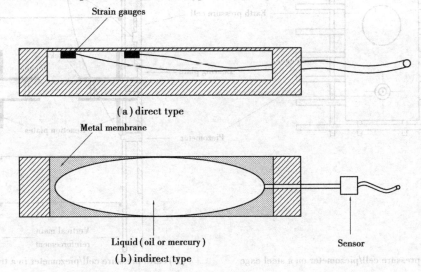

Figure 11.16　Configurations of the earth pressure cell

Figure 11.17 diagrams the installation of the earth pressure cell on the retaining wall. The installation is elucidated as follows:

①Assemble a steel cage and set it on a platform.

②Weld reaction plates onto the cage and bolt a hydraulic jack to the reaction plates.

③Add a bearing plate containing an earth pressure cell to one end of the hydraulic jack or two plates on each end. Extend the cable and hydraulic pipe along with the main reinforcement to the top of the cage and fix them.

④Examine whether there exists the phenomenon of soil collapsing from the trench wall at the positions where the earth pressure cells are to be installed. If not, place the cage with the earth pressure cells into the trench and measure the value of the earth pressure in bentonite fluid.

⑤Connect the hydraulic pipe of the jack with a manpowered pump and add pressure slowly to make the earth pressure cells move toward the faces of the trench walls. When the readings on the cells change, it follows that the earth pressure cells and the trench walls are contacting each other slightly. Keep pumping with the hydraulic pump and repeat preloading and unloading to improve the contacting condition. Lastly, add a light pre-load (of about 105% of the bentonite fluid pressure or the equivalent at-rest lateral earth pressure) and fix it [Figure 11.17(b)].

⑥When casting the concrete of the diaphragm wall using Tremie pipes, because the unit weight of concrete is heavier than that of bentonite, the trench walls may be pushed outward. In this situation, the readings of the earth pressure cells should be constantly taken. If the readings reveal signs of decreasing, more pumping is required to bring the earth pressure cells and the trench walls to full contact.

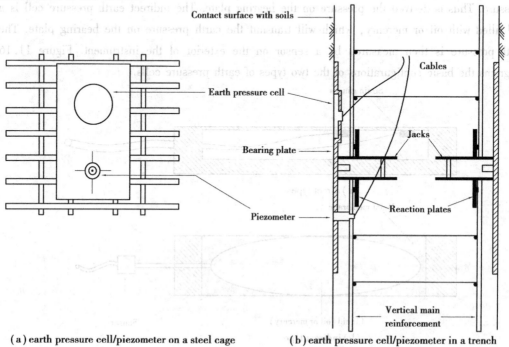

(a) earth pressure cell/piezometer on a steel cage          (b) earth pressure cell/piezometer in a trench

**Figure 11.17   Installation of the earth pressure cell/piezometer**

⑦Dismantle the hydraulic pump. The installation is then finished.

The contact between the earth pressure cells and the trench walls is a crucial point to the success of the installation. If preloading is too much, stress concentration, which will influence the accuracy of readings, will occur. If preloading is insufficient, the earth pressure cells may be enveloped by concrete during the Tremie casting, as shown in Figure 11.18. To ensure success, the installation of the earth pressure cells has to be carried out under the direction of experienced engineers.

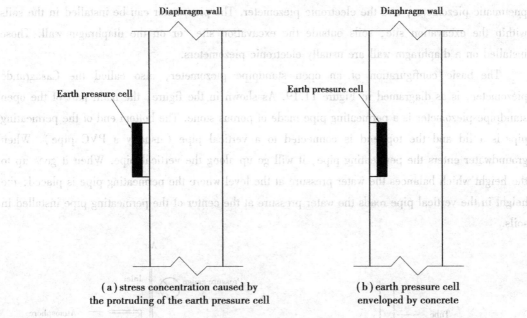

(a) stress concentration caused by the protruding of the earth pressure cell

(b) earth pressure cell enveloped by concrete

**Figure 11.18   Failures of the installation of an earth pressure cell**

The earth pressure measured by an earth pressure cell is the total earth pressure. To obtain the effective earth pressure, an electronic piezometer has to be added on the bearing plate next to the earth pressure cell, as shown in Figure 11.17(a).

According to the reason for its use, the earth pressure cell can be set on the active side, the passive side, or both. The same as with the inclinometer, rebar stress meter, strain gauge, and settlement points on the ground surface, the central section of the excavation site is the one where the earth pressure cell should be installed, considering the largest earth pressure on the section.

The maximum value of the earth pressure on the front of the retaining wall is the passive earth pressure, whereas the minimum on the back of the wall is the active earth pressure. Earth pressures on the wall usually fall between the two limiting values. An excavation design usually has taken these values into consideration. Thus, though measuring the earth pressures on the retaining wall is helpful to understand the bearing behavior of the wall and the surrounding soils, the results are not easily applied to the judgment of the excavation safety.

# 11.5 Measurement of Water Pressure and Ground-water Level

## 11.5.1 Water pressure

Some commonly used piezometers in excavations are the open standpipe piezometer, the pneumatic piezometer, and the electronic piezometer. The piezometer can be installed in the soils within the excavation site, soils outside the excavation site, or on the diaphragm wall. Those installed on a diaphragm wall are usually electronic piezometers.

The basic configuration of an open standpipe piezometer, also called the Casagrande piezometer, is as diagramed in Figure 11.19. As shown in the figure, the main part of the open standpipe piezometer is a permeating pipe made of porous stone. The bottom end of the permeating pipe is a lid and the top end is connected to a vertical pipe (usually a PVC pipe). When groundwater enters the permeating pipe, it will go up along the vertical pipe. When it goes up to the height which balances the water pressure at the level where the permeating pipe is placed, the height in the vertical pipe reads the water pressure at the center of the permeating pipe installed in soils.

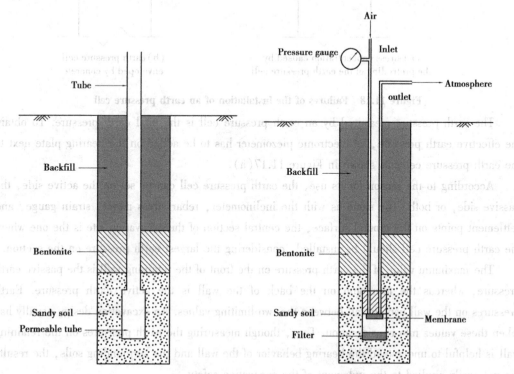

Figure 11.19  Schematic configuration
of an open standpipe piezometer

Figure 11.20  Schematic configuration
of a pneumatic piezometer

The height in the vertical pipe can be measured by using a water level indicator, which is a coaxial cable of negative and positive poles in which the groundwater can be seen as the electrolyte. Put the indicator into the vertical pipe and let it contact the water, and a low-current circuit would then be formed and a signal be generated to reveal the height of water in the vertical pipe. The open standpipe piezometer is simple in its principle and reliable in its result, with easy installation. It is required, however, to wait till the groundwater has fully flowed into the vertical pipe before the height is measured. As a result, measuring the water pressure in clayey soils usually takes a long time.

The pneumatic piezometer, also called the closed piezometer, is as diagramed in Figure 11.20, illustrating its basic configuration. When the air pressure in the air pipe is smaller than the water pressure, $p$, as shown in the figure, the air keeps flowing into the pipe, and the air pressure will grow. When the air pressure in the pipe is equal to or exceeds the water pressure, $p$, the membrane moves outward and the air current is facilitated; air will flow out of it and the air pressure will thus be decreased. The largest reading of the air pressures in the pipe is then the water pressure. Nitrogen and $CO_2$ are the most widely used gases. Using a pneumatic piezometer to measure the water pressure does not require water in great quantity to flow into the main part of the piezometer and thus does not take a long time to wait. It is suitable to be used in clayey as well as sandy soils.

The electronic piezometer in its basic design is similar to the earth pressure cell ( see Figure 11.16) except it isolates the bearing plate from the influence of earth pressure. Therefore, a permeable stone is placed in front of the bearing plate which is acted on by groundwater flowing through the permeable stone. Thus, the water pressure is measured. The electronic piezometer can be divided into the resistance and vibrating types. Figure 11.21 diagrams the basic configuration and the installation of the electronic piezometer. The strengths and shortcomings of the open standpipe piezometer, the pneumatic piezometer, and the electronic piezometer are listed in Table 11. 1.

The installation of a piezometer in soils is elucidated as follows:

①Bore a hole using a drill machine to 50 cm below the designed depth.

②Place the main part of the piezometer (no matter whether it is the open standpipe type, the pneumatic type, or the electronic type) in the bored hole and have the center of the piezometer located at the designed depth. Fill the bored hole with sandy soil. The sandy soil is used to help groundwater permeate into the main part of the piezometer and be measured.

③Seal the top (or both ends) of the sandy soil with bentonite to avoid groundwater from other elevations reaching the permeable pipe and influencing the measurement results. Above the bentonite is again filled with sandy soils to provide sufficient overburden weight to stabilize bentonite and sealing.

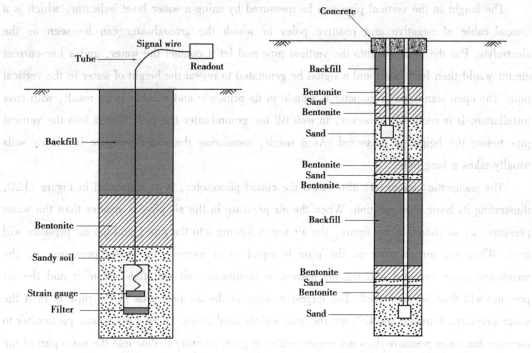

**Figure 11.21   Schematic configuration of an electronic piezometer**

**Figure 11.22   Two piezometers installed in the same borehole**

In the same bole can be placed more than one piezometer, all of them requiring sealing to avoid water pressure from other elevations, as shown in Figure 11.22. Depending on the monitoring aim, the piezometer can be embedded in clayey or sandy soils. Using one in clayey soils can derive the effective stress by measuring excess pore water pressure induced by soil deformation during excavation. Neither the excess pore water pressure nor the effective stress directly relates to excavation safety. They are, nevertheless, helpful to understand the deformation behaviors and stability characteristics. The piezometer aimed, as above elucidated, should be embedded in the central section of the excavation site.

The piezometer embedded in sandy or permeable soils can monitor the variation of water pressure, which is helpful for the diagnosis of the excavation safety. If the water pressure changes abruptly while the stress on the reinforcement of the retaining wall is abnormal, it must get extraordinary attention and an effective remedial measure should be adopted to safeguard the excavation safety.

**Table 11.1  Strengths and shortcomings of various types of piezometers**

| Type | Strengths | Shortcomings |
|------|-----------|--------------|
| Open standpipe piezometer | ①Reliable.<br>②Can measure the permeability of soils.<br>③Can take groundwater sample.<br>④Applicable to sandy layers.<br>⑤Durable. | ①Takes long time to read and is not suitable to be applied in clayey soils.<br>②Easily damaged during construction process.<br>③The shaft often obstructs construction.<br>④The filter stones in the permeable pipe are easily clogged. |
| Pneumatic piezometer | ①Does not take long time to read; applicable to sandy and clayey soils.<br>②Does not obstruct construction.<br>③Not easily damaged or destroyed. | ①The reading equipment is not as conveniently portable as that with the open standpipe piezometer.<br>②Requires better installation techniques.<br>③The membrane is susceptible to erosion in the long term and thereby the durability is affected. |
| Electronic piezometer | ①Does not take long time to read; applicable to sandy and clayey soils.<br>②Easy to read.<br>③Does not obstruct construction.<br>④Though usually not durable, some specially designed types of electronic piezometers are also good in durability.<br>⑤Fits automatic systems. | ①For long-term usage, durability and accuracy of readings have to be considered.<br>②More expensive. |

## 11.5.2  Groundwater level

The instrument for groundwater level is the observation well where a vertical pipe with many holes, enveloped with two layers of nylon net, is placed. The installation procedure of the vertical pipe is as elucidated in the following (see Figure 11.23):

①Bore to the designed depth using a drill machine.

②Fill the bottom of the bored hole with sand.

③Place the water observation pipe into the hole and fill the space with sand.

④Take some protective measures on the top of the vertical pipe. Thus is finished the installation of the observation pipe.

After groundwater flows into the pipe and the water level inside reaches the stable state, the water level in the pipe is then the groundwater level. As with the open standpipe piezometer, the water level in the pipe can be measured by using a water level indicator.

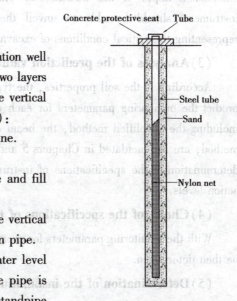

**Figure 11.23  Installation of an observation well**

The aim of the observation well is to monitor the change of the groundwater before or during excavation and to supply information for the determination of the excavation method, the design of water pumping, and the management of construction period.

# 11.6 Plan of Monitoring Systems

A complete plan of a monitoring system should include the following items: (1) determination of the monitoring parameters; (2) determination of the locations of the monitoring instruments or devices; (3) analyses of the prediction values of the parameters; (4) choice of the specifications of the monitoring instruments; (5) determination of the installation specifications of the instruments; (6) setting of the alert and action levels; and (7) determination of the measurement frequency. According to the sequence of execution, they are elucidated as follows:

## (1) Determination of the monitoring parameters

Excavation-related materials are soils, retaining walls, and struts. Excavation will produce physical quantities such as stress and strain on, and displacement of these materials, which can be measured by using instruments. Considering the cost, it is impossible to measure every physical quantity within the excavation influence range. Thus, suitable parameters have to be determined according to the excavation scale, geological conditions, and the situations of adjacent properties.

## (2) Determination of locations of the monitoring instruments

The measurement parameters determined, what follows is the selection of the locations and embedment depth of the instruments. The proper locations and embedment depths of the instruments should be able to unveil the excavation behaviors and the physical quantities representing the critical conditions of excavation.

## (3) Analyses of the prediction values of the parameters

According to the soil properties, the type of the retaining wall and the excavation conditions predict the monitoring parameters for each of the excavation stages. The methods of prediction, including the simplified method, the beam on elastic foundation method, and the finite element method, are as elucidated in Chapters 5 and 6. The prediction results are not only helpful to the determination of the specifications of instruments but also useful to the setting of the alert and action levels.

## (4) Choice of the specifications of the monitoring instruments

With the monitoring parameters for excavation stages, the specifications of the instruments can be then determined.

## (5) Determination of the installation specifications of the instruments

Whether the monitoring parameters are accurate affects the judgment of the excavation safety. Though the precisions of the instruments are important, whether they have been correctly installed also influences the measurement results. As a result, appropriate installation specifications have to be mapped out according to the characteristics of the instrument and the geological conditions.

### (6) Setting of the alert and action levels

The alert level is an important index for the judgment whether the excavation is under normal conditions. The action level is the value indicating the immediate necessity of measures to be taken. Before excavation, the alert and action levels for each excavation stage should be determined in advance.

### (7) Determination of the measurement frequency

The measurement frequency refers to the times of the measurement taken within a certain period. Basically, the measurement frequency has to be increased during the process of excavation.

# 11.7 Application of Monitoring Systems

In former sections, commonly used monitoring items and instruments have been introduced. The lateral deformations of the retaining structure and soils, the tilt and settlement of the building, the ground settlement, and the heave of the excavation bottom, the uplift of the central post, the strut load, the stress of the retaining structure, and the water pressure within the excavation zone are directly related to the safety of both excavations and adjacent properties. Figures 11.24 and 11.25 diagram the monitoring system of a MRT (Mass Rapid Transit) station in Singapore (Zhang et al.,2018).

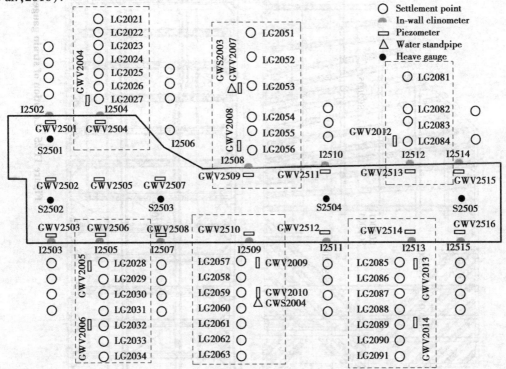

**Figure 11.24　Instrumentation layout of the monitoring system (Zhang et al.,2018)**

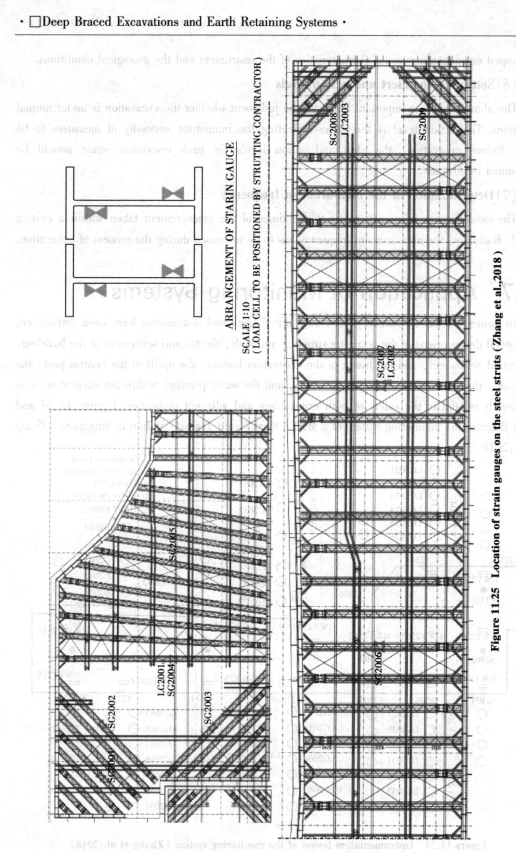

ARRANGEMENT OF STARIN GAUGE

SCALE 1:10
(LOAD CELL TO BE POSITIONED BY STRUTTING CONTRACTOR)

Figure 11.25  Location of strain gauges on the steel struts (Zhang et al,2018)

Each monitoring item can be set with the alert and action levels separately. These measured values, and alert and action levels all contribute to the understanding of the existing conditions of the excavation and serve as the basis to judge whether and when emergency measures are to be taken.

An alert level is an important indicator that tells whether the excavation is under abnormal conditions. An action level is the limit indicating the necessity for emergency measures. Under normal conditions, measured values will vary with the excavation depth. Generally speaking, when the measured value exceeds the alert level, it follows that the excavation is under a certain abnormality and engineers have to be cautious and make a judgment of the cause, though the excavation can continue. When the measured value exceeds the action level, it indicates the excavation is under dangerous conditions and has to be suspended right away to take necessary measures to restore the stability.

In fact, there are many objects of measurement related to the safety of an excavation. That some of the measured values grow beyond the action levels does not necessarily lead to the conclusion that the excavation is really under unsafe conditions. To our certainty, however, the higher measured values grow beyond the action levels, the closer the excavation is to failure. Engineers have to make their own judgment of the present conditions of safety on the basis of the relations between the measured values and the alert/action levels and the characteristics of the excavation.

The setting of alert and action levels is based on empirical experiences so far and no generally accepted standard formula is available. In engineering practice, the allowable value is usually taken as the action level and 70%-80% of the allowable value as the alert level. The allowable value is a generally accepted physical quantity, when taking the factor of safety into consideration. For example, the allowable settlement at LG2057 (see Figure 11.24) is 9 cm, the action level for the settlement is then 9 cm and the alert level 6.3 cm (70% action level).

# 12  Back Analysis for Excavation

## 12.1  Introduction

As the discussion in Chapter 6, finite element method is a powerful tool for geotechnical problem analysis. The normal procedure of the analysis can be demonstrated as 3 steps: (1) build mechanical models, including geometry, constitutive model of soils and rocks and initial boundary conditions; (2) determine the parameters of soils and rocks; (3) get the stress and deformation field. It can be called forward analysis.

However, the determination of parameters of soils and rocks usually has uncertainty, making challenges for prediction of geotechnical performance. Because of the limitation of time and cost, the geological survey before the construction may not be comprehensive enough to reveal real condition of the whole construction field. On the other hand, the results of laboratory tests may fail to fully represent the in situ soils and the complicated soil stratum conditions because of disturbance caused by sampling. Also, some parameters, like the small strain stiffness (Section 7.5.1), are difficult be determined via conventional tests. All of these factors lead FEM prediction to fail to match the observed response.

In this regard, observational method (Terzaghi and Peck, 1943; Peck, 1969b), a framework wherein construction and design procedures and details are adjusted based upon observations and measurements made as construction proceeds, was usually used to optimize the design parameters. In this framework, the excavation-induced response has been measured and the more reasonable soil (rock) parameters are supposed to be back-figured based on the observation, allowing more reliable prediction for following construction activities. The procedure is so-called back analysis.

Because braced excavations are generally carried out in stages, back-analysis to update key soil parameters (such as the normalized undrained shear strength and the normalized initial tangent modulus in an excavation in clays) is generally realized in multiple stages; before the first excavation stage, the wall and ground responses are predicted with field tests and/or laboratory data. After the first stage excavation is completed and wall and/or ground responses are measured, the key soil parameters can be updated with the observed responses to refine the knowledge of the soil parameters. With the updated soil parameters, the wall and/or ground responses in the subsequent excavation stages may be predicted with improved fidelity. This process can be repeated stage by stage until the final excavation depth is reached.

Figure 12.1 diagrams an example reported by Finno (2005). Before the optimization, the computed wall lateral displacements by using initial parameters were significantly larger than the observation at every construction stage. The maximum computed displacements were about twice the measured ones and the displacement profiles result in significant and unrealistic movements at the bottom. It indicates that the tests overestimate the soil stiffness properties. To optimize the soil stiffness parameters, back analysis was carried out. When parameters optimized based on Stage 1 observations were used, the improvement of the fit between the computed and measured response at every stage is significant. At Stage 5, the maximum computed displacement exceeds the measured data by only about 15%. Analyses were also made wherein parameters were recalibrated at every stage until the final construction stage and the fit between the computed and measured displacement is further improved.

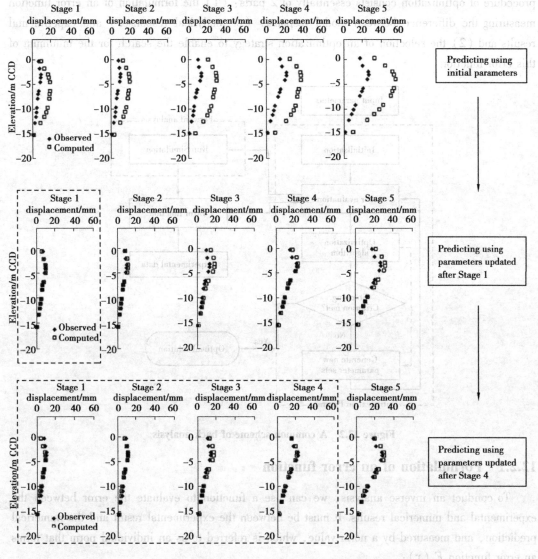

Figure 12.1  Optimizing analysis via updating parameters (Finno, 2005)

While back analysis is necessary, note that it cannot replace laboratory and filed tests, the back-figured soil parameters should be regarded as the equivalent soil parameters, not the real soil parameters. It only helps to update the design parameters to provide references for following construction activities.

Generally, back analysis methods used in excavation issues are not specialized. This chapter will only make introduction to some selected methods.

# 12.2 General Procedure of Back Analysis in Excavation Issues

The procedures of back analysis can be generally illustrated as Figure 12.2. The mathematical procedure of optimization consists essentially of 2 parts: (1) the formulation of an error function measuring the difference between model responses (or analytical responses) and experimental results and (2) the selection of an optimization strategy to enable the search for the minimum of this error function.

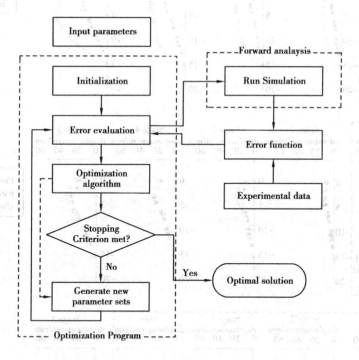

Figure 12.2  A common scheme of back analysis

## 12.2.1  Formulation of an error function

To conduct an inverse analysis, we can use a function to evaluate the error between the experimental and numerical results. It must be between the experimental result and the numerical prediction, and measured by a norm value, which is referred to as an individual norm that forms an error function $F_{\text{err}}(x)$:

$$F_{err}(\boldsymbol{x}) \rightarrow \min \qquad (12.1)$$

where $\boldsymbol{x}$ is a vector containing the parameters to be optimized. Bound constraints are introduced on these variables,

$$\boldsymbol{x}_{min} \leqslant \boldsymbol{x} \leqslant \boldsymbol{x}_{max} \qquad (12.2)$$

where $\boldsymbol{x}_{min}$ and $\boldsymbol{x}_{max}$ are, respectively, the lower and upper bounds of $\boldsymbol{x}$.

As the first step in the formulation of an error function, an expression for the individual norm has to be established. In general, the individual norm is based on Euclidean measures between discrete points, composed of the experimental and the numerical results. The simplest error function can take the following expression:

$$F_{err}(\boldsymbol{x}) = \frac{1}{N} \left( \sum_{i=1}^{N} |u_i - u_i^*| \right) \qquad (12.3)$$

where $N$ is the number of values, $u_i^*$ is the computed value, and $u_i$ is the value of observation.

Another formulation of the error function was introduced:

$$F_{err}(\boldsymbol{x}) = \frac{1}{N} \left[ \sum_{i=1}^{N} (u_i - u_i^*)^k \right]^{\frac{1}{k}} \qquad (12.4)$$

where $k$ is a non-null positive value with $k=1$ for the sum of error at every point and $k=2$ for the least square function.

However, Eqs. 12.3 and 12.4 present some disadvantages that the error is dependent on the type of test and the number of measurement points. To eliminate the scale effects on the fitness between the experimental and the simulated results, some normalized formula was developed.

Levasseur et al. (2008) adopted an advanced error function which can be adopted with 2 modifications of 100 percentage and add weight to each calculation point. The average difference between the measured and the simulated results is expressed in the form of the least square method:

$$F_{err}(\boldsymbol{x}) = \sqrt{\frac{1}{N} \sum_{i=1}^{N} w_i \left( \frac{u_i - u_i^*}{u_i^*} \right)^2} \times 100 \qquad (12.5)$$

where $w_i$ is weight for the calculation at point $i$.

## 12.2.2  Optimization strategy

Methodologies used for back analysis can be classified into two groups, i.e., (1) deterministic methods and (2) probabilistic methods. In a deterministic method, analysis model is usually believed or assumed accurate, and the purpose of back analysis is to find a set of parameters fitting the observation. In a probabilistic back analysis, however, it is recognized that the analysis model may not be perfectly accurate and numerous combinations of parameters may result in the same response. Table 12.1 lists several optimization techniques widely used in geotechnical engineering and some of them will be introduced later.

**Table 12.1  Optimization techniques used in geotechnical engineering**

| Deterministic method | References | Probabilistic method | References |
|---|---|---|---|
| Gradient-based algorithms | Levasseur (2008) | Least-squares method | Xu & Zheng (2001) |
| Nelder-Mead simplex | Nelder & Mead (1965) | Maximum likelihood method | Ledesma et al. (1996) |
| Genetic algorithms | Holland (1992) | | |
| Particle swarm optimizations | Kennedy & Eberhart (2011) | | |
| Simulated annealing | Busetti (2003) | | |
| Differential evolution algorithm | Price & Storn (1997a, b) | | |
| Artificial bee colony | Karaboga (2005) | | |
| Artificial neural network | Ghaboussi et al. (1998a, b) Hashash et al. (2002,2004) | Bayesian method | Fitzpatrick (1991) Honjo et al. (1994) |

## 12.2.3  Experimental and observed data

Note that the forward analysis is not only restricted to FEM, but also any framework for solving geotechnical problems, such as limit equilibrium method and simplified methods involving semi-empirical models. When the excavation is carried out in a heterogeneous or layered soil deposit, back analysis of soil parameters can be quite complex because parameters of each layer have to be treated as separate variables. In theory, it is not a problem to back figure the parameters in FEM. In theory, this is not a problem if the analysis model, which is a required component in any updating scheme based on field observations, allows for consideration of soil parameters separately in each of multiple layers, as in the case of finite-element modeling of an excavation in a layered soil deposit. However, back-analysis using FEM that involves multiple sets of soil parameters is computationally intensive, especially under updating framework with uncertainty.

As for the experimental data, displacements are the most accessible observation objects in excavation issues. Thus it is the most commonly practice to back figure parameter based on displacement data. In theory, the two concerns in deep excavation monitoring system, data of retaining wall displacements and surface settlements, could be used in back analysis. However, it is a much more complicated task to predict ground settlement because of various factors (such as small strain behaviors of soils), which makes wall displacements based back analysis more common.

# 12.3　Deterministic Method

## 12.3.1　Gradient-based method

　　The gradient method is probably one of the oldest optimization algorithms, going back to 1847 with the initial work of Cauchy. The gradient method is an algorithm for examining the directions defined by the gradient of a function at the current point. Based on the basic principle, different gradient-based methods have been developed, such as the steepest descent method, the conjugate gradient method, the Levenberg-Marquardt method, the Newton method and several Quasi Newton methods, the Davidon-Fletcher-Powell, and the Broyden-Fletcher-Goldfarb-Shanno methods.

　　A rapid convergence is the primary advantage of a gradient-based method. Clearly, the effective use of gradient information can significantly enhance the speed of convergence compared to a method that does not compute gradients. However, gradient-based methods have some limitations, being strongly dependent on user skills (e.g., the basic knowledge of typical values of parameters and the ability of selecting ranges of parameters), due to the need to choose the initial trial solutions. Also, they can easily fall into local minimums, mainly when the procedure is applied to multi-objective functions, as it is the case for material parameter identification with a nonlinear soil model. The requirement of derivative calculations makes these methods nontrivial to implement. Another potential weakness of the gradient-based methods is that they are relatively sensitive to difficulties such as noisy objective function spaces, inaccurate gradients, categorical variables, and topology optimization.

### 1) Steepest descent algorithm

　　To characterize the discrepancy between the experimental behavior and the modelled one, a scalar error function, $F_{err}$ can be defined by using the least-square method (Levasseur, 2008):

$$F_{err} = \left[ \frac{1}{N} \sum_{i=1}^{N} \frac{(u_i - u_i^*)}{\Delta u_i^2} \right]^{\frac{1}{2}} \quad (12.6)$$

where $N$ is the number of observation, $u_i$ is the $i$th observed value, $u_i^*$ is the computed value and $1/\Delta u_i$ is the weight of the discrepancy between $u_i$ and $u_i^*$. Basically, $1/\Delta u_i$ can be decomposed into two parts:

$$\Delta u_i = \varepsilon + \alpha u_i \quad (12.7)$$

where $\varepsilon$ is an absolute error and $\alpha$ a dimensionless relative error. Depending on the physical meaning of the experimental data, $\varepsilon$ and $\alpha$ allow to give more or less weight to measured points. If $\varepsilon$ is equal to zero, the points with weak measured values have more weight. The sensitivity of the parameters which influence these points is well taken into account by $F_{err}$. On the contrary, when $\alpha$ is equal to zero, all the points of the curve have the same weight. $\Delta u_i$ is directly linked to measurement errors, which is similar to the discrepancy norm usually used in the literature.

The error function $F_{err}(p)$ is defined as a scalar for each set of $N_p$ unknown parameters, which is noted as a vector $p$. The inverse problem is then solved as a minimization problem in the $N_p$-dimension space (the search space) restricted to authorized values of $p$ between $p_{min}$ and $p_{max}$ by using the steepest descent algorithm

Step 1: Given an a priori set

Step 2: The next set of unknown parameters $p_{i+1}$ is then chosen such that

$$p_i = p_{i+1} + xd_i \qquad (12.8)$$

Where $d_i$ is a vector indicating the downward direction

$$d_i = -\frac{F_{err}(p_i)}{\|\nabla F_{err}(p_i)\|^2} \nabla F_{err}(p_i) \qquad (12.9)$$

and $x$ is a non-dimensional scalar step of which the optimum value is determined by a quadratic estimation of $F_{err}$ in the $d_i$ direction.

Step 3: The optimization program starts again from Stage 1 until $F_{err}$ is less than the experimental measured error or until the norm of the incremental parameter vector, $\|p_{i+1}-p_i\|$ is less than the expected precision on the parameter values.

This algorithm corresponds to the traditional steepest descent method which is the simplest gradient method.

## 2) Gauss-Newton method

Finno used *UCODE* (Poeter and Hill 1998), a computer code designed to allow inverse modeling posed as a parameter estimation problem, to conduct the back analysis. The model independency allows the chosen numerical code to be used as a separate entity wherein modifications only involve model input values, allowing one to develop a procedure that can be easily employed in practice and in which the engineer will not be asked to use a particular finite element code or inversion algorithm. Rather, macros can be written in a Windows environment to couple *UCODE* with any finite element software.

With the results of a finite element prediction in hand, the computed results are compared with field observations in terms of an objective function. In *UCODE*, the weighted least-squares objective function $S(b)$ is expressed as

$$S(b) = [y - y'(b)]^T \omega [y - y'(b)] = e^T \omega \, e \qquad (12.10)$$

where
$b =$ vector containing values of the parameters to be estimated;
$y =$ vector of the observations being matched by the regression;

$y'(b)$ = vector of the computed values which correspond to observations;

$\omega$ = weight matrix wherein the weight of every observation is taken as the inverse of its error variance;

$e$ = vector of residuals.

A sensitivity matrix, $X$ is then computed by using a forward difference approximation based on the changes in the computed solution due to slight perturbations of the estimated parameter values. This step requires multiple runs of the finite element code. Regression analysis of this nonlinear problem is used to find the values of the parameters that result in a best fit between the computed and observed values. In *UCODE*, this fitting is accomplished with the modified Gauss-Newton method, the results of which allow the parameters to be updated by using

$$( C^{T}X^{T}r\omega X_{r}C + Im_{r} ) C^{-1}d_{r} = C^{1}X^{T}r \; \omega [ y - y'(b) ] \tag{12.11}$$

$$b_{r+1} = \rho_{r}d_{r} + b_{r} \tag{12.12}$$

where

$d_{r}$ = vector used to update the parameter estimates $b$;

$r$ = parameter estimation iteration number;

$X_{r}$ = sensitivity matrix $X_{ij} = \partial y_{i}/\partial b_{i}$ evaluated at parameter estimate $b_{r}$;

$C$ = diagonal scaling matrix with elements $c_{jj}$ equal to $1/( X^{T}r\omega X_{r})_{jj}$;

$I$ = identity matrix;

$m_{r}$ = Marquardt parameter (Marquardt 1963) used to improve regression performance;

$\rho_{r}$ = damping parameter, computed as the change in consecutive estimates of a parameter normalized by its initial value, but it is restricted to values less than 0.5.

In Finno's (2005) study, $m_{r}$ is initially set equal to 0 for each parameter estimation iteration $r$. For iterations in which the vector $d$ defines parameter changes that are unlikely to reduce the value of the objective function, as determined by the Cooley and Naff (1990) condition, $m_{r}$ is increased by 1.5 $m_{r}$(old) +0.001 until the condition is no longer met.

At a given iteration, after performing the modified Gauss-Newton optimization, decide whether the updated model is optimized according to either of two convergence criteria:

a. The maximum parameter change of a given iteration is less than a user-defined percentage of the value of the parameter at the previous iteration; or

b. The objective function, $S(b)$ changes less than a user-defined amount for three consecutive iterations.

After the model is optimized, the final set of input parameters is used to run the finite element model one last time and produce the "updated" prediction of future performance.

The relative importance of the input parameters being simultaneously estimated can be defined by using various parameter statistics (Hill, 1998). The statistics found most useful for this work are the composite scaled sensitivity, $ccs_{j}$, and correlation coefficient, $cor(i,j)$. The value of $ccs_{j}$

indicates the total amount of information provided by the observations for the estimation of parameter $j$, and is defined as

$$\text{ccs}_j = \left[ \frac{\sum_{j=1}^{N} \left( \left( \frac{\partial y_i'}{\partial b_i'} \right) b_j \omega_{ii}^{\frac{1}{2}} \right)^2 \Big|_b}{N} \right]^{\frac{1}{2}} \qquad (12.13)$$

where $y_i'$ = $i$th computed value; $b_j$ = $j$th estimated parameter; $\partial y_i'/\partial b_i'$ = sensitivity of the $i$th computed value with respect to the $j$th parameter; $\omega_{ii}$ = weight of the $i$th observation; and $N$ = number of observation.

The values of $\text{cor}(i, j)$ indicate the correlation between the $i$th and $j$th parameters, and are defined as

$$\text{cor}(i,j) = \frac{\text{cov}(i,j)}{\text{var}(i)^{\frac{1}{2}} \text{var}(j)^{\frac{1}{2}}} \qquad (12.14)$$

where $\text{cor}(i,j)$ = off-diagonal elements of the variance-covariance matrix $V(b') = s^2 (X^T r \omega X_r)^{-1}$ in which $s^2$ = model error variance; $\text{var}(i)$ and $\text{var}(j)$ refer to the diagonal elements of $V(b')$.

Inverse analysis algorithms allow the simultaneous calibration of multiple input parameters. However, identifying the important parameters to include in the inverse analysis can be problematic, and it is generally not possible to use a regression analysis to estimate every input parameter of a given simulation. The number and type of input parameters that one can expect to estimate simultaneously depend on a number of factors, including the soil models used, the stress conditions of the simulated system, available observations, and numerical implementation issues. Note that within the context of finite element simulations, individual entries within element stiffness matrices may depend on combinations of soil parameters, implying that more than one combination of these parameters can yield the same result. Consequently, these parameters will be correlated within the context of the optimization solution, even though the parameters may be independent of a geotechnical perspective.

The total number of input parameters can be reduced in three steps to the number of parameters that are likely to be optimized successfully by inverse analysis.

In Step 1, the number of relevant and uncorrelated parameters of the constitutive model chosen to simulate the soil behavior is determined. The number of parameters that can be estimated by inverse analysis depends upon the characteristics of the model, the type of observations available, and the stress conditions in the soil. Composite scaled sensitivity values (Eq. 12.13) can provide valuable information on the relative importance of the different input parameters of a given model. Parameter correlation coefficients (Eq. 12.14) can be used to evaluate which parameters are correlated and are, therefore, not likely to be estimated simultaneously by inverse analysis.

In Step 2, the number of soil layers to calibrate and the type of soil model used to simulate the layers determine the total number of relevant parameters of the simulation. An additional sensitivity analysis may be necessary to be checked for correlations between parameters relative to different layers.

Finally, in Step 3, the total number of observations available and computational time considerations may prompt a final reduction of the number of parameters to optimize simultaneously. A detailed example of this procedure is presented by (Calvello and Finno, 2004).

The relative fit improvement, RFI, which indicates by what percentage the optimized results improved compared to the predictions at the beginning of a certain stage, can be used to quantify the effectiveness of optimization procedure and the reliability of the predictions. RFI is defined as:

$$\mathrm{RFI}_i = \frac{S(b)_{\mathrm{in}_i} - S(b)_{\mathrm{fin}_i}}{S(b)_{\mathrm{in}_i}} \qquad (12.15)$$

where $S(b)_{\mathrm{in}\_i}$—initial value of the objective function is at stage $i$; and $S(b)_{\mathrm{fin}\_i}$—optimized value of the objective function is at the end of stage $i$.

The error variance $s^2$, a commonly used indicator of the overall magnitude of the weighted residuals, is computed as

$$s^2 = \frac{S(b)}{ND - NP} \qquad (12.16)$$

where $S(b)$ = objective function; $ND$ = number of observations; and $NP$ = number of estimated parameters.

## 3) quasi-Newton method

Tang et al. (2008) employed an objective function $f(x)$ defined as

$$f(x) = \frac{1}{N} \sum_{i=1}^{N} \left[ \frac{u_i(x)}{u_i^*} - 1 \right]^2 \qquad (12.17)$$

where $N$ = number of observations; $u_i^*$ is the measured values; $u_i(x)$ is the computed values; $x$ is the selected parameter vector.

Then, specify an initial set of target variables $x^0$, and then these variables are updated iteratively in the optimization analysis process. The iterative procedure used in this study can be expressed as

$$x^{k+1} = x^k + \alpha^k S^k \qquad (12.18)$$

where $k$ is the iteration number and $S$ is a search direction. The step size in terms of the scalar quantity $\alpha^k$ defines the moving distance in the search direction $S$. Use of Eq. 12.18 consists of two major components: the first is to determine the search direction $S^k$, and the second is to calculate a $\alpha^k$ with which the minimized $f(x)$ can be obtained in the determined search direction $S^k$. The second component is also called line search. The detailed procedures are described as follows:

(a) Compute the search direction $S^k$.

(b) Determine the optimum step size $\alpha^k$ so that $f(x^k+\alpha^k S^k)$ is the minimum value in the search direction.

(c) Let $x^{k+1}=x^k+\alpha^k S^k$ where $x^{k+1}$ are new and $x^k$ old parameter vectors, respectively.

(d) Repeat (1) to (3) until convergence is reached.

The difficulty in solving highly nonlinear Eq. 12.17 is expected, and thus this study employs quadratic Taylor series approximation at $k$th iteration to solve this equation, which can be expressed as

$$f(x) \approx f(x) + \nabla f(x)\delta_x + \frac{1}{2}\delta_x H(x^k)\delta_x \qquad (12.19)$$

where $\delta_x=x^{k+1}-x^k$ and $H(x^k)$ is the Hessian matrix, which denotes the matrix of second partial derivatives of the objective function with respect to the target variables.

By incorporating the stationary condition $\nabla f(x^k)=0$ into Eq. 12.19, the following equation can be obtained:

$$x^{k+1} = x^k - [H(x^k)]^{-1}\nabla f(x^k) \qquad (12.20)$$

By comparing Eq. 12.20 with Eq. 12.18, when $\alpha^*=1$, the $S^k$ can be expressed as

$$S^k =- [H(x^k)]^{-1}\nabla f(x^k) \qquad (12.21)$$

Eq. 12.21 is the well-known Newton method and can be used to determine the search direction $S^k$. In theory, only one step computation is required to reach the minimum of $f(x)$ when $f(x)$ is a quadratic. However, since $f(x)$ in Eq. 12.17 is not a quadratic, it is necessary to invert the matrix $H(x^k)$ to compute the second partial derivatives of $f(x)$, in which one iteration requires $n(n+1)/2$ differentiations of the objective function ($n$ is the number of variables). Note that such iterative process in the back analysis is cumbersome and time-consuming. Therefore, for upgrading the efficiency of computation, the quasi-Newton method is incorporated in the present study instead of the Newton method. To this end, the quasi-Newton method, BFGS (Broyden-Fletcher-Goldfarb-Shanno) method (Fletcher, 1980) is adopted and expressed as

$$H_{BFGS}^{k+1} = H + (1 + \frac{\gamma^T H\gamma}{\delta^T})\frac{\delta\delta^T}{\delta^T\gamma} - (\frac{\delta\gamma^T H + H\gamma\delta^T}{\delta^T\gamma}) \qquad (12.22)$$

where $\gamma^k=\nabla f(x^{k+1})-\nabla f(x^k)$ and $\delta^k=x^{k+1}-x^k=\alpha^k S^k$. Note that, for clarity, the sign of the $k$th iteration on the right hand side of this equation is omitted.

In addition, the combination of the DSC (Davies-Swann-Campey) method (Himmelblau, 1986) and the Powell's (1962) quadratic interpolation (PQI) method is employed in this study to determine the optimum step size. Details of this method can be referred to Himmelblau (1986).

Next, the principle of the convergence criterion for determining when the process of the back analysis should be terminated is based on whether one of the three components of the convergence criterion is reached at two successive iterations. The three components are expressed as

$$|f(x^{k+1}) - f(x^k)| \leqslant \varepsilon_a \qquad (12.23a)$$

$$\frac{|f(x^{k+1}) - f(x^k)|}{f(x^k)} \leqslant \varepsilon_b \qquad (12.23b)$$

$$\frac{|x^{k+1} - x^k|}{x^k} \leq \varepsilon_c \tag{12.23c}$$

where $\varepsilon_a$, $\varepsilon_b$ and $\varepsilon_c$ are tolerances.

The first component of the convergence criterion (Eq. 12.23a) is based on the consideration that if the absolute value of change of the objective function obtained between two successive iterations is less than the specified (user-defined) tolerance. The percentage of the change of the objective function is employed as the second component (Eq. 12.23b) to judge if the convergence is reached. Finally, the percentage of change of target parameters is used as the third component (Eq. 12.23c). When one of three components of the convergence criterion is satisfied at two successive iterations, the optimization analysis is terminated because the next effective search direction cannot be found. Conversely, if none of the three components is satisfied, the optimized search direction can be determined and the analysis proceeds to search the step size.

The PQI method is incorporated in the developed NOT to effectively compute the optimized step size. Specifically, the convergence criterion of the step size consists of two components:

$$\frac{|f(\alpha^{l+1}) - f(\alpha^l)|}{f(\alpha^l)} \leq \varepsilon_d \tag{12.24a}$$

$$\frac{|\alpha^{l+1} - \alpha^l|}{\alpha^l} \leq \varepsilon_e \tag{12.24b}$$

where $l$ is iteration number of line search; $\alpha^l$ and $\alpha^{l+1}$ are the optimum step sizes computed at $l$th and $(l+1)$th iterations, respectively; $f(\alpha^l)$ and $f(\alpha^{l+1})$ are the objective function at $l$th and $(l+1)$th iterations accordingly. Tang et al. (2008) adopted $\varepsilon_a = \varepsilon_b = \varepsilon_c = 0.01$ and $\varepsilon_d = \varepsilon_e = 0.001$ as the tolerances of converge.

To enhance convergence rate of the optimization analysis, three auxiliary techniques are employed.

## (1) Scaling

The convergence rate decreases when orders of magnitude of variables are not identical. Vanderplaats (1984) suggested that each variable can be multiplied by a scaling factor and then solved by the optimization method. As illustrated in Figure 12.3(a), many iterations are required to search the minimum value for the elliptic shape of contours of the objective function defined by the unscaled variables. However, the objective function with scaled variables exhibits a circular shape of contours, and only one iteration is required to obtain the minimum value [see Figure 12.3(b)].The mathematical formulas are expressed as follows:

$$\bar{x} = Dx \tag{12.25}$$

$$\overline{\nabla}f(\bar{x}) = D^{-1} \nabla f(x) \tag{12.26}$$

where $x$ and $\bar{x}$ are the original and scaled vectors of variables, respectively. $\nabla f(x)$ and $\overline{\nabla}f(\bar{x})$ are the original and scaled vectors of gradient, respectively. $D = [1/|x_1|], [1/|x_2|], \cdots [1/|x_n|]$, $x_1, x_2, \cdots, x_n$ are values of variables for each iteration.

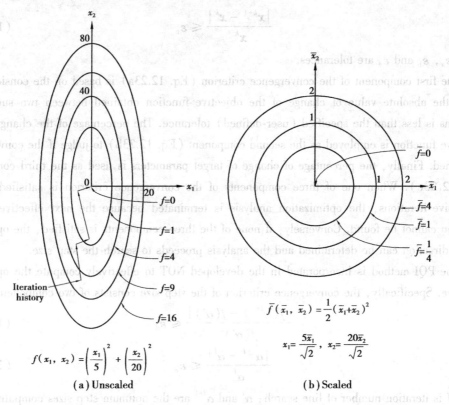

$$f(x_1, x_2) = \left(\frac{x_1}{5}\right)^2 + \left(\frac{x_2}{20}\right)^2$$

(a) Unscaled

$$\bar{f}(\bar{x}_1, \bar{x}_2) = \frac{1}{2}(\bar{x}_1 + \bar{x}_2)^2$$

$$\bar{x}_1 = \frac{5\bar{x}_1}{\sqrt{2}}, \quad x_2 = \frac{20\bar{x}_2}{\sqrt{2}}$$

(b) Scaled

**Figure 12.3　Contours of objective function with scaled and unscaled variable**
(Tang et al. ,2008)

### (2) Restarting

Since the quasi-Newton method employs the current Hessian matrix to update the matrix at the next iteration, both the iteration process and digital computers may cause the problem of error accumulation. Search direction, as obtained from this method, may no longer be the direction of minimum value after a great number of iterations. To deal with this problem, a negative gradient ascent based on a new point after a certain number of iterations is used as the new search direction, instead of using the previously calculated direction. This action is called "restarting". In this study, restarting is activated when the minimum value of the search direction does not cause the value of the objective function to decrease.

### (3) Assurance

Although the PQI method is an efficient way to obtain the optimum step size in a certain search direction, numerical instability may occasionally occur during the iteration process because the objective function formed by various variables is assumed to be a unimodality function. In reality, the type of the function is unknown before iteration begins. Thus, it is not easy to obtain the minimum objective function through the interpolation method. Therefore, this study employs the DSC method to determine three equal step sizes, which can generate a concave curve formed by corresponding objective functions at the beginning of searching the step size. The optimum step size can then be determined by using the PQI method and thus a higher stability and accuracy of results of optimization analyses can be obtained.

Since soil parameters are implicitly included in the objective function, it is difficult to determine the gradient vector of the objective function. To this end, this study adopts the FDM to obtain the gradient vector by

$$\frac{\partial f}{\partial x_j} = \lim_{\delta x \to 0} \frac{f(x_1, \cdots, x_j + \delta x_j, \cdots, x_n) - f(x_1, \cdots, x_j, \cdots, x_n)}{\delta x_j} \quad (12.27)$$

where $x_j$ is the $j$th variable at variable vector.

For the FDM, the value of $\delta x_j$ essentially has a significant effect on the analysis result. An improper value of $\delta x_j$ usually causes a large numerical error. Dennis and Schnabel (1983) suggested that $\delta x_j$ can be determined by

$$\delta x_j = \sqrt{\varepsilon_m} x_j \quad (12.28)$$

where $\varepsilon_m$ is the machine error. Tang et al. (2007) adopt $\delta x_j = 9.77 \times 10^{-4} x_j$ so that $\delta x_j$ is varied with each iteration.

## 4) Limitation of gradient-based method

Tang et al. (2008) studied the applicability of the gradient-based methods. It is found that gradient-based method performs well when only one parameter is adopted in the back analysis. However, the further examination indicates the methods are not capable of optimizing a set of parameters.

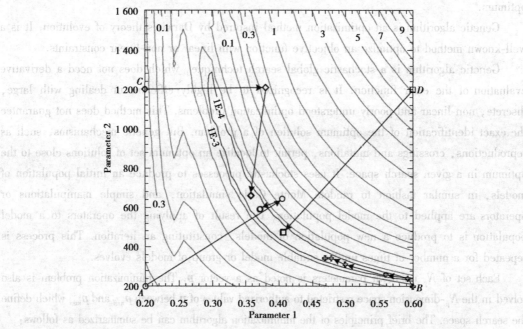

**Figure 12.4  Contours of the objective function and identification iteration paths simultaneously using two target parameters (after Tang et al. ,2008)**

Figure 12.4 shows the identification iteration paths for the four initial values and the contour of the objective function of a hypothetical case, where the exact value of the parameters is (0.35, 700).

The four identification iteration paths fall into the "parabolic-pattern valley", where the value of the objective function is less than $10^{-4}$. Theoretically, any point within the valley can be regarded as a solution when the criterion of objective function equal to $10^{-4}$ is adopted, causing the back analysis does not yield a unique solution under different initial values. Thus, it should be noted that those optimized soil parameters are one of the possible solutions although the unique solution cannot be yielded.

## 12.3.2 Genetic algorithm

The gradient method the uniqueness of the solution of the inverse problem, however, modelling errors or in situ measurement uncertainties are not sufficiently taken into account by inverse analysis processes to be sure about solution accuracy. There is not one unique and exact solution but rather an infinity of approximate solutions around an optimum. The aim of the parameter identification should better be the identification of this infinity of approximate solutions rather than the one with the lower error function whose definition is arbitrary.

The genetic algorithm (GA) is robust and highly efficient method, which is able to solve complex optimization problems, but does not guarantee an exact identification of the optimum solution. However, it does permit the localization of an optimum set of solutions close to this optimum.

Genetic algorithm is an optimization method inspired by Darwin's theory of evolution. It is a well-known method to optimize an objective function with linear or non-linear constraints.

Genetic algorithm is a stochastic global search technique, which does not need a derivative evaluation of the error function. It is recognized to be highly efficient in dealing with large, discrete, non-linear and poorly understood optimization problems. This method does not guarantee the exact identification of the optimum solution of a problem. But genetic mechanisms, such as reproductions, crossings and mutations, permit to localize an optimum set of solutions close to the optimum in a given search space. It uses stochastic processes to produce an initial population of models, in similar fashion to random Monte Carlo simulation, and simple manipulations or operators are applied to the model population. The result of applying the operators to a model population is to produce a new population of models, constituting an iteration. This process is repeated for a number of times until a suitable model or group of models evolves.

Each set of $N_p$ unknown parameters is noted as a vector $\boldsymbol{p}$. The minimization problem is also solved in the $N_p$-dimension space restricted to authorized values of $\boldsymbol{p}$ between $\boldsymbol{p}_{min}$ and $\boldsymbol{p}_{max}$ which define the search space. The brief principles of the minimization algorithm can be summarized as follows:

### (1) Encoding, individual and population

Each mechanical parameter is binary encoded and represents a gene. The concatenation of several genes forms an individual. Each individual defines a point of the search space. A group of $N_i$ individuals represents a population of the $i$th generation.

## (2) Generation of an initial population

Group of $N_i$ individuals is randomly chosen in the search space. The scalar error function for each individual of a population is evaluated. Mechanisms of selection, reproduction, and mutation are used to make the population evolve to the best individuals in the search space.

## (3) Selection

Depending on their fitness (minimal cost of scalar error function), only the best $N_i/3$ individuals are preserved for the constitution of the next population. They are called parents. This 'elitist' selection is known to be more efficient for unimodal function optimizations.

## (4) Reproduction and crossing

The parents are randomly selected by pairs and crossed over into $N_{coup}$ points to generate new pairs of offsprings (Table 12.2). To improve the algorithm efficiency, the $N_{coup}$ number is chosen equal to the number of sought parameters. Crossing process is repeated until $2N_i/3$ offspring are created. These new offspring are called children.

**Table 12.2 Illustration of the reproduction between a pair of parents to generate a new pair of offspring in genetic algorithm optimization method**

| Parents | | | | |
|---|---|---|---|---|
| Parent A | **1100** | 110 | **10011101** | **11100** |
| Crossover points | — | — | — | |
| Prarent B | 0110 | 001 | 1111011 | 00111 |
| Offspring | | | | |
| Offspring A | **1100** | 001 | **10011101** | 00111 |
| Crossover points | — | | | |
| Offspring A | 0110 | **110** | 01111011 | **11100** |

## (5) Mutation and generation of a new population

Putting together parents and children create a new population of $N_i$ individuals. To limit the convergence problems and to diversify the population, some of the individuals are randomly mutated (Table 12.3).

## (6) Test of convergence

The three previous stages are repeated until the error function average (or standard deviation) of the parent part of the population is less than a given error.

**Table 12.3 Illustration of the mutation of an individual in genetic algorithm optimization method**

| Offspring | | | |
|---|---|---|---|
| Offspring | 1100 | 0 | 10011101 |
| Selected gene | | — | |
| | | | |
| Offspring after mutation | | | |
| Mutated offspring | 1100 | 1 | 10011101 |
| Mutated gene | | — | |

# 12.4 Probabilistic Methods

## 12.4.1 Maximum likelihood method

Ledesma et al. (1996) illustrated a maximum likelihood to perform back analysis. The formulation allows the introduction of the error structure of the observations and gives a minimum bound of the variances of the parameters identified. The formulation results in a minimization problem of an "objective function", which can be solved by means of any suitable optimization technique.

### 1) Basic formulation

It is assumed that a relation between state variables, $x$ and parameters, $p$, has been defined by means of a model, $M$ (generally non-linear) which is considered fixed: $x = M(p)$. The information available includes some measurements, that is, a set of measure set variable, $x^*$, and some prior information on the parameters to be estimated, $p^*$.

In the maximum likelihood approach, the best estimation of the system parameters is found by maximizing the likelihood of a hypothesis, $L$. Likelihood is defined as proportional to the joint probability of errors in measuring state variables and in the prior information of the parameters:

$$L = kP(x,p) = kP(x)P(p) \tag{12.29}$$

where measured state variables and parameters have been considered as independent.

Assume that the model is correct and differences between measured values, $x^*$, and the values computed using the model, $x$, are due to the error measurement process. Also, differences between the prior information on the parameters, $p^*$, and the parameters to be estimated, $p$ are due to the error in that prior information. Therefore, when Eq. 12.29 is used, we are maximizing the probability of reproducing the errors we have obtained in the measurement process and in the generation of prior information.

If probability distributions are supposed to be multivariate Gaussian, then

$$P(x) = \frac{1}{\sqrt{((2\pi)^m |C_x|)}} \exp\left[ -\frac{1}{2}(x^* - x)^T (C_x)^{-1}(x^* - x)\right] \quad (12.30a)$$

$$P(p) = \frac{1}{\sqrt{((2\pi)^n |C_p^0|)}} \exp\left[ -\frac{1}{2}(p^* - p)^T(C_p^0)^{-1}(p^* - p)\right] \quad (12.30b)$$

where $(x^* - x)$ is the vector of differences between an computed values measured by using a fixed model, $(p^* - p)$ is the vector of differences between prior information and parameters to be estimated, $C_x$ is the measurements covariance matrix, which represents the structure of the error measurements, $C_p^0$ is the *a prior* parameters covariance matrix, which represents the error structure of the available prior information, $m$ is the number of measurements, $n$ is number of parameters, $(\ )^T$ is used to indicate a transposed matrix, and $| |$ is the determinant symbol.

Note that in (　), $L = L(p)$, i.e. likelihood depends only on parameters, because the relationship $x = M(p)$ is introduced in the probability density function (Eq. 12.30a). Maximizing $L$ is equivalent to minimize the function $S = -2 \ln L(p)$, that is

$$S = (x^* - x)^T(C_x)^{-1}(x^* - x) + (p^* - p)^T(C_p^0)^{-1}(p^* - p) + \ln|C_x| + \ln|C_p^0| + m\ln(2\pi) +$$
$$n\ln(2\pi) - 2\ln(k) \quad (12.31)$$

where the last three terms are constant and can be disregarded in the minimization process.

Eq. 12.31 shows that the function to be minimized, i.e. objective function, depends on the error structure of measurements and prior information through the covariance matrices which are usually difficult to define. Generally, the information available will not be sufficient to specify all the elements of the covariance matrices and some terms will have to be fixed. To do that, it is convenient to separate measurements and prior information in groups with independent covariance matrices. For instance, if $m$ measurements have been obtained from $r$ independent instruments and $n$ parameters can be divided in $s$ groups with individual *a priori* covariance matrices, then the objective function, $S$, becomes

$$S(p) = \sum_{i=1}^r (x^* - x)^T (C_x)_i^{-1}(x^* - x) + \sum_{j=1}^s (p^* - p)^T (C_p^0)_j^{-1}(p^* - p) +$$
$$\sum_{i=1}^r \ln|(C_x)_i| + \sum_{j=1}^s \ln|(C_p^0)_j| \quad (12.32)$$

It should be pointed out that when no prior information is available, only the first and third terms in Eq. 12.32 must be considered. Moreover, if measurements are independent and all of them have the same variance, we obtain $C_x = \sigma^2 I$ where $I$ is the identity matrix. If the value of $\sigma^2$ is fixed in the process, only the first term in Eq. 12.32 is relevant and classical least-squares criterion is obtained from that equation.

## 2) Variance estimation

It is convenient to express each individual covariance matrix as

$$(C_x)_i = \sigma_i^2 (E_x)_i \qquad (12.33a)$$

$$(C_p^0)_i = \sigma_i^2 (E_p^0)_j \qquad (12.33b)$$

where $\sigma^2$ plays the role of a scale factor which represents the global variance of the data, whereas the $E_x$ and $E_p^0$ matrices represent the error structure associated with that particular type of data.

Generally, the error structure is constant, that is, it depends only on the measurement instrument or on the procedure used to obtain the prior information on the parameters.

The global variances $\sigma^2$ could be determined from the standard error of the measurement device, but in general there are uncontrolled factors that influence that value, for instance, operator skill or equipment conditions. Therefore, those values are difficult to determine in practice, not only for the measurement covariance matrices but usually also for the prior information covariance matrices. Hence, it is convenient to consider those global variances as additional parameters to be identified. This is consistent with the maximum likelihood approach. Introducing Eqs. 12.33a and b into the objective function Eq. 12.32 leads to

$$S = \sum_{i=1}^{r} \sigma_i^{-2} (x^* - x)^{\mathrm{T}} (E_x)_i^{-1} (x^* - x) + \sum_{j=1}^{s} \sigma_i^{-2} (p^* - p)^{\mathrm{T}} (E_p^0)_j^{-1} (p^* - p) + \sum_{i=1}^{r} m_i \ln \sigma_i^2 +$$

$$\sum_{j=1}^{s} n_i \ln \sigma_j^2 + \sum_{i=1}^{r} \ln |(E_x)_i| + \sum_{j=1}^{s} \ln |(E_p^0)_j| \qquad (12.34)$$

where $m_i$ and $n_i$ are the dimensions of the individual covariance matrices. The last two terms are now constant (assuming fixed the error structure and only variable the global error $\sigma^2$) and will not be taken into account in the minimization process.

Any procedure to minimize function Eq. 12.34 can be adopted. Note that the objective function now depends both on the parameters and on the variances. However, the minimization problem can be simplified uncoupling the estimation of parameters from the estimation of variances. This is convenient from a practical point of view, as conventional formulations are geared to the identification of parameters only. The minimization is then performed in an iterative two-step procedure: particular values of the variances are first selected and the minimization process is carried out to identify only parameters. Variance values are varied according to an independent optimization procedure until the global minimum is obtained.

To show the validity of this approach, let us consider first a set of $\mu_i = \sigma^{*2}/\sigma_i^2$ fixed, where $\sigma^{*2}$ is any variance taken as reference. This is equivalent to performing a constrained minimization of Eq. 12.34. The extended Lagrangian of function Eq. 12.34 is

$$\sigma^{*2} - \mu_k \sigma_k^2 = 0, k = 1, \cdots, (r + s) \qquad (12.35)$$

i.e.

$$\mathrm{Lag} = S + \sum_{k}^{r+s} l_k (\sigma^{*2} - \mu_k \sigma_k^2) \qquad (12.36)$$

where $l_k$ is a Lagrange multiplier. After deriving Eq 12.36 with respect to the variances, imposing the minimum condition and eliminating $l_i$, from the equations, the following equations are obtained:

$$\sigma^{*2} = \frac{J'}{m+n} \qquad (12.37a)$$

$$\sigma_k^2 = \frac{J'}{\mu_k(m+n)} \qquad (12.37b)$$

where

$$J' = \sum_{i=1}^{r} \mu_i (x^* - x)^{\mathrm{T}} (E_x)_i^{-1}(x^* - x) + \sum_{j=1}^{s} \mu_j (p^* - p)^{\mathrm{T}} (E_p^0)_j^{-1}(p^* - p) \quad (12.38)$$

Substituting Eqs. 12.37a and b in Eq. 12.34 leads to a condition for $S$ in the minimum:

$$S = m + n + \sum_{i=1}^{r} m_i \ln \frac{J'}{\mu_i(m+n)} + \sum_{j=1}^{s} n_i \ln \frac{J'}{\mu_j(m+n)} \qquad (12.39)$$

Note that in Eq. 12.39 the dependence of parameters is through $J'$. Moreover, as Eq. 12.39 is a monotonic function of $J'$, a minimum of $J'$ will minimize $S$ as well.

The general problem was defined as finding a set of parameters, $\hat{p}$ and variances $\hat{\mu}$ that minimized Eq. 12.34. Therefore, if $\hat{\mu}$ is found using an independent algorithm, the parameters $\hat{p}$ that minimize $S(p, \mu)$ will also minimize for those $\hat{\mu}$ values. This result allows to uncouple the minimization procedure: i.e. by minimizing $J'$, different sets of $\hat{p}$ for different values of $\hat{\mu}$ are found. The values $(\hat{p}, \hat{\mu})$ that minimize $S$ are obtained via a direct search algorithm.

## 3) Parameters reliability

The maximum likelihood formulation provides a statistical framework within which information on the reliability of the parameters estimated can be obtained. For instance, the covariance matrix of the estimated parameters (*a posteriori* covariance matrix) is (Tarantola, 1987)

$$C_p = [(C_p^0)^{-1} + A^{\mathrm{T}}(C_x)^{-1}]^{-1} \qquad (12.40)$$

where $A$ is the sensitivity matrix ($m \times n$):

$$A = \frac{\partial x}{\partial p} \qquad (12.41)$$

Eq. 12.40 takes into account both the prior information and the measurement error to estimate the final covariance of the estimated parameters. It should be pointed out that $C_p$, computed from Eq. 12.40 is a minimum bound of the parameters variance, due to the linearization of the model implicit in Eq. 12.41.

## 4) Optimization procedure

The mathematical problem to be solved in order to perform the estimation of parameters is an unconstrained minimization of Eq. 12.34. Following the procedure described above, the simpler Eq. 12.38 can be used instead. Using different values of variances, associated values of $J'$ are obtained by minimization of Eq. 12.38. The global minimum of $J'$ will give the parameters and variances finally estimated.

There are a wide range of algorithms available to find the minimum, but Gauss-Newton method is convenient for this kind of objective functions. In case of convergence difficulties, the extension of that method due to Levenberg-Marquardt algorithm (Marquardt, 1963) usually gives good results. As the function to minimize is in general non-linear with respect to the parameters, the procedure works iteratively in the parameters' space. Hence, starting from a point in that space, the parameters correction is found by means of

$$\Delta p = \Delta p_0 + [A^T (C_x)^{-1}A + (C_p^0)^{-1} + \lambda I]^{-1} A^T (C_x)^{-1}[\Delta x - A\Delta p^0] \quad (12.42)$$

where $\Delta p = p^* - p$, $\Delta x = x^* - x$ and $\lambda$ is a scalar which controls the convergence process. If $\lambda \to 0$, the Gauss-Newton method is obtained. As $\lambda$ increases, Ap tends towards the maximum gradient direction. Near the minimum $\lambda$ of 0 (or a very small value). No general rule exists to decide the value of $\lambda$ to be used in each iteration. Usually, if $J'$ becomes smaller in an iteration, $\lambda$ is decreased, reaching zero close to the minimum. However, if $J'$ increases after using Eq. 12.42, the value of $\lambda$ is also increased in order to approach the gradient direction. In the examples presented in this book, the following criteria has been used: $\lambda = 10$ as initial value, and, $\lambda_{n+1} = 10 \lambda_n$ if $J'_{n+1} > J'_n$; otherwise $\lambda_{n+1} = \lambda_n/10$, where $n$ is the iteration number.

## 12.4.2 Bayesian methods

Juang et al. (2013) proposed a Bayesian framework using field observations for back-analysis and updating of soil parameters in a multistage braced excavation. The framework was adapted for the KJHH model (Kuang et al., 2007a), a semi-empirical equation for predicting the maximum wall deflection or maximum settlement. Both the updating methodology using one and two type response (wall deflection or maximum settlement) were given.

### 1) Updating soil parameters using one type of response observation

As for updating use of only one type of response, the implementation starts with expressing the model as follows:

$$y = c \cdot \delta(\theta) \quad (12.43)$$

where $y$ = predicted response; $\delta(\theta)$ = prediction model; $\theta$ = vector of the soil parameters; and $c$ = model bias factor, which represents the model uncertainty.

Based on Eq. 12.43, the likelihood that the prediction ($y$) is equal to the observation ($Y_{obs}$) can be expressed as a conditional probability density function (PDF) of $\theta$:

$$L(\theta | y = Y_{obs}) = N[Y_{obs}/\delta(\theta)] \quad (12.44)$$

where $L(\theta | y = Y_{obs})$ = likelihood; and the notation $N$ = normal PDF that is function of $[Y_{obs}/\delta(\theta)]$. It is noted that at a given $Y_{obs}$, the term $N[Y_{obs}/\delta(\theta)]$ is a function of $\theta$ only. Recalling that $c = y/\delta(\theta)$, this normal PDF can be characterized with a mean of $\mu_c$ and a standard deviation $\sigma_c$. In a Bayesian framework, given a prior PDF, $f(\theta)$, the posterior PDF of can be obtained as follows:

$$f(\theta | y = Y_{obs}) = m_1 N[Y_{obs}/\delta(\theta)] f(\theta) \quad (12.45)$$

where $m_1$ is a normalization factor that guarantees a unity for the cumulative probability over the entire range of $\theta$.

## 2) Updating soil parameters using two types of response observation

To update the soil parameters with the observation of both wall deflection and surface settlement, the likelihood equal to the corresponding observations is a conditional PDF of $\theta$:

$$L(\theta|y_1 = Y_{obs1}, y_2 = Y_{obs2}) = N_2[Y_{obs1}/\delta_1(\theta), Y_{obs2}/\delta_2(\theta)] \tag{12.46}$$

$N_2$ = PDF of a bivariate normal distribution with a mean vector $[\mu] = [\mu_1, \mu_2]$ and a covariance matrix of

$$[\sigma] = \begin{bmatrix} \sigma_1^2 & \sigma_{12}^2 \\ \sigma_{21}^2 & \sigma_2^2 \end{bmatrix} \tag{12.47}$$

where $\sigma_{12}^2 = \upsilon_{21}^2 = \rho\sigma_1\sigma_2$, and $\rho$ = correlation coefficient between the two model bias factor $c_1$ and $c_2$. The preceding formulation is simply an extension of the formulation presented in Eq. 12.44 from using one type of observation to using two types of observations. Similarly, the posterior PDF of $\mu$ updated with two types of observations can be obtained as follows:

$$f(\theta|y_1 = Y_{obs1}, y_2 = Y_{obs2}) = m_2 N_2[Y_{obs1}/\delta_1(\theta), Y_{obs2}/\delta_2(\theta)] f(\theta) \tag{12.48}$$

where $m_1$ is a normalization factor that guarantees a unity for the cumulative probability over the entire range of $\theta$.

The posterior distribution may be obtained through optimization or sampling techniques. Markov Chain Monte Carlo (MCMC) simulation method is an efficient sampling technique that can yield samples of a posterior distribution. MCMC performs a random walk within the domain defined by the uncertain soil parameters according to their prior distributions. At each random walk, if the likelihood of model predictions matching the observations is increased, then the candidate point is accepted. Otherwise, the candidate is rejected. One advantage of MCMC is that the computation of the normalization factor may be avoided, which is generally difficult for multiple dimensional problems.

## 3) Procedure of Markov chain Monte-Carlo simulation using the Metropolis-Hastings algorithm

Juang et al. (2013) adopted the Metropolis-Hastings algorithm (Metropolis et al., 1953; Hastings, 1970) for its efficiency to implement MCMC sampling of the key parameter $\theta$ for its posterior PDF. The procedure can be summarized as follows:

①At Stage $k=1$, determine the first point $\theta_1$ in the Markov chain. This first instance $\theta_1$ may be obtained by random selection from the prior distribution or may simply be assigned the mean value.

②At next Stage $k$ ($k$ starts from 2), randomly generate a new $\theta_p$ from a proposal distribution $T(\theta_p|\theta_{k-1})$, which is assumed to be a multivariate normal distribution where the mean is set to be the current point $\theta_{k-1}$ in the Markov chain and the covariance matrix is equal to $sC_\theta$, where $s$ is a scaling factor and $C_\theta$ is the covariance matrix of the prior distribution of $\theta$. The multivariate normal distribution is chosen for its good convergence properties in the Bayesian inference.

③Generate a random number $\theta$ from a uniform distribution $U(0,1)$

④Compute the ratio of densities $r$:

$$r = \frac{q(\theta_p \,|\, y = y_{obs})}{q(\theta_{k-1} \,|\, y = y_{obs})} \leqslant 1 \tag{12.49}$$

where $q(\theta_p \,|\, y = y_{obs})$ is the unnormalized posterior PDF, which is essentially Eq. 12.45 or 12.49 without the normalization factor. Note that $q(\theta \,|\, y = y_{obs})$ is $q(\theta)$ evaluated at $y = y_{obs}$.

⑤Determine whether $\theta_p$ is acceptable (and thus yields a new point in the Markov chain) with the following acceptance rule: if $u \leqslant r$, then up is acceptable and set $\theta_k = \theta_p$; otherwise, set $\theta_k = \theta_{k-1}$. Then go back to Step 2.

⑥Repeat Steps 2-5 until the target number of samples (i.e., Markov chain length) is reached.

The Metropolis-Hastings algorithm randomly samples from the posterior distribution. Typically, initial samples are not completely valid because the Markov chain has not stabilized. These initial samples may be discarded as burn-in samples. Several factors influence the efficiency of sampling posterior distribution with the MCMC approach, such as the proposal distribution, Markov chain length, and number of burn-in samples. Therefore, the construction of a Markov chain is problem specific and needs to be examined case by case. It should be noted that other MCMC algorithms may be used in this procedure as long as efficient Markov chains can be achieved.

Results from Juang et al. (2013) suggested that the Bayesian updating is not much affected by the assumed prior distributions and the levels of the coefficient of variation of the soil parameters. Thus, while prior knowledge is important, the Bayesian updating with observations through stages of excavation can reduce the influence of this prior knowledge, and converged results can be obtained even if the prior knowledge is imperfect.

### 12.4.3 First-order reliability Method

Zhang (2015a) provides a framework using spreadsheet to updating parameters, based on First-Order Reliability Method (FORM).

The second moment reliability index $\beta$ is defined by Hasofer and Lind (1974), and the matrix formulation (Veneziano 1974; Ditlevsen 1981) is given

$$\beta = \min_{x \in F} \sqrt{\left[\frac{x_i - \mu_i}{\sigma_i}\right]^{\mathrm{T}} [\boldsymbol{R}]^{-1} \left[\frac{x_i - \mu_i}{\sigma_i}\right]} \tag{12.50}$$

where $x_i$ is random variables, $\mu_i$ is the mean value, $\boldsymbol{R}$ = correlation matrix, $\sigma_i$ is standard deviation, $F$ = failure domain.

The point denoted by the $x_i$ values which minimize the square root of the quadratic form in

Eq. 12.50 and satisfies $x \in F$ is the design point—the point of tangency of an expanding dispersion ellipsoid with the limit state surface which separates safe combinations of parametric values from unsafe combinations.

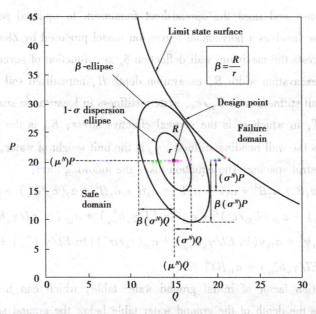

**Figure 12.5　Equivalent dispersion ellipses and reliability index in the original space of the basic random variables (after Low and Tang ,1997)**

As a multivariate normal dispersion ellipsoid expands from the mean-value point, its expanding surfaces are contours of decreasing probability values, according to the following established probability density function of the multivariate normal distribution:

$$f(x) = \frac{1}{(2\pi)^{\frac{n}{2}} |C|^{\frac{1}{2}}} \exp\left\{-\frac{1}{2}\left[\frac{x_i - \mu_i}{\sigma_i}\right]^{\mathrm{T}} [R]^{-1} \left[\frac{x_i - \mu_i}{\sigma_i}\right]\right\}$$

$$= \frac{1}{(2\pi)^{\frac{n}{2}} |C|^{\frac{1}{2}}} \exp\left[-\frac{1}{2}\beta^2\right] \tag{12.51}$$

where $\beta$ is defined by Eq. 12.50, without the "min". Hence, to minimize $\beta$ in the previous multivariate normal distribution is to maximize the value of the multivariate normal probability density function, and to find the smallest dispersion ellipsoid tangent to the limit state surface is equivalent to finding the most probable failure point (the design point).

For normal variates, the design point, the first point of contact between the expanding dispersion ellipsoid and the limit state surface in Figure 12.5, is the most probable failure point with respect to the safe mean-value point at the centre of the expanding ellipsoid. The reliability index is the axis ratio $(R/r)$ of the ellipse that touches the limit state surface and the standard deviation dispersion ellipse. By geometrical properties of ellipses, this codirectional axis ratio is the same along any "radial" direction.

For calculating $\beta$ in an efficient way, Low and Tang (1997, 2007) developed procedures using Mircosoft Excel's Solver tool. The procedures have been widely used for geotechnical problems.

Zhang (2015a) let the limit state surface equation be the deviation between the observation and predicted response and used the spreadsheet framework to updated soil parameters. The prediction of response involves a polynomial regression model proposed by Zhang (2015b).

The model suggests the maximum wall deflection $\delta_h$ as a function of seven input parameters: soil unit weight $\gamma$, excavation width $B$, excavation depth $H$, normalized soil shear strength ratio $c_u/\sigma_v'$, normalized soil stiffness ratio $E_{50}/c_u$, system stiffness in logarithmic scale $\ln EI/\gamma_w h_{avg}^4$ and soft clay thickness $T$, in which $\sigma_v'$ is the vertical effective stress, $E_{50}$ is the secant soil stiffness modulus, $EI$ denotes the wall bending stiffness, $\gamma_w$ is the unit weight of water, and $h_{avg}^4$ represents the average vertical strut spacing. The equation takes the following form:

$$\delta_h = \mu_w [a_0 + a_1 B + a_2 B^2 + a_3 T + a_4 T^2 + a_5 H + a_6 H^2 + a_7 (c_u/\sigma_v') + a_8 (c_u/\sigma_v')^2 +$$
$$a_9 (E_{50}/c_u) + a_{10} (E_{50}/c_u)^2 + a_{11} (\ln EI/\gamma_w h_{avg}^4) + a_{12} (\ln EI/\gamma_w h_{avg}^4)^2 +$$
$$a_{13} \gamma + a_{14} \gamma^2 + a_{15} \gamma (\ln EI/\gamma_w h_{avg}^4) + a_{16} (c_u/\sigma_v')(\ln EI/\gamma_w h_{avg}^4) +$$
$$a_{17} H (\ln EI/\gamma_w h_{avg}^4) + a_{18} HT] \qquad (12.52)$$

where $\mu_w$ is medication factor of initial ground water table, which can be approximated as $\mu_w = 1 - 0.1l$, and $l$ is the depth of the ground water table below the ground surface (in meters) and $l \leqslant 2$.

The differences, defined as $g(x_i)$, between the PR predicted maximum wall deflection $\delta_{h,PR}$ and the field measurement $\delta_{h,M}$ for each excavation stage are minimized though $g(x_i) = \delta_{h,PR} - \delta_{h,M}$ under condition of

$$d = \sqrt{x_1'^2 + x_2'^2 + x_n'^2} = (x'^T x)^{1/2} \qquad (12.53)$$

where a primed dimensionless variable $x_i'^2$ is defined as $x_i'^2 = (x_i - \mu_{x_i})/\sigma_{x_i}$, in which each soil variable $x_i$ is defined in terms of its mean $\mu_{x_i}$ and its standard deviation $\sigma_{x_i}$. $x'$ is the vector of the $x_i'$s and the superscript $T$ indicates the transpose of a matrix or vector, $x_i^*$ is the design point which minimizes $d$ and satisfies $g(x_i) = 0$. The problem is then a constrained minimization that can be summarized as the requirement to minimize $d$ subject to the constraint that $g(x_i) = 0$ is satisfied.

Generally speaking, the constrained minimization can be performed through a spreadsheet or mathematical software such as MATLAB. In this study, the design point $x_i^*$ and thus the minimal $d$ value were obtained by using the EXCEL spreadsheet's built-in optimization routine SOLVER.

Figure 12.6 shows the example spreadsheet setup for deriving the updated soil parameters through minimizing the difference between the actual measured wall deflection $\delta_{h,M}$ and the estimated value $\delta_{h,PR}$ obtained using this PR model.

The spreadsheet-based procedure for estimating the posterior distribution can be summarized as follows:

Step 1: The semi-empirical PR model is firstly implemented in the spreadsheet, as shown in cells J3 : K4 in Figure 12.6. For the excavation case histories presented later, $\gamma$, $B$, $H$ for each excavation stage, $\ln(EI/\gamma_w h_{avg}^4)$, and $T$ are simply treated as constants since they are either deterministic geometrical properties or can be determined fairly precisely. These five constant inputs are input in cells D7 : D11. For simplicity, the distribution types of the random parameters $c_u/\sigma_v'$ and $E_{50}/c_u$ are assumed as normal. The prior mean and standard deviation (SD) of input parameters $c_u/\sigma_v'$ and $E_{50}/c_u$ are input into cells D3 : D4 and E3 : E4, respectively; the COVs for the two soil parameters are input into cells H3 : H4. Cells L3 : L4 contain the measured wall deflection corresponding to excavation to depth $H_i$. The $x_i'$ vector in Cells O3 : O4 contains equations for $(x_i-\mu_{x_i})/\sigma_{x_i}$ as defined for Eq. 12.53.

Step 2: In this step, the posterior mean of the two input parameters are obtained by minimizing the difference between $\delta_{h,M}$ and $\delta_{h,PR}$, which is input into Cells P3 : P4. The design point $x_i^*$ value in Cells (I3 : I4) was obtained by using the EXCEL spreadsheet's built-in optimization routine SOLVER to minimize the cell, by changing the $x_i^*$ values, under the constraint that the performance function $g(x_i)=0$. Prior to invoking the SOLVER search algorithm, the $x_i^*$ values were set equal to the original mean values (150, 0.25) of the two variables. Iterative numerical derivatives and directional search for the design point $x_i^*$ are automatically carried out in the spreadsheet environment with "Automatic scaling", "Quadratic for Estimate", "Central for Derivatives", and "Newton for search" set as default options. It can be observed from Figure 12.6 that, based on the measured $\delta_{h,M}$ or 27.0 mm at an excavation depth $H_i$ of 7.0 m and the prior mean values of (163.9, 0.303) of the two soil parameters, the updated soil parameters are (164.3, 0.305) and the corresponding forward prediction of the wall deflection at the subsequent stage (where $H_{i+1}=9.5$m) is 35.2 mm (Cell I 11).

Step 3: These procedures are repeated stage by stage till the end of excavation (where $H$ equals to the final excavation depth) and final pair of $c_u/\sigma_v'$ and $E_{50}/c_u$ values and be determined.

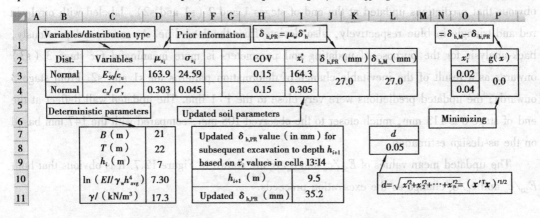

Figure 12.6　Example spreadsheet for updating of soil parameters (Zhang, 2015a)

A well-documented case history, TEC (Taipei Enterprise Center), is taken as an example of the application of this back analysis method. Table 12.4 summarizes the inputs of this case and Table 12.5 lists the measured wall deflection at each excavation stages.

**Table 12.4　Initial parameters of TEC case**

| $\gamma$ | $B$ | $\ln\left(\dfrac{EI}{\gamma_w h_{avg}^4}\right)$ | $T$ | $E_{50}/c_u$ | $c_u/\sigma_v'$ | $\delta_{h,M}$ |
|---|---|---|---|---|---|---|
| 19.0 | 35 | 7.30 | 27 | 200 | 0.34 | 108 |

**Table 12.5　Observed wall deflection of TEC case**

| | Stage 1 | Stage 2 | Stage 3 | Stage 4 | Stage 5 | Stage 6 | Stage 7 |
|---|---|---|---|---|---|---|---|
| Excavation depth $H$ (m) | 2.8 | 4.9 | 8.6 | 11.8 | 15.2 | 17.3 | 19.7 |
| Measured wall deflection $\delta_{h,M}$ (mm) | 19 | 30 | 43 | 67 | 81 | 95 | 108 |

The mean values of these predictions versus the field observations were plotted in Figure 12.7, in which the predictions made prior to excavation, herein referred to as the as-design predictions, are denoted with the square symbol. These are the predictions made from Eq. 12.53, without any field measurements and involving no updating of the soil parameters. The as-design predictions using the initially assumed prior information deviate from the 1∶1 line (the predicted deflections versus the observed values). Here the observed wall deflections are those measured at the end of stages 1-7. The predicted maximum wall deflections are those predicted prior to stages 1-6, respectively. It should be noted that prior to each excavation stage, the wall deflection predictions are made for various excavation depths that are anticipated at all subsequent stages. That is, prior to stage 3, predictions are made at depths corresponding to end of stages 3-7, respectively. It is obvious that predictions updated at the end of stages 1 and 2 (s1 and s2), labeled with circle in red and triangle in blue respectively, also deviate from the 1∶1 line. As mentioned previously, back-analysis for the purpose of updating soil parameters is more meaningful for stage 3 (s3) onwards as a result of the inevitable change of deformation patterns for s1 and s2. From stage 3 onwards, the updated predictions were very close to the 1∶1 line. The updated wall deflect at the end of stage 6 is 113 mm, much closer to the observed 108 mm, compared with the 147 mm based on the as-design estimation.

The updated mean values of $E_{50}/c_u$ and $c_u/\sigma_v'$ are shown in Figure 12.7. It is obvious that both $E_{50}/c_u$ and $c_u/\sigma_v'$ change as the excavation proceeds.

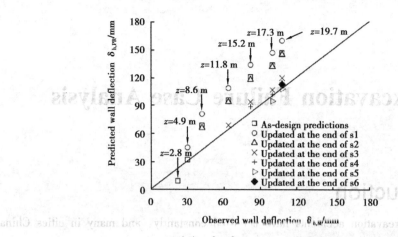

(a) updated maximum wall deflection

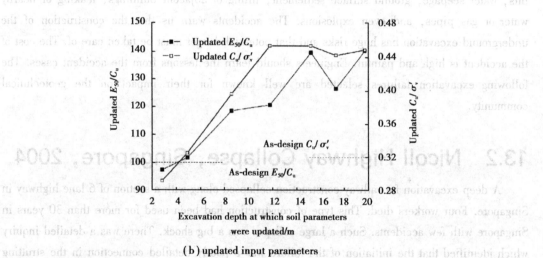

(b) updated input parameters

**Figure 12.7   Updating at various excavation depths for TEC case (Zhang, 2015a)**

# 13    Excavation Failure Case Analysis

## 13.1    Introduction

In recent years, excavation accidents have occurred constantly, and many in cities China such as Shanghai, Guangzhou and Hangzhou. These accidents include the collapse of foundation pits, water seepage, ground surface settlement, tilting of adjacent buildings, leaking of nearby water or gas pipes, and even explosions. The accidents warn us that the construction of the underground excavation has huge risks and that potential danger must be taken care of. The cost of the accident is high and painful. Engineers should learn the lessons from the accident cases. The following excavation failures selected are well known for their impact on the geotechnical community.

## 13.2    Nicoll Highway Collapse, Singapore, 2004

A deep excavation for railway construction collapsed along with a section of 6 lane highway in Singapore. Four workers died. This type of construction had been used for more than 30 years in Singapore with few accidents. Such a large collapse was a big shock. There was a detailed inquiry which identified that the initiation of the failure was a poorly detailed connection in the strutting system and the total collapse was due to the inability of the system to withstand the failure of one row of strutting. Moreover, there were many other contributory factors that were identified, such as an incorrect use of a computer program to design the earth retaining system, erroneous back analyses, deficient monitoring, failure to observe stop work limits and many more including an over-riding culture of complacency. This chapter briefly summarizes the events, the findings of the Committee of Inquiry and the remedial actions, and sets out some of the lessons we have learned.

### 13.2.1    Case Description

Around 3:30 pm on April 20, 2004, the 33.5 m deep cut and cover excavation required for the construction of tunnels between Nicoll Highway and Boulevard Stations (Figures 13. 1) collapsed. The collapse was catastrophic, resulting in four fatalities and a lot of damage (Figure 13.2), which delayed the completion of the CCL1 stage by about four years.

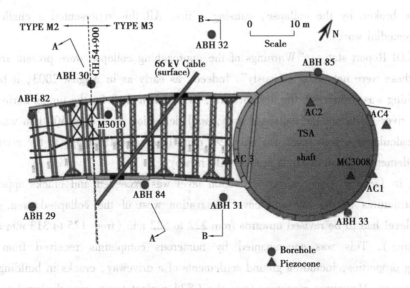

**Figure 13.1  Plan showing location of diaphragm wall panels, 9th level strutting system and site investigation (Whittle A.J. and Davies R.V)**

**Figure 13.2  The surrounding ground deformation of the collapsed excavation**

The Singapore Government appointed a Committee of Inquiry (COI) to ascertain the causes and circumstances of the incident. The COI Report (2005), submitted to the Government in May 2005 and made public on the Internet, is the main source of information for this chapter.

When the incident occurred, the surrounding ground collapsed into the excavation area (Figure 13.2). The area of the collapsed zone was approximately 100 m×130 m and 30 m deep. Its edge was only about 10 m away from the closest building-Golden Mile Complex. The Nicoll Highway and the approach slab before the abutment of the Merdeka Bridge over the Kallang River were also damaged. A storm drain located south of the cut and cover tunnel was the main drainage outlet, conveying water to the Kallang River. When the collapse occurred, the river water rushed into the collapse area. Several key utilities, including power electric mains, gas mains, and water mains were broken by the collapse, causing a fire. All this represented a challenge for the following remedial works.

The COI Report states: "Warnings of the approaching collapse were present from an early stage but these were not taken seriously". Indeed, as early as in August 2003, it became clear that something was wrong with the design. First, in a launch shaft for the tunnel boring machines, across the river east from the collapsed area, wall deflection exceeding 500 mm was measured, while the calculated design level was only 190 mm. This caused damage to the retaining walls, ground settlements and cracks at a cricket field nearby.

Next, in January 2004, design deflection level was exceeded and cracks appeared in the retaining structures in the cut and cover excavation west of the collapsed area. The design deflection level had to be revised upwards from 222 to 522 mm (from 125 to 313 mm closer to the collapse area). This was accompanied by numerous complaints received from owners of neighboring properties, including ground settlements of a driveway, cracks in buildings and other building damage. Manpower resources from the C824 project team were deployed to handle the repair works. Finally, in the collapsed area itself, the inclinometer I-104 indicated that the original design level of 145 mm in the southern diaphragm wall was exceeded as early as February 2004 (Figure 13.3). At that moment installation of the sixth strut level was in progress.

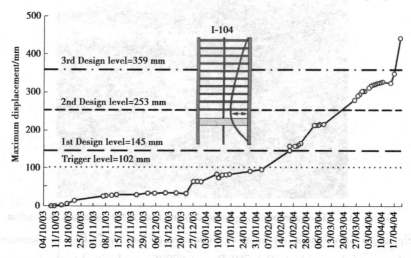

**Figure 13.3 Displacement of the southern wall in time (After the COI Report, 2005): measurements for inclinometer I-104 (Puzrin et. al., 2010)**

The COI concluded, which in our computer-dominated age sounds like a paradox: a wrong use of numerical modeling in geotechnical design, together with some structural errors, were the main causes of the Nicoll Highway collapse.

## 13.2.2 Lessons Learned

The final report of the Committee of Inquiry into the collapse of the Nicoll Highway was submitted on May 11, 2005. Faced with the monumental task of piecing together information from

193 witnesses and volumes of documents, the COI elucidated the crucial errors and persons responsible for the collapse of Nicoll Highway. In the COI's words: "Warnings of the approaching collapse were present from an early stage but these were not taken seriously". It concluded that "the death of four persons was the direct result of the collapse" and that "the Nicoll Highway collapse could have been prevented". The following recommendations were made by the Committee of Inquiry to avoid similar incidents in the future.

(1) **Effective risk management**

The potential for major accidents, whether due to the construction process or deficiencies in design, must be recognized and expeditiously controlled. It is inappropriate to leave the control of risk wholly to contractors. Owner's and builder's management must seek a balance between production pressures and quality and safety.

(2) **Robustness of design**

A robust design is essential. This robustness is provided by identifying the hazards and checking that the proposed design can adequately withstand them. The design should have sufficient redundancy to prevent a catastrophic collapse in the event of a failure of any particular element. Temporary works for deep excavation should be given the same respect as permanent works.

(3) **Numerical modeling in geotechnical design**

Numerical analysis should supplement and not supplant sound engineering judgment and practice. Those who perform geotechnical numerical analysis must have a fundamental knowledge of soil mechanics and a clear understanding of numerical modeling.

(4) **Back analysis**

A proper back analysis should be done with an understanding of why the design is not performing as originally predicted, and not just to increase the design levels. The input parameters (e.g., soil properties) should be adjusted to allow for the displacements and forces to fit the measurements. If these adjusted input parameters fall outside the meaningful range, the analysis model is probably wrong, and should not be used for further predictions.

# 13.3  Xianghu Metro Station Collapse, Hangzhou, China, 2008

Xianghu Station is the starting point of Hangzhou Metro Line 1. This station was constructed by using the bottom-up construction method in eight excavations from north to south. The collapsed position is the second excavation on the northern side (N2 excavation), which is near the junction of Fengqing Road and Xiangxi Road as shown in Figure 13.4. The excavation occupies a planned area of about 2 313 $m^2$, with approximately 21.5 m in width and 107.6 m in length. The maximum digging depth is up to 16.2 m. The accident happened at the No. 2 foundation pit as shown in Figure 13.5. The foundation pit is 107.8 m long and 21.05 m wide, and the depth is 15.7-16.3 m.

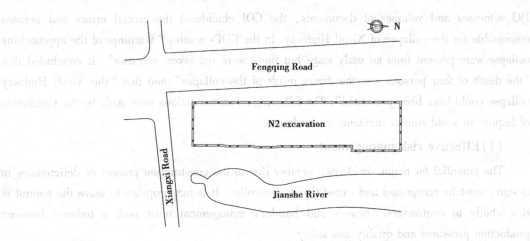

Figure 13.4　Plan view of the N2 excavation

At around 15:15 on November 15, 2008, a large-scale collapse accident occurred on the west side of the North 2 foundation pit. The collapsed area was 75 m long, about 20 m wide and 15 m deep. More than 11 vehicles in progress fell into the collapsed area. At least 50 people were buried in the collapse, causing 21 persons dead and 4 injured, and the direct economic losses was more than 49.62 million yuan. This is the most serious accident in the history of China's subway construction.

On November 17th, 2008, there was a crack on the ground at the southern end of the collapsed foundation pit. In order to facilitate the rescue work, the houses near the accident were completely knocked down.

Figure 13.5　Some site scenes of the accident

## 13.3.1　Case analysis

Through field investigation and preliminary analysis, the reasons for the excavation collapse were found to be the misuse of the soil parameters, over excavation, incorrect installation of steel struts, invalid monitoring data, and inadequate ground improvement.

The soil condition of Hangzhou is unnormal. According to field investigation, the accident was in silty clay. As a main road, Fengqing Road beside the foundation pit has a large traffic and a large load, which has a significant influence on the load-bearing wall on the west side of the foundation pit, causing the entire foundation pit to collapse; furthermore, a rare continuous rainfall in Hangzhou in October increased the instability of sand.

The whole design did not consider the influence of the surrounding environment. The demanded demolition and transfer of nearby house was not carried out. The construction method should be closely integrated with the stratum and the surrounding environment; however, no good construction method has been developed in this project. Since a cheaper one has been chosen, which caused a serious damage to the environment, and many problems, it ends up with great losses.

In order to save money and time, the construction was in a rush, the construction period and cost are unreasonable—these are the main reasons that affected the safety and quality. During the project, there is a collapsed sewage pipe leaking for a long time, which causes the formation of the slipping surface; consequently it may cause the collapse with the help of continuous rain.

There are some other causes for the collapse:

## 1) Direct causes

### (1) Over-excavation

The design document stipulated that when the foundation pit was excavated to 0.5 m below the support design elevation, the excavation must be stopped, and the support should be set in time, and no over-excavation was allowed.

However, during the excavation, when the last strut was supposed to install, some sections have been dug to the bottom of the foundation pit, and the strut installation and the construction of concrete floor are not in time.

### (2) Weak links in the support system

The supporting system was diaphragm wall with steel pipe strut; however, the design report did not include a detailed description about the connection of the supporting steel pipe strut and the diaphragm wall, or an example of the joint of strut. There were no corresponding technical requirements put forward regarding to the joint, or welding requirements for steel strut and diaphragm wall embedded parts. No welding was performed.

The excessive lateral displacement of sectional wall together with large axial forces and severely eccentric of the struts resulted in instability of the support system.

### (3) **Monitoring failed**

Before November 15th, the maximum settlement of the ground had reached 316 mm, and the maximum displacement of the inclinator was at the depth of 18 m is 43.7 mm. On November 13th, the maximum lateral displacement of the inclinator had reached 65 mm. All of them had already exceeded the alarm value, but with no alarm.

## 2) Indirect causes

①The original plan of the designer was reinforcing the foundation pit by strip-shaped mixing pile group at the bottom; however, the contractor suggested consolidating the base by dewatering instead of using reinforcement measures.

The design report proposed:

a. The groundwater in the foundation pit be reduced to 1 m below the lowest point of the structural internal members and 3 m below the bottom of the foundation pit;

b. The dewatering test should be carried out before excavation, and on-site dewatering should be done 4 weeks in advance;

c. When the ground subsidence exceeds the alarm value, the dewatering should be stopped and reported in time.

But none of these were fully implemented in the actual construction.

②The supervisor did not stop the serious problems that was against the design codes and specifications.

③The results of physical and mechanical properties test were reliable. However, the design of the supporting system had not taken into consideration the comprehensive design and reasonable recommendation according to the characteristics of the local soft soil, which caused the selected parameters too large and reduced the factor of safety of the foundation pit system.

## 13.3.2　Lesson learned

After field investigation and preliminary analysis, five reasons of excavation collapse were concluded: (1) using the average shear strength value derived from consolidated quick sheartests and not the standard value from consolidated undrained tests as the strength parameters; (2) canceling the fourth level struts; (3) poor connection of pipe struts with coupling beams and the diaphragm wall; (4) insufficient monitoring instruments and ignoring the unusual performance of the excavation; and (5) using artesian wells for consolidation instead of jet grouting.

①This collapse accident once again proves that excavation of soils should be based on the principles of layered and zoned construction, symmetrical and balanced excavation, early cast of basal slabs, and over-excavation is strictly prohibited, especially in soft soil areas. This project was not implemented in accordance with these principles, which is one of the direct causes of the accident.

②When using retaining wall with inner support structure in thick soft soil area, we must consider stability problems in the deep area, including wall settlement, uplift and "toe kick" damage; it should be calculated and verified by a variety of methods. If the requirements cannot be met, the insert depth of walls should be deepened or the passive area should be reinforced. The design of this project was not firm enough, and passive area reinforcement was not carried out.

③If the inner support is made of steel truss or steel pipe strut, a requirement of joint structure and welding should be proposed properly to avoid eccentricity and instability; and the first inner support should be strengthened by reinforced concrete.

④The selection of the shear strength parameters should be matched with the calculation purpose and the safety factor. For the calculation of earth pressure, the related calculation of "toe-kick" and "overturning" should choose the results of consolidation quick shear test or triaxial CU test; the calculation of basal heave, sliding and overall stability should choose the results of quick direct shear teat, triaxial unconsolidated undrained test or cross plate shear strength teat when soft clay plas an important role. The design should be based on comprehensive consideration, not just consider the average but also the maximum and minimum data.

⑤The monitoring should be carried out strictly, independently and without interference. When the alarm value is reached, it must be timely reported and must not be concealed. The supervision department should be familiar with geotechnical engineering and should strictly stop major violations.

# 13.4　Guangzhou Haizhu City Square Foundation Pit Collapse

The foundation pit of Zhuhai City Plaza collapsed around 12:20 on July 21st, 2005. The pit had been put into construction since October 2002, but suspended several times during the course until it collapsed on July 21, 2005, which lasted 2 years and 9 months.

After the collapse on the south side of the foundation pit, the strut at southeast corner failed, causing the east 20 m deep retaining wall unsupported, and posing a threat to the safety of the surrounding environment (as shown in Figure 13.6). The foundation piles of the Seafarer Hotel at the south of collapsed pit were broken and displaced, and the cap was suspended in the air, resulting in the building collapse near the foundation pit (as shown in Figure 13.7); the foundation pile of the residential building No. 1 near the pit was exposed and failed, and soil under the caps were hollowed out (as shown in Figure 13.8).

**Figure 13.6   Collapse on the south side of the foundation pit**

**Figure 13.7   The collapse of the building near the pit**

**Figure 13.8   The ground surface settlement and
cap failure caused by pit collapse**

## 13.4.1 Case analysis

The main technical reasons for this collapse are

①It can be concluded from the geological exploration data that the site was consist of rock layer with argillaceous siltstone mingled with interlayers of strongly weathered soil within the depth of excavation; during the excavation, it was found that the rock mass on the south side was inclined to the pit, and there was water seepage and mud flowing in the strongly weathered interlayer; according to the deformation monitoring data, it had shown obvious signs of the collapse on the south side of the foundation pit. However, none of these problems had drawn due attention, and no effective measures had been taken, which is the main cause of the accident.

②The initial design depth of the foundation pit was 17.0 m, but the actual excavation depth was 20.3 m. The 3.3 m deep over-excavation not only caused the insufficient of the anchoring force, but also made the 20 m deep retaining wall a suspended wall which cannot be adjusted, resulting in a potential danger.

③The time of intermittent construction of the foundation pit was 2 years and 9 months. It was far longer than the period of one year according to the code concerning temporary construction, which led to the softening of the soil layers because of seepage, and the reduction of anchoring force of anchor (cable) because of the corrosion of steel components.

④When the foundation pit collapsed, there was still one backhoe excavator, one crane and one 16 m$^2$ dump truck on the south side of the foundation pit near the Seafaring hotel. When the foundation pit was in danger, these construction loads were also one of the causes of the collapse of the foundation pit.

## 13.4.2 Lesson learned

The following situation should be strictly avoid:

①Over-excavation: The original design depth was 17 meters, but the actual pit had been excavated to 20.3 meters, causing the retaining wall to be hanging wall;

②Overtime: The service period of the foundation pit support structure was one year, and the actual excavation had been nearly three years;

③Overloading: overloading at the top of pit such as dump truck, crane and backhoe excavator;

④Ignoring geological response: The rock layer was shallow and inclined into the pit. The designers used the software to design and checked the foundation pit, and ignored the fact that the rock layer was inclined 25°. It was found that the displacement of the wall on the south side was too large, so this part of the wall was reinforced with anchor cables but the range was too small which only covered 20-30 m along the west side of the south. Constructors believed that the insufficient reinforcement wes caused by unreasonable design.

# 13.5　Shanghai Metro Line 4 Seepage

　　In the early morning of July 1st, 2003, a great amount of water and quicksand burst into the section of cross-river tunnel of Shanghai Rail Transit Line 4, which was a connection passage used to connect the upper and lower lines (as shown in Figure 13.9). It caused about 270 m tunnel to be damaged and the ground settlement reached a maximum 7 m in the surrounding area, which caused three buildings heavily inclined. Even worse, the adjacent flood wall of Huangpu River collapsed and caused piping (as shown in Figure 13.10). Fortunately, due to the timely alarm, all the personnel in the tunnel and nearby building were safely evacuated without causing casualties. However, the accident caused direct economic losses of around 150 million.

**Figure 13.9　Aerial view of accident site**

**Figure 13.10　The collapsed flood wall of huangpu river**

　　Due to the influence of ground settlement, the 30-meter-long cut-off wall at Dongjiadu Road section was sunk and cracked (as shown in Figure 13.11), which caused threat to flood control. The city rescue headquarters mobilized the armed police force to rush to the scene to take measures such as using sandbags and plugging, and timely controlled the danger. In addition, in order to protect the buildings in the surrounding areas, technical measures were adopted such as grouting to minimize the surging of underground quicksand. In order to balance the water and earth pressure inside and outside the tunnel, a method of closing the tunnel opening and injecting water is adopted. A large number of grouting filling and reinforcement were carried out for the roads and important buildings, large settlements on the ground were backfilled, and all the damaged buildings were demolished (as shown in Figure 13.12).

**Figure 13.11   Ground settlement**

**Figure 13.12   Heavily inclined buildings**

## 13.5.1   Cause analysis

The excavation sequence of the shaft and the bypass channel is wrong. The failure of the refrigeration equipment caused the rising of temperature and the underground pressurized water caused sandblasting. These three disadvantages eventually led to the accident. For projects with high construction risks like this, it would be an important cause of the accident if there is lack of emergency response plan, or the on-site management and supervision are out of control.

### 1)Excavation sequence error

There was a large shaft above the tunnel, and other two small shafts 8-9 m below the large shaft were excavated to connect the already formed tunnel. According to the construction practice, the bypass channel should be dug first and then the shaft. However, the construction sequence had been changed, which was very likely to cause collapse. When the accident occurred, one small shaft had been dug and the other shaft was excavated about 2 m.

## 2) The incorrect application of freezing method

Before the accident, the contractor, the project department of Shanghai Branch of China Coal Mine Engineering Co., Ltd., arbitrarily adjusted the original construction design while the expert group determined that the adjustment did not strictly follow the relevant provisions of the freezing construction process, resulting in weak layers of the frozen soil in the adjacent channel construction. The adjusted scheme reduced the average temperature requirement for frozen soil, from $-10$ ℃ in the original scheme to $-8$ ℃; the number of vertical freezing tubes at the bypass channel decreased from 24 in the original scheme to 22, while for the original 7 vertical freezing pipes of 25 meters deep, 4 of them were shortened to 14.25 meters, and 3 shortened to 16 meters, resulting in weak frozen soil in the intersection below the waistline of the bypass channel and the under tunnel. A row of 6 frozen inclined holes with a hole spacing of 1 meter was adopted. Although the length of the frozen hole was increased, the number was too small and the spacing was too large, so that the freezing effect was insufficient to resist the water and soil pressure of the corresponding part.

## 3) On-site management dereliction

On the morning of June 28th, the small refrigerator of the down line of the tunnel broke down, and the cooling was stopped for 7.5 hours. At about 2 p.m., the construction crew installed a hydrological observation hole in the down-line tunnel and found that there was always pressure water leaking. Although some measures such as blocking the tunneling surface with wood were adopted, the effect was not good. At 3 o'clock on the morning of the 29th, the water pressure measured at the water valve was close to the confined water pressure of the seventh external layer. The danger was first revealed, but no one at the scene reported the situation to the contractor or supervisor, resulting in a gradual increase in the risk.

It was in such a dangerous situation that at 0:00 on July 1st, although the deputy manager was aware of the serious potential dangers in the frozen soil layer of the connection passage, and the project had been suspended, he still determined to arrange the construction personnel to partially remove the wood sealing at the excavation face besides the frozen soil, and used the wind chisel to cut the hole with a diameter of 0.2 m to prepare the concrete conveying pipe. It was from this hole that the water emerged, and the water sand continuously rushed in from the lower right corner of the tunneling face and the side wall, so that the sealing became invalid and eventually caused an accident.

## 13.5.2　Repair plan

In August 2004, the repair work of Dongjiadu section of Line 4 was restarted. After comparing the method of the in-situ repair and rerouting scheme, the in-situ repair scheme was finally adopted. After overcoming a series of technical problems such as the 65 m ultra-deep underground continuous wall, 38 m ultra-deep underground obstacle clearing, ultra-deep foundation reinforcement in complex strata, ultra-deep confined water dewatering, 41 m ultra-deep foundation pit excavation, tunnel pumping cleanup, large section freezing dug, the Dongjiadu section of Shanghai Line 4 was re-opened on July 9th, 2007.

Both east and west bulkhead were built by using freezing method as cut-off wall. After the debris in the tunnel were removed, R.C. bulkhead was built to protect the unaffected tunnels from the cut and cover recovery works (as shown in Figure 13.13).

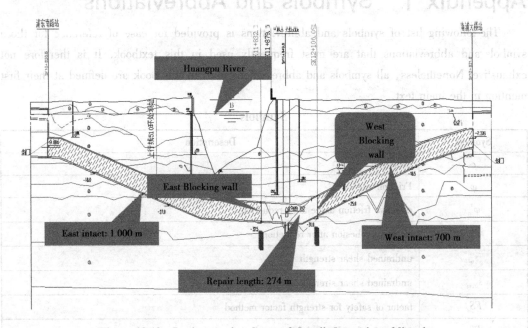

Figure 13.13　In-situ repair scheme of dongjiadu section of line 4

# 13.6　Other Cases

*International Conferences on Case Histories in Geotechnical Engineering* aim to share knowledge gained from geotechnical projects and Missouri University of Science and Technology provides many volumes of the conference proceedings. We can find some case histories involving failure in deep excavation, such as

Boones and Westland (2004);

Rosenvinge and Teodore (2013);

Rahimi et al. (1993).

# Appendix

## Appendix I    Symbols and Abbreviations

The following list of symbols and abbreviations is provided for ease of reference for those symbols and abbreviations that are most frequently used in this textbook. It is therefore not exhaustive. Nonetheless, all symbols and abbreviations used in this book are defined at their first mention in the main text.

### Symbols

| Symbols | Description |
|---------|-------------|
| $c$ | Cohesion |
| $\varphi$ | Friction angle |
| $\varphi'_m$ | effective friction angle after reduction |
| $c'_m$ | effective cohesion after reduction |
| $s_u$ | undrained shear strength |
| $s_{u,m}$ | undrained shear strength after reduction |
| $FS_s$ | factor of safety for strength factor method |
| $FS_l$ | factor of safety for load factor method |
| $FS_d$ | factor of safety for dimension factor method |
| $q_s$ | surcharge |
| $F_b$ | factor of safety against basal heave |
| $N_c$ | Skempton's bearing capacity factor |
| $M_d$ | driving moment |
| $M_r$ | resistant moment |
| $M_s$ | allowable bending moment of the retaining wall |
| $F_p$ | factor of safety against push-in |
| $K_a$ | coefficient of active earth pressure |
| $K_p$ | coefficient of passive earth pressure |

continued

| Symbols | Description |
|---------|-------------|
| $\sigma_a$ | total active earth pressure (horizontal) acting on the retaining wall |
| $\sigma_p$ | total passive earth pressure (horizontal) acting on the retaining wall |
| $\sigma_a'$ | effective active earth pressure acting on the retaining wall |
| $\sigma_p'$ | effective passive earth pressure acting on the retaining wall |
| $K$ | lateral earth pressure coefficient |
| $K_0$ | coefficient of at-rest earth pressure |
| $K_0^{nc}$ | $K_0$-value for normal consolidation (default $K_0^{nc} = 1-\sin\varphi$) |
| $\sigma_y$ | the vertical stress at a point in the ground |
| $\sigma_x$ | the horizontal stress at a point in the ground |
| $\sigma_t$ | tension cut-off and tensile strength |
| $\varphi_{cv}$ | the critical state friction angle |
| $\psi$ | dilatancy angle |
| $\psi_m$ | the mobilized dilatancy angle |
| $M$ | the stress ratio at failure |
| $\varepsilon_v$ | the volumetric strain |
| $p_v'$ | effective overburden pressure |
| $\sigma_h'$ | horizontal (lateral) effective stress |
| $p'$ | The mean effective stress |
| $\gamma$ | bulk density |
| $\gamma'$ | effective unit weight of soil |
| $\gamma_m$ | the moisture unit weight of soil |
| $\gamma_{unsat}$ | unit weight above phreatic level |
| $\gamma_{sat}$ | unit weight below phreatic level |
| $\gamma_w$ | unit weight of water |
| $\gamma_{ave}$ | average unit weight of the soil |
| $H_e$ | excavation depth |
| $c_u$ | undrained shear strength of soil |
| $c_u^*$ | index capable of quantifying both the effects of $c_u$ profile and value |
| $\Delta t$ | change in strut temperature |
| $K_{ur}$ | elastic bulk modulus |
| $E$ | elastic Young's modulus |

continued

| Symbols | Description |
|---|---|
| $E_i$ | initial tangent modulus |
| $E_u$ | secant modulus, $E_u \approx 60\% \, E_i$ |
| $E_s$ | representative soil stiffness |
| $E_{50}$ | the triaxial loading stiffness |
| $E_{ur}$ | the triaxial unloading stiffness |
| $E_{oed}$ | the oedometer loading stiffness |
| $E_{ref}$ | reference Young's modulus |
| $E_{50}^{ref}$ | secant stiffness in standard drained triaxial test |
| $E_{oed}^{ref}$ | tangent stiffness for primary oedometer loading |
| $E_{ur}^{ref}$ | unloading/reloading stiffness from drained triaxial test |
| $p^{ref}$ | reference pressure |
| $m$ | Power for stress level dependency of stiffness |
| $S_H$ | horizontal strut spacing |
| $S_V$ | vertical strut spacing |
| $h_{avg}$ | average vertical strut spacing |
| $\alpha$ | wall stiffness coefficient |
| $S$ | system stiffness |
| $\nu$ | Poisson's ratio |
| $v_{ur}$ | Poissons ratio for unloading-reloading |
| $I$ | moment of inertia |
| $G$ | shear modulus |
| $G^*$ | rate of increase in soil shear modulus with depth |
| $G_0^{ref}$ | reference shear modulus at very small strains |
| $N_{pre}$ | load on the strut before strut failure |
| $N_{post}$ | load on the strut after strut failure |
| $N_{fail}$ | load on the failed strut before failure |
| $R_{inter}$ | Interface reduction factor |
| $f_c'$ | 28th-day compressive strength of concrete |
| $\Delta$ | displacement flexibility ($kN/m^4$) |
| $M_{wall}$ | wall bending moment |
| $M_{max}$ | maximum bending moment |

continued

| Symbols | Description |
|---|---|
| $\delta_{max}$ | flexural stress |
| $\sigma_a$ | allowable stress of the steel |
| $\lambda$ | short-term magnified factor of the allowable stress |
| $\rho_s$ | dimensionless system stiffness |
| $\delta_v$ | ground surface settlement |
| $\delta_{v(max)}$ | maximum ground surface settlement |
| $\delta_{H(max)}$ | maximum wall displacement |
| $N_b$ | the stability number of soil, is defined as $\gamma H_e/s_u$ |
| $N_{cb}$ | the critical stability number against basal heave |
| $E_{50}/c_u$ | soil stiffness |
| $C_u/\sigma_v'$ | soil shear strength ratio |
| $\{\sigma\}$ | stress matrix |
| $\{\varepsilon\}$ | strain matrix |
| $[C]$ | stress-strain relational matrix |
| $\varepsilon_v$ | volumetric strain |
| $[K]$ | global stiffness matrix |
| $V$ | volume |
| $\Delta V$ | change of the volume |
| $V_p$ | compression wave velocity |
| $V_s$ | shear wave velocity |
| $\tau$ | shear stress |
| $\tau_{max}$ | the shear stress at failure |
| $\gamma_r$ | the threshold shear strain |
| $\gamma_{0.7}$ | threshold shear strain at which $G_s=0.722G_0$ |
| $T_{skin,start,max}$ | skin resistance at the top of the embedded beam |
| $T_{skin,end,max}$ | skin resistance at the bottom of the embedded beam |
| $\lambda^*$ | modified compression index |
| $\lambda$ | Cam-clay compression index |
| $\kappa$ | the standard Cam-clay swelling index |
| $\kappa^*$ | modified swelling index |
| $\sigma_3'$ | the minor principal stress |

continued

| Symbols | Description |
|---|---|
| $R_f$ | failure ratio |
| $C_c$ | compression index |
| $C_s$ | swelling index or reloading index |
| $e_{init}$ | initial void ratio |
| $e_{min}$ | minimum void ratio |
| $G_0$ | the initial or very small-strain shear modulus |
| $G_s$ | the secant shear modulus |
| $G_t$ | lower cut-off of the tangent shear modulus |
| $G_{ur}$ | unloading reloading stiffness |
| $\gamma_{cut\text{-}off}$ | cut-off shear strain |
| $\gamma_{hist}$ | scalar valued shear strain |
| $\Delta \underline{e}$ | the actual deviatoric strain increment |
| $\bar{\bar{H}}$ | a symmetric tensor representing the deviatoric strain history of the material |
| $\varepsilon_q$ | the second deviatoric strain invariant |
| $k$ | coefficient of permeability |
| $W(u)$ | well function |
| $i_e$ | entry hydraulic gradient of groundwater flowing into a well |
| $e_0$ | initial void ratio |
| $R_w$ | radius of the imaginary well |
| $C_c$ | coefficient of compressibility |
| $C_s$ | the coefficient of swelling |
| $\tau_t$ | shear strength of the treated soil |
| $\tau_s$ | shear strength of the untreated soil |
| $F_{err}(x)$ | error function |
| $u_i$ | obeseved value |
| $u_i^*$ | computed value |
| $w_i$ | calculation weight at a point |
| $ccs_j$ | composite scaled sensitivity |
| $ND$ | number of observations |
| $NP$ | number of estimated parameters |

continued

| Symbols | Description |
|---|---|
| $H(x^k)$ | Hessian matrix |
| $\beta$ | reliability index |
| $EI$ | wall stiffness ($kN \cdot m^2/m$) |
| $EA$ | compressive stiffness ($kN/m$) |

**Abbreviations**

| Abbreviations | Description |
|:---:|:---|
| APD | apparent pressure diagrams |
| DPL | distributed prop load |
| FEL | final excavation level |
| TEC | Taipei Enterprise Center project |
| BW | basement wall method |
| ICOS | Impresa Construzioni Opere Specializzate method |
| MHL | Masago Hydraulic Long bucket method |
| PIP | packed-in-place |
| ACI | American Concrete Institute code |
| PSR | plane strain ratio |
| MARS | multivariate adaptive regression splines |
| LR | logarithmic regression |
| BF | basis functions |
| FoS | factor of safety |
| CST | constant strain triangular elements |
| LST | linear strain elements |
| OCR | over consolidation ratio |
| POP | pre-overburden pressure |
| LE | linear elastic model |
| MC | Mohr-Coulomb model |
| HS | hardening soil model |
| SS | soft soil model |
| SSC | soft soil creep model |
| MCC | Modified Cam-Clay model |
| PI | Plasticity index |
| GA | Genetic algorithm |
| FORM | First-Order Reliability Method |

# Appendix II  Database of Propped and Anchored Deep Excavation

| No. | Location | Soil at dredge level | $S_u$/ kPa | $H$/m | $h$/m | $h/H$ | $s$ | FoS | $EI$/ $(N \cdot m^2)$ | Wall type | Construction method | Support type | $\delta_h$/ mm | $\delta_v$/ mm | Reference |
|---|---|---|---|---|---|---|---|---|---|---|---|---|---|---|---|
| 1 | Oslo Enerhaughen, Norway | Soft clay | 20 | 8 | 17 | 1 | 2.5 | 1.34 | 45,000 | Sheet | No data (NA) | Multi-prop | 40 | 106 | NGI (1962a), Long (2001) |
| 2 | Oslo Telecom, Norway | Soft clay | 20 | 8.5 | 10 | 1 | 2.25 | 0.9 | 35,850 | Sheet | NA | Multi-prop | 80 | 93 | NGI (1962b), Long (2001) |
| 3 | Oslo Gronland 2, Norway | Soft clay | 25 | 11.5 | 26 | 1 | 3.75 | 1.3 | 73,800 | Sheet | NA | Multi-prop | 100 | 178 | NGI (1962c), Long (2001) |
| 4 | Oslo Vaterland 3, Norway | Soft clay | 34 | 12 | 26 | 1 | 2 | 1.26 | 73,800 | Sheet | NA | Multi-prop | 125 | 114 | NGI (1962f), Long (2001) |
| 5 | Oslo Gunnerus, Norway | Soft clay | 35 | 10.5 | 18 | 1 | 2 | 1.21 | 82,350 | Sheet | NA | Multi-prop | 320 | 600 | Aas (1984), Long (2001) |
| 6 | Vasteras, Sweden | Soft clay | 30 | 6.3 | 12.5 | 1 | 2.1 | 1.5 | 17,000 | Sheet | NA | Multi-anchor | 100 | 175 | Broms and Stille (1976), Long (2001) |
| 7 | Gothenburg, Sweden | Soft clay | 17 | 4 | 10 | 1 | 2 | 1.6 | 17,000 | Sheet | NA | Single-anchor | 400 | 410 | Broms and Stille (1976), Long (2001) |
| 8 | Chicago Subway, USA | Soft clay | 35 | 19 | 11 | 1 | 3 | 0.96 | 50,000 | Sheet | NA | Multi-prop | 60 | NA | Flaate (1966), Long (2001) |
| 9 | Chicago A, USA | Soft clay | 35 | 9.4 | NA | 1 | 3.05 | 1.67 | 55,250 | Sheet | NA | Multi-prop | 64 | NA | Gill and Lucas (1990), Long (2001) |
| 10 | Chicago C, USA | Soft clay | 35 | 8.8 | NA | 1 | 1.98 | 1.18 | 55,250 | Sheet | NA | Multi-prop | 56 | NA | Gill and Lucas (1990), Long (2001) |
| 11 | Chicago D, USA | Firm clay | 70 | 5.7 | 11 | 1 | 2.44 | 1.55 | 55,250 | Sheet | NA | Multi-prop | 13 | NA | Gill and Lucas (1990), Long (2001) |

continued

| No. | Location | Soil at dredge level | $S_u$/kPa | $H$/m | $h$/m | $h/H$ | $s$ | $FoS$ | $EI$/(N·m²) | Wall type | Construction method | Support type | $\delta_h$/mm | $\delta_v$/mm | Reference |
|---|---|---|---|---|---|---|---|---|---|---|---|---|---|---|---|
| 12 | Chicago E, USA | Firm clay | 70 | 10.67 | 11 | 1 | 3.04 | 1.55 | 1,106,000 | Diaphragm | NA | Multi-prop | 38 | NA | Gill and Lucas (1990), Long (2001) |
| 13 | Chicago F, USA | Firm clay | 70 | 10.67 | 11 | 1 | 3.81 | 1.25 | 55,250 | Sheet | NA | Multi-prop | 89 | NA | Gill and Lucas (1990), Long (2001) |
| 14 | Chicago G, USA | Firm clay | 70 | 12.34 | 11 | 0.89 | 3.96 | 1.41 | 69,000 | Sheet | NA | Multi-prop | 178 | NA | Gill and Lucas (1990), Long (2001) |
| 15 | Chicago H, USA | Firm clay | 70 | 10.67 | 11 | 1 | 3.05 | 1.4 | 566,340 | Diaphragm | NA | Multi-prop | 69 | NA | Gill and Lucas (1990), Long (2001) |
| 16 | HDR-4 Chicago, USA | Firm clay | 30 | 12.2 | 16 | 1 | 2.4 | 1.1 | 161,000 | Sheet | NA | Multi-prop | 190 | 250 | Finno et al. (1989), Long (2001) |
| 17 | Washington, USA | Soft clay | 30 | 9.1 | 18 | 1 | 2.2 | 0.81 | 50,160 | Soldier | NA | Multi-prop | 254 | NA | Swanson and Larson (1990), Long (2001) |
| 18 | Bowlin Point, N.Y., USA | Stiff clay | 40 | 9.8 | 9.8 | 1 | 1.96 | 2.4 | 50,000 | Sheet | NA | Multi-prop | 80 | NA | Murphy et al. (1975), Long (2001) |
| 19 | Davidson 1, San Francisco, USA | Soft clay | 10 | 9.1 | 21 | 1 | 3 | 0.83 | 72,500 | Sheet | NA | Multi-prop | 254 | NA | Clough and Reed (1984), Long (2001) |
| 20 | Islais 2, San Francisco, USA | Soft clay | 13 | 9.1 | 21 | 1 | 3 | 1.22 | 55,250 | Sheet | NA | Multi-prop | 254 | NA | Mana and Clough (1981), Clough and Reed (1984), Long (2001) |
| 21 | Embarcadero III | Soft clay | 30 | 13.7 | 27 | 1 | 3.4 | 0.99 | 80,000 | Sheet | NA | Multi-prop | 38 | NA | O'Rourke (1992), Long (2001) |
| 22 | Levi Strauss San Francisco, USA | Soft clay | 35 | 14 | 19.5 | 1 | 2.74 | 1.3 | 80,000 | Sheet | NA | Multi-prop | 150 | NA | Tait and Taylor (1975), Long (2001) |

| No. | Location | Soil | | | | | | | Wall type | | Support | | | Reference |
|---|---|---|---|---|---|---|---|---|---|---|---|---|---|---|
| 23 | SNBB San Francisco, USA | Soft clay | 35 | 14 | 19.5 | 1 | 2.74 | 1.3 | 4,528,466 | Diaphragm | NA | Multi-prop | 22 | NA | Tait and Taylor (1975), Long (2001) |
| 24 | 3rd Har Tun Boston, USA | Soft clay | 35 | 15.8 | 30 | 1 | 2.63 | 1.2 | 72,500 | Sheet | NA | Multi-prop | 150 | NA | Cacoilo et al. (1998), Long (2001) |
| 25 | H'Fok B Singapore | Soft clay | 15 | 7.3 | 30 | 1 | 1.83 | 0.87 | 75,700 | Sheet | NA | Multi-prop | 235 | 250 | Davies and Walsh (1983), Long (2001) |
| 26 | Tokyo Airport, Japan | Soft clay | 35 | 11 | 15 | 1 | 2.75 | 1.64 | 172,000 | Sheet | NA | Multi-prop | 300 | NA | Tanaka (1996), Long (2001) |
| 27 | Mexico City, Mexico | Soft clay | 25 | 9 | 20 | 1 | 1.8 | 0.95 | 50,640 | Sheet | NA | Multi-prop | 155 | NA | Rodriguez and Flamand (1969), Long (2001) |
| 28 | Shanghai, China | Soft clay | 8 | 33.5 | NA | 1 | 2.5 | 2 | 567,000 | Diaphragm | NA | Multi-prop | 40 | 15 | Tan and Wei (2012) |
| 29 | Shanghai, China | Soft clay | 8 | 33.6 | NA | 1 | 2.5 | 2 | 567,000 | Diaphragm | NA | Multi-prop | 42 | 21 | Tan and Wei (2012) |
| 30 | Boston, USA | Soft clay | 35 | 13.4 | NA | 1 | 3.35 | 1.5 | 1,200,000 | Diaphragm | NA | Multi-prop | 97 | NA | Sen et al. (2004) |
| 31 | UOB Singapore | Soft clay | 30 | 13 | 30 | 1 | 2.6 | >3 | 4,320,000 | Diaphragm | NA | Multi-prop | 56 | 130 | Wallace et al. (1992), Long (2001) |
| 32 | H'Fok A Singapore | Soft clay | 15 | 7.3 | 19 | 1 | 1.83 | >3 | 75,700 | Sheet | NA | Multi-prop | 60 | NA | Davies and Walsh (1983), Long (2001) |
| 33 | CTC Singapore | Soft clay | 20 | 12 | 37 | 1 | 2 | >3 | 57,440 | Sheet | NA | Multi-prop | 188 | 150 | Lee et al. (1985), Long (2001) |
| 34 | Somerset Singapore | Peats/silts | 15 | 15.2 | 10 | 1 | 3.8 | >3 | 540,000 | Diaphragm | NA | Multi-prop | 20 | NA | Leonard et al. (1987), Long (2001) |
| 35 | MOE I2 Singapore | Soft clay | 18 | 6.8 | 24 | 1 | 1.7 | >3 | 45,436 | Sheet | NA | Multi-prop | 330 | NA | Tan et al. (1985), Long (2001) |
| 36 | MOE I9 Singapore | Soft clay | 18 | 6.4 | 12 | 1 | 1.6 | >3 | 45,436 | Sheet | NA | Multi-prop | 100 | NA | Tan et al. (1985), Long (2001) |
| 37 | Singapore Bugis, Singapore | Soft clay | 40 | 18.3 | 30 | 1 | 2.29 | >3 | 4,320,000 | Diaphragm | NA | Multi-prop | 160 | NA | Hulme et al. (1989), Long (2001) |
| 38 | Singapore CBD, Singapore | Soft clay | 13 | 15 | 17 | 1 | 2.5 | >3 | 70,000 | Sheet | NA | Multi-prop | 145 | 100 | Broms et al. (1986), Long (2001) |

continued

| No. | Location | Soil at dredge level | $S_u$/kPa | $H$/m | $h$/m | $h/H$ | $s$ | $FoS$ | $EI/$ (N · m²) | Wall type | Construction method | Support type | $\delta_h/$mm | $\delta_v/$mm | Reference |
|---|---|---|---|---|---|---|---|---|---|---|---|---|---|---|---|
| 39 | Singapore Parking, Singapore | Soft clay | 35 | 9.5 | 12 | | 5.9 | >3 | 540,000 | Diaphragm | NA | Single-prop | 70 | NA | Vuillemin and Wong (1991), Long (2001) |
| 40 | Taiwan Airline, China | Firm clay | 45 | 9.6 | 9 | 0.94 | 2.4 | >3 | 18,850 | Contiguous | NA | Multi-prop | 22 | 22 | Ou et al. (1993), Long (2001) |
| 41 | Taiwan Power, China | Firm clay | 120 | 16.2 | 15 | 0.93 | 3.24 | >3 | 540,000 | Diaphragm | NA | Multi-prop | 80 | 56 | Ou et al. (1993), Long (2001) |
| 42 | Taiwan Quen M, China | Firm clay | 47 | 10.7 | 8 | 0.75 | 2.68 | >3 | 540,000 | Diaphragm | NA | Multi-prop | 70 | 35 | Ou et al. (1993), Long (2001) |
| 43 | Taiwan Tax, China | Soft clay | 38 | 7.65 | 8 | 1 | 1.91 | >3 | 40,000 | Sheet | NA | Multi-prop | 69 | 41 | Ou et al. (1993), Long (2001) |
| 44 | Taiwan Formosa, China | Soft clay | 35 | 18.45 | 20 | 1 | 2.64 | >3 | 1,280,000 | Diaphragm | NA | Multi-prop | 60 | 42 | Ou et al. (1993), Long (2001) |
| 45 | Taiwan Cathay, China | Firm clay | 90 | 21 | 12 | 0.57 | 2.63 | >3 | 857,500 | Diaphragm | NA | Multi-prop | 62 | 31 | Ou et al. (1993), Long (2001) |
| 46 | Bangkok A, Thailand | Soft clay | 35 | 9.8 | 15 | 1 | 3.1 | >3 | 1,378,420 | Diaphragm | NA | Multi-prop | 50 | NA | Balasubramaniam et al. (1994), Long (2001) |
| 47 | Bangkok C, Thailand | Soft clay | 35 | 18.5 | 12 | 1 | 4.6 | >3 | 1,378,420 | Diaphragm | NA | Multi-prop | 30 | NA | Balasubramaniam et al. (1994), Long (2001) |
| 48 | Bangkok E, Thailand | Soft clay | 35 | 7.2 | 12 | 1 | 1.8 | >3 | 50,000 | Sheet | NA | Multi-prop | 220 | NA | Balasubramaniam et al. (1994), Long (2001) |
| 49 | Oslo Vaterland 1, Norway | Soft clay | 25 | 11 | 16 | 1 | 2 | >3 | 73,800 | Sheet | NA | Multi-prop | 220 | 270 | NGI (1962d), Long (2001) |
| 50 | Oslo Vaterland 2, Norway | Soft clay | 20 | 11 | 16 | 1 | 2 | >3 | 73,800 | Sheet | NA | Multi-prop | 140 | 260 | NGI (1962e), Long (2001) |

| No. | Location | Soil | | | | | | | | Wall | | Support | | | Reference |
|---|---|---|---|---|---|---|---|---|---|---|---|---|---|---|---|
| 51 | Oslo Studenterlu, Norway | Soft clay | 40 | 16 | 37 | 1 | 5.3 | >3 | 2,500,000 | Diaphragm | NA | Multi-prop | 42 | 65 | Karlsrud (1981, 1983, 1986), Long (2001) |
| 52 | Oslo Jerbanetorget, Norway | Soft clay | 20 | 10 | 35 | 1 | 5 | >3 | 2,500,000 | Diaphragm | NA | Multi-prop | 20 | NA | Karlsrud (1981, 1983), Long (2001) |
| 53 | Oslo Bank of Norway, Norway | Soft clay | 20 | 16 | 18 | 1 | 3.2 | >3 | 2,500,000 | Diaphragm | NA | Mult-prop | 16 | 62 | Roti and Friis (1985), Long (2001) |
| 54 | Eastbourne 1, UK | Soft clay | 35 | 11 | 15 | 1.00 | 10.00 | >3 | 2,500,000 | Diaphragm | NA | Single-prop | 61 | NA | Fernie and Suckling (1996), Long (2001) |
| 55 | Eastbourne 2, UK | Soft clay | 35 | 14 | 15 | 1.00 | 13.00 | >3 | 2,500,000 | Diaphragm | NA | Single-prop | 15 | NA | Fernie and Suckling (1996), Long (2001) |
| 56 | Pietrafitta, Italy | Soft-hard clay | 40 | 5.5 | 20 | 1.00 | 7.80 | >3 | 42,000 | Sheet | NA | Multi-prop | 71 | NA | Rampello et al. (1992), Long (2001) |
| 57 | Chicago, USA | Soft clay | 35 | 13.4 | 15 | 1.00 | 4.46 | >3 | 1,055,000 | Sheet | NA | Multi-prop | 150 | NA | Fernie and Suckling (1996), Long (2001) |
| 58 | Inland Steel Chicago, USA | Soft clay | 35 | 11 | 19 | 1.00 | 2.00 | >3 | 50,000 | Sheet | NA | Multi-prop | 55 | NA | Flaate (1966), Long (2001) |
| 59 | Osaka A, Japan | Soft clay | 35 | 20.6 | 25 | 1.00 | 3.00 | >3 | 2,500,000 | Diaphragm | NA | Multi-prop | 78 | NA | Tamano et al. (1996), Long (2001) |
| 60 | Lake zone, Mexico | Soft clay | 25 | 15.7 | 20 | 1.00 | 2.62 | >3 | 2,500,000 | Diaphragm | NA | Multi-prop | 135 | NA | Auvinet and Organista (1998), Long (2001) |
| 61 | Shanghi-Jin Mao, China | Soft-hard clay | 10 | 19.65 | 36 | 1.00 | 3.93 | >3 | 2,500,000 | Diaphragm | NA | Multi-prop | 81 | NA | Zhao et al. (1999), Long (2001) |
| 62 | Shanghi-Heng Long, China | Soft-hard clay | 10 | 18.2 | 29 | 1.00 | 3.64 | >3 | 2,500,000 | Diaphragm | NA | Multi-prop | 99 | NA | Zhao et al. (1999), Long (2001) |
| 63 | Shanghi, China | Soft-hard clay | 10 | 17.85 | 24 | 1.00 | 3.57 | >3 | 2,500,000 | Diaphragm | NA | Multi-prop | 129 | NA | Onishi and Sugawara (1999), Long (2001) |
| 64 | River Wall M'Boro, UK | Soft clay | 35 | 9.5 | 12 | 1.00 | 4.25 | >3 | 177,660 | Sheet | NA | Single-anchor | 125 | NA | Baggett and Buttling (1977), Long (2001) |
| 65 | Detroit, USA | Soft clay | 35 | 7 | 10 | 1.00 | 11.20 | >3 | 83,400 | Sheet | NA | Single-anchor | 45 | NA | Fernie and Suckling (1996), Long (2001) |

continued

| No. | Location | Soil at dredge level | $S_u$/ kPa | $H$/m | $h$/m | $h/H$ | $q_s$ | FoS | $EI$/ $(N \cdot m^2)$ | Wall type | Construction method | Support type | $\delta_h$/ mm | $\delta_v$/ mm | Reference |
|---|---|---|---|---|---|---|---|---|---|---|---|---|---|---|---|
| 66 | TP, Bogota, Colombia | Soft clay/silt | 15 | 16 | 34 | 1.00 | 3.75 | >3 | 540,000 | Diaphragm | NA | Multi-anchor | 125 | 1000 | Maldonado (1998), Long (2001) |
| 67 | Newton Singapore | Soft clay | 18 | 14.5 | 12 | 1.00 | 3.63 | >3 | 2,500,000 | Diaphragm | Top-down | Undifferentiated multi-support | 110 | 220 | Nicholson (1987), Long (2001) |
| 68 | Taiwan Chi Ching, China | Soft clays | 30 | 13.9 | 15 | 1.00 | 2.78 | >3 | 857,500 | Diaphragm | Top-down | Undifferentiated multi-support | 65 | 65 | Ou et al. (1993), Long (2001) |
| 69 | Taiwan Far East, China | Firm clay | 60 | 20 | 24 | 0.60 | 3.33 | >3 | 857,500 | Diaphragm | Top-down | Undifferentiated multi-support | 135 | 67 | Ou et al. (1993), Long (2001) |
| 70 | Oslo Christiana, Norway | Soft clay | 35 | 9.6 | 23 | 1.00 | 3.00 | >3 | 483,600 | Sheet | Top-down | Undifferentiated multi-support | 48 | 100 | Finstad (1991), Long (2001) |
| 71 | Shanghai, China | Soft clay | 30 | 14.5 | 50 | 1.00 | 3.00 | >3 | 2,500,000 | Diaphragm | Top-down | Multi-prop | 35 | 45 | Ng et al. (2012) |
| 72 | Illinois, USA | Soft clay | 35 | 7.5 | 10.1 | 1.00 | 5.00 | >3 | 625,000 | Sheet | Top-down | Multi-prop | 40 | 190 | Blackburn and Finno (2007) |
| 73 | NTUH, Taipei, China | Soft Soil | 36 | 15.7 | 18 | 1.00 | 2.40 | >3 | 680,000 | Diaphragm | Top-down | Multi-anchor | 80 | NA | Liao and Hsieh (2002) |
| 74 | TCC, Taipei, China | Soft Soil | 35 | 12.5 | 22 | 1.00 | 2.00 | >3 | 340,000 | Diaphragm | Top-down | Multi-anchor | 32 | NA | Liao and Hsieh (2002) |
| 75 | TCAC, Taipei, China | Soft Soil | 35 | 20 | 24 | 1.00 | 2.12 | >3 | 4,200,000 | Diaphragm | Top-down | Multi-anchor | 50 | NA | Liao and Hsieh (2002) |
| 76 | Shanghai, China | Soft clay | 40 | 33.6 | NA | 1.00 | 2.50 | >3 | 1,215,000 | Diaphragm | Bottom-up | Multi-prop | 27 | 6 | Tan and Wei (2012) |

| No. | Location | Soil | | | | | | | Wall | Method | Support | | | Reference |
|---|---|---|---|---|---|---|---|---|---|---|---|---|---|---|
| 77 | Shanghai, China | Soft clay | 40 | 33.6 | NA | 1.00 | 2.50 | >3 | 1,215,000 | Diaphragm | Bottom-up | Multi-prop | 31 | NA | Tan and Wei (2012) |
| 78 | Shanghai, China | Very soft clay | 20 | 9 | NA | 1.00 | 4.50 | >3 | 60,000 | Deep soil mix | Top-down | Multi-prop | 10 | 16 | Shao et al. (2005) |
| 79 | Chicago, USA | Soft-medium stiff clay | 36 | 12.8 | 12.8 | 1 | 4.5 | >3 | 131,220 | Sheet | Top-down | Multi-anchor | 90 | 74 | Finno and Roboski (2005) |
| 80 | Chicago, USA | Soft-medium stiff clay | 46 | 10.8 | 10.5 | 0.97 | 4.5 | >3 | 131,220 | Sheet | Top-down | Multi-anchor | 45 | NA | Finno and Roboski (2005) |
| 81 | Case 2 | Soft clay | 35 | 10 | 42 | 1 | 2.5 | >3 | 447,000 | Diaphragm | Bottom-up | Multi-prop | 22 | NA | Kung et al. (2007a) |
| 82 | Hsinkuang, Pakistan | Soft clay | 35 | 16 | 55 | 1 | 2.9 | >3 | 709,000 | Diaphragm | Bottom-up | Multi-prop | 83 | NA | Kung et al. (2007a) |
| 83 | Taipei Gas, China | Soft clay | 35 | 18.1 | 46 | 1 | 2.5 | >3 | 2,067,000 | Diaphragm | Bottom-up | Multi-prop | 76 | NA | Kung et al (2007a) |
| 84 | Tzuchyang | Soft clay | 35 | 13.6 | 50 | 1 | 3.2 | >3 | 709,000 | Diaphragm | Top-down | Multi-prop | 53 | NA | Kung et al. (2007a) |
| 85 | Case 10 | Soft clay | 35 | 12.3 | 40.5 | 1 | 2.7 | >3 | 447,000 | Diaphragm | Bottom-up | Multi-prop | 59 | NA | Kung et al. (2007a) |
| 86 | Baisern | Soft clay | 35 | 12.3 | 37 | 1 | 3.2 | >3 | 447,000 | Diaphragm | Bottom-up | Multi-prop | 39 | NA | Kung et al. (2007a) |
| 87 | MRT-1 | Soft clay | 35 | 16.8 | 45 | 1 | 3.2 | >3 | 1,059,000 | Diaphragm | Bottom-up | Multi-prop | 30 | NA | Kung et al. (2007a) |
| 88 | MRT-2 | Soft clay | 35 | 16.4 | 52 | 1 | 3.1 | >3 | 1,059,000 | Diaphragm | Bottom-up | Multi-prop | 41 | NA | Kung et al. (2007a) |
| 89 | MRT-3 | Soft clay | 35 | 12.4 | 47.9 | 1 | 3.4 | >3 | 2,067,000 | Diaphragm | Top-down | Multi-prop | 22 | NA | Kung et al. (2007a) |
| 90 | MRT-4 | Soft clay | 35 | 16.2 | 45 | 1 | 2.4 | >3 | 2,067,000 | Diaphragm | Top-down | Multi-prop | 49 | NA | Kung et al. (2007a) |

continued

| No. | Location | Soil at dredge level | $S_u$/ kPa | $H$/m | $h$/m | $h/H$ | $s$ | $F_oS$ | $EI$/ (N · m²) | Wall type | Construction method | Support type | $\delta_h$/ mm | $\delta_v$/ mm | Reference |
|---|---|---|---|---|---|---|---|---|---|---|---|---|---|---|---|
| 91 | Subway 8 | Soft clay | 35 | 28.8 | 50 | 1 | 2.6 | >3 | 3,572,000 | Diaphragm | Bottom-up | Multi-prop | 69 | NA | Kung et al. (2007a) |
| 92 | Subway 9 | Soft clay | 35 | 26.4 | 50 | 1 | 3.1 | >3 | 3,572,000 | Diaphragm | Bottom-up | Multi-prop | 55 | NA | Kung et al. (2007a) |
| 93 | Subway 10 | Soft clay | 35 | 21.7 | 50 | 1 | 2.6 | >3 | 3,572,000 | Diaphragm | Bottom-up | Multi-prop | 41 | NA | Kung et al. (2007a) |
| 94 | Sinyi, China | Soft clay | 35 | 12.3 | 46 | 1 | 2.6 | >3 | 447,000 | Diaphragm | Bottom-up | Multi-prop | 46 | NA | Kung et al. (2007a, b) |
| 95 | Taiwan Sugar, China | Soft clay | 35 | 13.2 | 45 | 1 | 3.2 | >3 | 1,059,000 | Diaphragm | Bottom-up | Multi-prop | 58 | NA | Kung et al. (2007a, b) |
| 96 | Tai Kai | Soft clay | 35 | 12.6 | 48 | 1 | 2.7 | >3 | 447,000 | Diaphragm | Bottom-up | Multi-prop | 60 | NA | Kung et al. (2007a, b) |
| 97 | Case 12 | Soft clay | 35 | 13.7 | 36 | 1 | 3.1 | >3 | 709,000 | Diaphragm | Bottom-up | Multi-prop | 48 | NA | Kung et al. (2007a, b) |
| 98 | Subway 1 | Soft clay | 35 | 14.5 | 46 | 1 | 2.5 | >3 | 1,059,000 | Diaphragm | Bottom-up | Multi-prop | 35 | NA | Kung et al. (2007a, b) |
| 99 | Subway 2 | Soft clay | 35 | 19.4 | 45.1 | 1 | 2.8 | >3 | 2,751,000 | Diaphragm | Bottom-up | Multi-prop | 60 | NA | Kung et al. (2007a, b) |
| 100 | Subway 3 | Soft clay | 35 | 19.4 | 50 | 1 | 2.4 | >3 | 2,751,000 | Diaphragm | Bottom-up | Multi-prop | 62 | NA | Kung et al. (2007a, b) |
| 101 | Subway 4 | Soft clay | 35 | 16.2 | 50 | 1 | 3.4 | >3 | 2,067,000 | Diaphragm | Top-down | Multi-prop | 47 | NA | Kung et al. (2007a) |
| 102 | Subway 5 | Soft clay | 35 | 15.5 | 45.8 | 1 | 2.4 | >3 | 3,572,000 | Diaphragm | Bottom-up | Multi-prop | 36 | NA | Kung et al. (2007a) |

| No. | Case | Soil | | | | | | | Wall | Method | Support | | | Reference |
|---|---|---|---|---|---|---|---|---|---|---|---|---|---|---|
| 103 | Subway 6 | Soft clay | 35 | 12.7 | 45.8 | 1 | 3.1 | >3 | 3,572,000 | Diaphragm | Bottom-up | Multi-prop | 29 | NA | Kung et al. (2007a) |
| 104 | Subway 7 | Soft clay | 35 | 19.9 | 45.8 | 1 | 3.1 | >3 | 3,572,000 | Diaphragm | Bottom-up | Multi-prop | 52 | NA | Kung et al. (2007a, b) |
| 105 | Syed Alwi, Singapore | Soft clay | 35 | 7.8 | 16 | 1 | 3.5 | >3 | 447,000 | Diaphragm | Bottom-up | Multi-prop | 48 | NA | Lim et al. (2003), Kung et al. (2007a) |
| 106 | Lavender, Singapore | Soft clay | 35 | 15.7 | 20.5 | 1 | 2.6 | >3 | 2,067,000 | Diaphragm | Bottom-up | Multi-prop | 31 | NA | Lim et al. (2003), Kung et al. (2007a) |
| 107 | BIF, Boston, USA | Marine clay | 50 | 19.4 | 15 | 0.77 | 3.2 | >3 | 60,000 | Deep soil mix | NA | Multi-prop | 41 | NA | O'Rourke and McGinn (2006) |
| 108 | Gu Bei Station, Shanghai, China | Soft clay | 26 | 14.5 | 50 | 1 | 3.5 | >3 | 1,280,000 | Diaphragm | Top-down | Multi-prop | 20 | 39 | Hong et al. (2015) |
| 109 | Block 37, Chicago, USA | Soft clay | 35 | 15 | NA | 1 | 3.75 | >3 | 180,000 | Diaphragm | Top-down | Multi-prop | 45 | 30 | Mu and Huang (2016) |
| 110 | Shanghai, China | Soft clay | 35 | 22.8 | NA | 1 | 3.8 | >3 | 1,200,000 | Diaphragm | Bottom-up | Multi-prop | 80 | 8 | Shi et al. (2015) |
| 111 | Shanghai, China | Soft clay | 35 | 15.5 | NA | 1 | 2.7 | >3 | 600,000 | Diaphragm | Top-down | Multi-prop | 40 | 20 | Liu et al. (2005) |
| 112 | Salerno, Italy | Sand and gravel | 150 | 8 | 1 | 0.13 | 4 | >3 | 540,000 | Diaphragm | Top-down | Multi-prop | 28 | 11 | Bilotta et al. (2004) |
| 113 | Longwood, Boston, USA | Stiff clay | 60 | 23 | 1 | 0.04 | 3.1 | >3 | 1,200,000 | Diaphragm | Top-down | Multi-anchor | 20 | 4 | Konstantakos et al. (2004) |
| 114 | Longwood, Boston, USA | Stiff clay | 60 | 23 | 1 | 0.04 | 3.1 | >3 | 1,200,000 | Diaphragm | Top-down | Multi-anchor | 65 | 12 | Konstantakos et al. (2004) |
| 115 | Longwood, Boston, USA | Stiff clay | 60 | 23 | 1 | 0.04 | 3.1 | >3 | 1,200,000 | Diaphragm | Top-down | Multi-anchor | 65 | 21 | Konstantakos et al. (2004) |
| 116 | Marmaris, Turkey | Silty sand | 65 | 6 | 3.5 | 0.58 | 2 | >3 | 289,000 | Secant | Bottom-up | Multi-anchor | 29 | NA | Nalcakan et al. (2004) |
| 117 | Marmaris, Turkey | Silty sand | 65 | 6 | 3.5 | 0.58 | 2 | >3 | 289,000 | Secant | Bottom-up | Multi-anchor | 49 | NA | Nalackan et al. (2004) |
| 118 | Shanghai, China | Soft clay | 35 | 12.2 | NA | 1 | 3 | >3 | 1,200,000 | Diaphragm | Top-down | Undifferentiated multi-support | 40 | 25 | Dong et al. (2013) |

continued

| No. | Location | Soil at dredge level | $S_t$/ kPa | $H$/m | $h$/m | $h/H$ | $s$ | FoS | $EI$/ $(N \cdot m^2)$ | Wall type | Construction method | Support type | $\delta_h$/ mm | $\delta_v$/ mm | Reference |
|---|---|---|---|---|---|---|---|---|---|---|---|---|---|---|---|
| 119 | Constance, Germany | Soft clay | 20 | 9.9 | 25 | 1 | 5 | >3 | 91,140 | Sheet | Bottom-up | Multi-prop | 113 | 160 | Becker et al. (2008) |
| 120 | Nangang, Taipei, China | Soft clay | 20 | 11.6 | 21 | 1 | 2.9 | >3 | 5,570,000 | Diaphragm | Bottom-up | Multi-prop | 114 | 83 | Chen et al. (2014) |
| 121 | Shihuduan, Taipei, China | Soft clay | 35 | 12.1 | NA | 1 | 3 | >3 | 460,844 | Diaphragm | Bottom-up | Undifferentiated multi-support | 94 | 68 | Chen et al. (2014) |
| 122 | Hepingsihang, Taipei, China | Soft clay | 35 | 12.1 | NA | 1 | 3 | >3 | 460,844 | Diaphragm | Bottom-up | Undifferentiated multi-support | 71 | NA | Chen et al. (2014) |
| 123 | Jyru Road, China | Soft clay | 35 | 9.4 | NA | 1 | 2.4 | >3 | 460,844 | Diaphragm | Bottom-up | Undifferentiated multi-support | 54 | 43 | Chen et al. (2014) |
| 124 | Neihu Road, Taipei, China | Soft clay | 35 | 12.3 | NA | 1 | 3.1 | >3 | 687,908 | Diaphragm | Bottom-up | Undifferentiated multi-support | 38 | NA | Chen et al. (2014) |
| 125 | Jianbei, Taipei, China | Soft clay | 35 | 16 | NA | 1 | 3.2 | >3 | 687,908 | Diaphragm | Bottom-up | Undifferentiated multi-support | 52 | NA | Chen et al. (2014) |
| 126 | Duenbei, Taipei, China | Soft clay | 35 | 12 | NA | 1 | 4 | >3 | 460,844 | Diaphragm | Top-down | Undifferentiated multi-support | 28 | 15 | Chen et al. (2014) |
| 127 | CPC, Taipei, China | Soft clay | 35 | 31.7 | NA | 1 | 4 | >3 | 4,534,547 | Diaphragm | Top-down | Undifferentiated multi-support | 192 | NA | Hwang and Moh (2007), Chen et al. (2014) |
| 128 | CPC, Taipei, China | Soft clay | 35 | 31.7 | NA | 1 | 4 | >3 | 1,454,766 | Diaphragm | Top-down | Undifferentiated multi-support | 123 | NA | Hwang and Moh (2007), Chen et al. (2014) |
| 129 | Taipei 101, China | Soft clay | 35 | 21.7 | NA | 1 | 4.3 | >3 | 2,321,688 | Diaphragm | Top-down | Undifferentiated multi-support | 106 | NA | Lin and Woo (2007), Chen et al. (2014) |
| 130 | Taipei 101, China | Soft clay | 35 | 21.7 | NA | 1 | 3.6 | >3 | 2,321,688 | Diaphragm | Top-down | Undifferentiated multi-support | 54 | NA | Lin and Woo (2007), Chen et al. (2014) |

| No. | Project | Soil type | | | | | | | | | | Support | | | Reference |
|---|---|---|---|---|---|---|---|---|---|---|---|---|---|---|---|
| 131 | UPIB, Taipei, China | Soft clay | 35 | 32.6 | NA | 1 | 4.1 | >3 | 4,534,547 | Diaphragm | Top-down | Undifferentiated multi-support | 74 | 30 | Ou et al. (2006), Chen et al. (2014) |
| 132 | Jungshan, Taipei, China | Soft clay | 35 | 26.6 | NA | 1 | 3.8 | >3 | 2,321,688 | Diaphragm | Top-down | Undifferentiated multi-support | 122 | NA | Hwang and Moh (2007), Chen et al. (2014) |
| 133 | Kuntsevo Plaza, Russia | Sand and clay | 35 | 25 | NA | 1 | 4.2 | >3 | 1,200,000 | Diaphragm | Bottom-up | Multi-anchor | 8 | 13 | Mothersille et al. (2015) |
| 134 | Shanghai, China | Soft soil | 35 | 15 | 40 | 1 | 3.7 | >3 | 1,200,000 | Diaphragm | Top-down | Multi-prop | 70 | 136 | Xu et al. (2013) |
| 135 | Shanghai, China | Soft soil | 35 | 14.5 | 40 | 1 | 3.7 | >3 | 1,200,000 | Diaphragm | Top-down | Multi-prop | 21 | 16 | Xu et al. (2013) |
| 136 | Shanghai, China | Soft soil | 35 | 15.2 | 40 | 1 | 3.7 | >3 | 1,200,000 | Diaphragm | Top-down | Multi-prop | 21 | 16 | Xu et al. (2013) |
| 137 | Brescia, Italy | Clay and sand | 75 | 18.5 | 1 | 0.05 | 2.6 | >3 | 1,200,000 | Diaphragm | NA | Multi-anchor | 40 | 38 | Sanzeni et al. (2013) |
| 138 | YBCC, Shanghai, China | Soft clay | 35 | 15.9 | NA | 1 | 4.4 | >3 | 680,000 | Diaphragm | Top-down | Multi-prop | 30.7 | NA | Jia et al. (2008) |
| 139 | Bangkok, Thailand | Soft clay | 18 | 5.6 | NA | 1 | 2.9 | >3 | 1,200,000 | Diaphragm | Top-down | single-prop | 38 | NA | Thasnanipan et al. (2004) |
| 140 | Patras, Greece | Silty sand and clay | 125 | 7.5 | 2 | 0.27 | 3.8 | >3 | 1,200,000 | Diaphragm | Bottom-up | single-anchor | 20 | NA | Zekkos et al. (2004) |
| 141 | Nangang, Taipei, China | Soft clay | 20 | 11.6 | 7 | 1 | 2.9 | >3 | 5,570,000 | Diaphragm | Bottom-up | Multi-prop | 217 | 178 | Chen et al. (2014) |
| 142 | CFSC, Norwich, UK | Sand and gravel | 50 | 12 | 10 | 0.83 | 6 | >3 | 340,000 | Contiguous | Top-down | Berm and support | 20 | NA | Bowden and Lees (2010) |
| 143 | CFSC, Norwich, UK | Sand and gravel | 150 | 12 | 2 | 0.17 | 6 | >3 | 340,000 | Contiguous | Top-down | Berm and support | 20 | NA | Bowden and Lees (2010) |

continued

| No. | Location | Soil at dredge level | $S_u$/kPa | $H$/m | $h$/m | $h/H$ | $s$ | FoS | $EI$/(N·m²) | Wall type | Construction method | Support type | $\delta_h$/mm | $\delta_v$/mm | Reference |
|---|---|---|---|---|---|---|---|---|---|---|---|---|---|---|---|
| 144 | CFSC, Norwich, UK | Sand and gravel | 50 | 12 | 10 | 0.83 | 3 | >3 | 340,000 | Contiguous | Top-down | Multi-prop | 8 | NA | Bowden and Lees (2010) |
| 145 | CFSC, Norwich, UK | Sand and gravel | 150 | 12 | 2 | 0.17 | 3 | >3 | 340,000 | Contiguous | Top-down | Multi-prop | 8 | NA | Bowden and Lees (2010) |
| 146 | CFSC, Norwich, UK | Sand and gravel | 50 | 12 | 10 | 0.83 | 4 | >3 | 340,000 | Contiguous | Top-down | Multi-prop | 6 | NA | Bowden and Lees (2010) |
| 147 | CFSC, Norwich, UK | Sand and gravel | 150 | 12 | 2 | 0.17 | 4 | >3 | 340,000 | Contiguous | Top-down | Multi-prop | 6 | NA | Bowden and Lees (2010) |
| 148 | John Rog. Quay, Dublin, Ireland | Soft silt | 35 | 7 | 12.5 | 1 | 7 | >3 | 690,135 | Secant | NA | single-anchor | 24 | NA | Long et al. (2012a) |
| 149 | Portmarnoch, Dublin, Ireland | Soft sandy silt | 18 | 4 | 6 | 1 | 4 | >3 | 191,000 | Secant | NA | single-anchor | 3 | NA | Long et al. (2012a) |
| 150 | Singapore River, Singapore | Stiff clay | 200 | 29 | 20 | 0.69 | 3.3 | >3 | 1,280,000 | Diaphragm | NA | Multi-prop | 79 | NA | Gronin et al. (1991), Long (2001) |
| 151 | Singapore Havelock, Singapore | Stiff clay | 200 | 16.5 | 16.4 | 0.99 | 4 | >3 | 1,280,000 | Diaphragm | NA | Multi-prop | 40 | NA | Gronin et al. (1991), Long (2001) |
| 152 | Singapore CBD, Singapore | Stiff clay | 70 | 14.7 | 14 | 0.95 | 2.1 | >3 | 60,000 | Sheet | NA | Multi-prop | 280 | 100 | Broms et al. (1986), Long (2001) |
| 153 | Singapore CE II, Singapore | Stiff clay | 125 | 10 | NA | 0.9 | 4 | >3 | 1,280,000 | Diaphragm | NA | Multi-prop | 22 | 15 | Wong et al. (1997), Long (2001) |

| No. | Project | Soil | | | | | | | | | | | | | Reference |
|---|---|---|---|---|---|---|---|---|---|---|---|---|---|---|---|
| 154 | Singapore CE II , Singapore | Stiff clay | 125 | 10 | NA | 0.9 | 4 | >3 | 73,500 | Sheet | NA | Multi-prop | 32 | 30 | Wong et al. (1997) , Long (2001) |
| 155 | Singapore CE II , Singapore | Stiff clay | 125 | 11 | NA | 0.9 | 4 | >3 | 73,500 | Sheet | NA | Multi-prop | 48 | 25 | Wong et al. (1997) , Long (2001) |
| 156 | Singapore CE II , Singapore | Stiff clay | 125 | 14 | NA | 0.9 | 4 | >3 | 100,000 | Contiguous | NA | Multi-prop | 37 | 18 | Wong et al. (1997) , Long (2001) |
| 157 | Singapore CE II , Singapore | Stiff clay | 125 | 15 | NA | 0.9 | 4 | >3 | 685,000 | Sheet | NA | Multi-prop | 88 | 75 | Wong et al. (1997) , Long (2001) |
| 158 | Singapore CE II , Singapore | Stiff clay | 125 | 17 | NA | 0.9 | 4 | >3 | 100,000 | Contiguous | NA | Multi-prop | 73 | 58 | Wong et al. (1997) , Long (2001) |
| 159 | Singapore CE II , Singapore | Stiff clay | 125 | 17 | NA | 0.9 | 4 | >3 | 73,500 | Sheet | NA | Multi-prop | 75 | 25 | Wong et al. (1997) , Long (2001) |
| 160 | Singapore CE II , Singapore | Stiff clay | 125 | 17.5 | NA | 0.9 | 4 | >3 | 685,000 | Sheet | NA | Multi-prop | 43 | NA | Wong et al. (1997) , Long (2001) |
| 161 | Singapore CE II , Singapore | Stiff clay | 125 | 18.5 | NA | 0.9 | 4 | >3 | 100,000 | Contiguous | NA | Multi-prop | 73 | NA | Wong et al. (1997) , Long (2001) |
| 162 | Singapore CE II , Singapore | Stiff clay | 125 | 19 | NA | 0.9 | 4 | >3 | 685,000 | Sheet | NA | Multi-prop | 50 | NA | Wong et al. (1997) , Long (2001) |
| 163 | Singapore Mstory , Singapore | Clayey sand | 500 | 17.3 | 17.2 | 0.99 | 2.88 | >3 | 2,500,000 | Diaphragm | NA | Multi-prop | 50 | NA | Lee et al. (1998) , Long (2001) |
| 164 | Singapore interchange , Singapore | Stiff clay | 500 | 20 | 15 | 0.75 | 4 | >3 | 1,338,750 | Contiguous | NA | Multi-prop | 40 | NA | Vuillemin and Wong (1991) , Long (2001) |
| 165 | Singapore canal , Singapore | Stiff clay | 140 | 6.5 | 6 | 0.92 | 5 | >3 | 50,000 | Sheet | NA | Single-prop | 85 | 120 | Wong and Chua (1999) , Long (2001) |
| 166 | Bangkok B , Thailand | Stiff clay | 140 | 15.5 | 15 | 0.97 | 5.1 | >3 | 1,378,420 | Diaphragm | NA | Multi-prop | 45 | NA | Balasubramaniam et al. (1994) , Long (2001) |
| 167 | Bangkok D , Thailand | Stiff clay | 140 | 16 | 14 | 0.88 | 3.2 | >3 | 2,500,000 | Diaphragm | NA | Multi-prop | 25 | NA | Balasubramaniam et al. (1994) , Long (2001) |
| 168 | Japan 1 , Japan | Stiff clay | 140 | 13.75 | 12 | 0.87 | 6.88 | >3 | 63,450 | Sheet | NA | Multi-prop | 120 | NA | Fernie and Suckling (1996) , Long (2001) |
| 169 | Sheung Wan HK CDG , Honk Kong , China | | 350 | 30 | 19 | 0.63 | 2.73 | >3 | 4,320,000 | Diaphragm | NA | Multi-prop | 20 | 23 | Fraser (1992) , Long (2001) |

continued

| No. | Location | Soil at dredge level | $S_u$/kPa | $H$/m | $h$/m | $h/H$ | $s$ | $FoS$ | $EI$/(N·m²) | Wall type | Construction method | Support type | $\delta_h$/mm | $\delta_v$/mm | Reference |
|---|---|---|---|---|---|---|---|---|---|---|---|---|---|---|---|
| 170 | Singapore CE II, Singapore | Stiff clay | 125 | 12 | NA | 0.9 | 4 | >3 | 1,280,000 | Diaphragm | NA | Multi-anchor | 15 | 17 | Wong et al. (1997), Long (2001) |
| 171 | Quai Gloria France | Sands | 150 | 12 | 9 | 0.75 | 10 | >3 | 1,378,420 | Diaphragm | NA | Single-anchor | 39 | NA | Vezinhet et al. (1989), Long (2001) |
| 172 | Hartford, Conn., USA | Dense gravels | 150 | 7 | 13.5 | 0.75 | 5 | >3 | 26,250 | Soldier | NA | Single-anchor | 20 | NA | Murphy et al. (1975), Long (2001) |
| 173 | C11A1 Boston, USA | Glacial till | 95 | 20 | 8 | 0.4 | 2.9 | >3 | 70,100 | Soldier | Top-down | Multi-prop | 40 | 33 | Hashash et al. (2008) |
| 174 | C15A1 Boston, USA | Medium stiff clay | 95 | 21.3 | NA | 0.7 | 5.2 | >3 | 730,000 | Soldier | Top-down | Multi-prop | 23 | 34 | Hashash et al. (2008) |
| 175 | C17A2 Boston, USA | Medium stiff clay | 95 | 17 | NA | 0.7 | 5.7 | >3 | 1,000,000 | Soldier | Top-down | Multi-prop | 39 | 22 | Hashash et al. (2008) |
| 176 | FXWS, Chicago, USA | Medium stiff clay | 95 | 12.2 | NA | 0.7 | 4 | >3 | 1,020,000 | Secant | Top-down | Multi-prop | 38 | 41 | Finno and Bryson (2002) |
| 177 | Kings Inn St., Dublin, Ireland | Dense gravel | 188 | 5.7 | 3.9 | 0.68 | 5.7 | >3 | 381,700 | Secant | NA | Single-prop | 6 | NA | Brangan (2007), Long et al. (2012b) |
| 178 | Croydon, UK | London clay | 85 | 11.4 | 2 | 0.18 | 5 | >3 | 500,000 | Secant | NA | Multi-prop | 20 | NA | Brooks and Spence (1993), Long (2001) |
| 179 | Holborn, UK | London clay | 120 | 11 | 3.5 | 0.32 | 8.65 | >3 | 1,169,950 | Secant | NA | Single-prop | 12 | NA | Ward (1992), Long (2001) |
| 180 | Minster Court, UK | London clay | 120 | 9 | 5 | 0.56 | 7.3 | >3 | 1,280,000 | Diaphragm | NA | Berm and support | 17 | NA | Tse and Nicholson (1992), Long (2001) |

| No. | Site | Soil | | | | | | | | | | | | Reference |
|---|---|---|---|---|---|---|---|---|---|---|---|---|---|---|
| 181 | Britannic Hse, UK | London clay | 120 | 14 | 1 | 0.07 | 4.67 | >3 | 1,280,000 | Diaphragm | NA | Berm and support | 60 | 34 | Cole and Burland (1972), Long (2001) |
| 182 | Chelsea, UK | London clay | 100 | 13 | 4 | 0.31 | 4.33 | >3 | 312,500 | Diaphragm | NA | Berm and support | 27 | NA | Corbett et al. (1978), Long (2001) |
| 183 | Walthamstow, UK | London clay | 140 | 7.9 | 1.4 | 0.18 | 7.4 | >3 | 8,437,500 | Secant | NA | Multi-prop | 18 | 20 | Watson and Carder (1994), Long (2001) |
| 184 | Barbican, UK | London clay | 180 | 16 | 2 | 0.13 | 4 | >3 | 8,437,500 | Diaphragm | NA | Multi-prop | 10 | NA | Stevens et al. (1977), Long (2001) |
| 185 | Charing Cross, UK | London clay | 180 | 11 | 5 | 0.45 | 2.75 | >3 | 1,280,000 | Diaphragm | NA | Multi-prop | 35 | 20 | Wood and Perrin (1984), Long (2001) |
| 186 | John Lewis KUT, UK | London clay | 140 | 12 | 2.5 | 0.21 | 3.8 | >3 | 1,280,000 | Diaphragm | NA | Berm and support | 20 | NA | Long (2001) |
| 187 | Victoria Emb, UK | London clay | 140 | 18 | 6 | 0.33 | 6 | >3 | 2,717,000 | Secant | NA | Berm and support | 34 | 28 | St. John et al. (1992), Long (2001) |
| 188 | London, UK | London clay | 140 | 8 | 2 | 0.25 | 5.2 | >3 | 312,500 | Diaphragm | NA | Single-prop | 3 | NA | Fernie and Suckling (1996), Long (2001) |
| 189 | Guildhall, UK | London clay | 140 | 6.5 | 2 | 0.31 | 3 | >3 | 312,500 | Diaphragm | NA | Single-prop | 9 | NA | Fernie and Suckling (1996), Long (2001) |
| 190 | Vauxhall, UK | London clay | 140 | 14.5 | 2 | 0.14 | 3.63 | >3 | 540,000 | Diaphragm | NA | Multi-prop | 22 | NA | Fernie and Suckling (1996), Long (2001) |
| 191 | Chingford, UK | London clay | 140 | 8 | 2 | 0.25 | 9.2 | >3 | 2,291,750 | Secant | NA | Single-prop | 21 | NA | Fernie and Suckling (1996), Long (2001) |
| 192 | Mark Lane, UK | London clay | 140 | 7 | 2 | 0.29 | 3.5 | >3 | 250,000 | Contiguous | NA | Multi-prop | 8 | NA | Fernie and Suckling (1996), Long (2001) |
| 193 | Mark Lane, UK | London clay | 140 | 5.5 | 2 | 0.36 | 6.3 | >3 | 250,000 | Contiguous | NA | Single-prop | 14 | NA | Fernie and Suckling (1996), Long (2001) |
| 194 | JLE III, UK | London clay | 140 | 8 | 2 | 0.25 | 8.2 | >3 | 540,000 | Diaphragm | NA | Single-prop | 10 | NA | Fernie and Suckling (1996), Long (2001) |
| 195 | Malden Way, UK | London clay | 80 | 7.5 | 2 | 0.27 | 5.5 | >3 | 2,544,700 | Contiguous | NA | Single-prop | 36 | NA | Symons and Carder (1991), Long (2001) |
| 196 | Bermondsey, UK | W1R beds | 300 | 19.5 | 0 | 0 | 6 | >3 | 2,500,000 | Diaphragm | NA | Multi-prop | 13 | NA | Dawson et al. (1996), Long (2001) |

continued

| No. | Location | Soil at dredge level | $S_t$/ kPa | $H$/m | $h$/m | $h/H$ | $s$ | $FoS$ | $EI$/ $(N \cdot m^2)$ | Wall type | Construction method | Support type | $\delta_h$/ mm | $\delta_v$/ mm | Reference |
|---|---|---|---|---|---|---|---|---|---|---|---|---|---|---|---|
| 197 | Canada Water, UK | W1R beds | 300 | 17 | 7 | 0.41 | 5.7 | >3 | 805,158 | Secant | NA | Multi-prop | 15 | NA | Powrie and Batten (1997), Long (2001) |
| 198 | Humber Bridge, UK | Kimmer clay | 250 | 24.5 | NA | 0 | 4.92 | >3 | 1,280,000 | Diaphragm | NA | Multi-prop | 21 | NA | Busbridge (1974), Long (2001) |
| 199 | Cambridge, UK | Gault clay | 120 | 10 | 2 | 0.2 | 3.3 | >3 | 540,000 | Diaphragm | NA | Multi-prop | 13 | 10 | Ng and Lings (1995), Long (2001) |
| 200 | Channel Tunnel, UK | Gault clay | 140 | 6.5 | 0 | 0 | 4 | >3 | 73,500 | Sheet | NA | Multi-prop | 45 | NA | Young and Ho (1994), Long (2001) |
| 201 | Lyon, France | Stiff clay | 80 | 8 | 0 | 0 | 3.2 | >3 | 540,000 | Diaphragm | NA | Multi-prop | 14 | NA | Kastner and Ferrand (1992), Long (2001) |
| 202 | Dublin-Jervis, Ireland | Glacial till | 501 | 9.7 | 3 | 0.31 | 8.5 | >3 | 1,254,800 | Secant | NA | Single-prop | 3 | 0 | Long (1997, 2001) |
| 203 | Dublin-Clarend, Ireland | Glacial till | 501 | 6.2 | 1 | 0.16 | 5 | >3 | 3,895 | Soldier | NA | Single-prop | 7 | 0 | Long (1997, 2001) |
| 204 | Dublin-M&S, Irealnd | Glacial till | 501 | 7.2 | 3 | 0.42 | 6 | >3 | 58,500 | Sheet | NA | Single-prop | 5 | 2 | Long (1997, 2001) |
| 205 | MBTA, Boston, USA | Stiff clay | 140 | 15.2 | 0 | 0 | 3.36 | >3 | 1,908,880 | Diaphragm | NA | Multi-prop | 25 | 13 | Becker and Haley (1990), Long (2001) |
| 206 | Oakland, USA | Stiff clay | 140 | 19 | NA | 0 | 3.36 | >3 | 63,500 | Sheet | NA | Multi-prop | 25 | 13 | Peck (1969a), Long (2001) |
| 207 | Houston, USA | Stiff clay | 140 | 18.3 | 3 | 0.16 | 6.1 | >3 | 704,900 | NA | NA | Multi-prop | 18 | NA | Peck (1969a), Long (2001) |
| 208 | Seattle, USA | Stiff clay | 300 | 23.8 | 0 | 0 | 2.64 | >3 | 63,500 | Sheet | NA | Multi-prop | 114 | NA | Peck (1969a), Long (2001) |

| No. | Location | Soil | | | | | | | | Wall type | | Support | | | Reference |
|---|---|---|---|---|---|---|---|---|---|---|---|---|---|---|---|
| 209 | West. Station, Seattle, USA | Stiff clay | 140 | 15.2 | 0 | 0 | 3.8 | >3 | 1,780,000 | Contiguous | NA | Multi-prop | 5 | 5 | Borst et al. (1990), Long (2001) |
| 210 | Pion. Square, Seattle, USA | Stiff clay | 140 | 21.9 | 0 | 0 | 5.5 | >3 | 1,126,000 | Secant | NA | Multi-prop | 14 | 5 | Borst et al. (1990), Long (2001) |
| 211 | Washington, USA | Hard clay | 300 | 17 | 4 | 0.24 | 11 | >3 | 1,535,850 | Secant | NA | Multi-prop | 26 | NA | O'Rourke (1992), Long (2001) |
| 212 | Washington, USA | Hard clay | 300 | 25 | 4 | 0.16 | 6.25 | >3 | 63,500 | Sheet | NA | Multi-prop | 28 | NA | O'Rourke (1992), Long (2001) |
| 213 | Washington, USA | Stiff clay | 140 | 15 | 4.5 | 0.3 | 7.5 | >3 | 20,000 | Contiguous | NA | Multi-prop | 14 | 7 | Eisenstein and Medeiroz (1983), Long (2001) |
| 214 | Houston-Exxon, USA | Stiff clay | 130 | 16.2 | 0 | 0 | 6.5 | >3 | 7,425 | Soldier | NA | Multi-prop | 9 | NA | Ulrich (1989a), Long (2001) |
| 215 | Houston-1Shell, USA | Stiff clay | 120 | 18 | 3 | 0.17 | 6 | >3 | 841,550 | Contiguous | NA | Multi-prop | 25 | NA | Ulrich (1989a), Long (2001) |
| 216 | Houston-Coker, USA | Stiff clay | 375 | 17.1 | 0 | 0 | 3.5 | >3 | 103,600 | Sheet | NA | Multi-prop | 28 | NA | Ulrich (1989a), Long (2001) |
| 217 | Tiong Bahru, Singapore | Stiff clay | 300 | 15.3 | 1.5 | 0.1 | 4.6 | >3 | 24,900 | Soldier | NA | Multi-prop | 25 | NA | Leonard et al. (1987), Long (2001) |
| 218 | Singapore CE II, Singapore | Stiff clay | 175 | 9 | 1 | 0.11 | 4 | >3 | 26,250 | Soldier | NA | Multi-prop | 30 | 10 | Wong et al. (1997), Long (2001) |
| 219 | Singapore CE II, Singapore | Stiff clay | 175 | 9 | 1 | 0.11 | 4 | >3 | 26,250 | Soldier | NA | Multi-prop | 30 | NA | Wong et al. (1997), Long (2001) |
| 220 | Singapore CE II, Singapore | Stiff clay | 175 | 9 | 1 | 0.11 | 4 | >3 | 26,250 | Soldier | NA | Multi-prop | 35 | NA | Wong et al. (1997), Long (2001) |
| 221 | Singapore CE II, Singapore | Stiff clay | 175 | 10.2 | 1 | 0.1 | 4 | >3 | 26,250 | Soldier | NA | Multi-prop | 10 | 5 | Wong et al. (1997), Long (2001) |
| 222 | Singapore CE II, Singapore | Stiff clay | 175 | 10.5 | 1 | 0.1 | 4 | >3 | 26,250 | Soldier | NA | Multi-prop | 20 | 9 | Wong et al. (1997), Long (2001) |
| 223 | Singapore CE II, Singapore | Stiff clay | 175 | 11.5 | 1 | 0.09 | 4 | >3 | 26,250 | Soldier | NA | Multi-prop | 17 | 10 | Wong et al. (1997), Long (2001) |
| 224 | Singapore CE II, Singapore | Stiff clay | 175 | 11.5 | 1 | 0.09 | 4 | >3 | 26,250 | Soldier | NA | Multi-prop | 10 | 15 | Wong et al. (1997), Long (2001) |

| No. | Location | Soil at dredge level | $S_u$/kPa | $H$/m | $h$/m | $h/H$ | $s$ | $FoS$ | $EI$/$(N \cdot m^2)$ | Wall type | Construction method | Support type | $\delta_h$/mm | $\delta_v$/mm | Reference |
|---|---|---|---|---|---|---|---|---|---|---|---|---|---|---|---|
| 225 | Singapore CE II, Singapore | Stiff clay | 175 | 10 | 1 | 0.1 | 4 | >3 | 73,500 | Sheet | NA | Multi-prop | 8 | 7 | Wong et al. (1997), Long (2001) |
| 226 | Singapore CE II, Singapore | Stiff clay | 175 | 10 | 1 | 0.1 | 4 | >3 | 73,500 | Sheet | NA | Multi-prop | 9 | 12 | Wong et al. (1997), Long (2001) |
| 227 | Singapore CE II, Singapore | Stiff clay | 175 | 10 | 1 | 0.1 | 4 | >3 | 73,500 | Sheet | NA | Multi-prop | 12 | 15 | Wong et al. (1997), Long (2001) |
| 228 | Singapore CE II, Singapore | Stiff clay | 175 | 10 | 1 | 0.1 | 4 | >3 | 73,500 | Sheet | NA | Multi-prop | 37 | 20 | Wong et al. (1997), Long (2001) |
| 229 | Singapore CE II, Singapore | Stiff clay | 175 | 11 | 1 | 0.09 | 4 | >3 | 73,500 | Sheet | NA | Multi-prop | 20 | 15 | Wong et al. (1997), Long (2001) |
| 230 | Singapore CE II, Singapore | Stiff clay | 175 | 11.5 | 1 | 0.09 | 4 | >3 | 73,500 | Sheet | NA | Multi-prop | 23 | 10 | Wong et al. (1997), Long (2001) |
| 231 | Singapore CE II, Singapore | Stiff clay | 175 | 11.5 | 1 | 0.09 | 4 | >3 | 73,500 | Sheet | NA | Multi-prop | 42 | 12 | Wong et al. (1997), Long (2001) |
| 232 | Singapore CE II, Singapore | Stiff clay | 175 | 12 | 1 | 0.08 | 4 | >3 | 73,500 | Sheet | NA | Multi-prop | 15 | 25 | Wong et al. (1997), Long (2001) |
| 233 | Singapore CE II, Singapore | Stiff clay | 175 | 17 | 1 | 0.06 | 4 | >3 | 73,500 | Sheet | NA | Multi-prop | 17 | 15 | Wong et al. (1997), Long (2001) |
| 234 | Singapore CE II, Singapore | Stiff clay | 175 | 17 | 1 | 0.06 | 4 | >3 | 73,500 | Sheet | NA | Multi-prop | 22 | 27 | Wong et al. (1997), Long (2001) |
| 235 | Singapore CE II, Singapore | Stiff clay | 175 | 11 | 1 | 0.09 | 4 | >3 | 100,000 | Contiguous | NA | Multi-prop | 7 | 12 | Wong et al. (1997), Long (2001) |
| 236 | Singapore CE II, Singapore | Stiff clay | 175 | 11 | 1 | 0.09 | 4 | >3 | 100,000 | Contiguous | NA | Multi-prop | 7 | 10 | Wong et al. (1997), Long (2001) |

| | Location | Soil type | | | | | | | | Wall type | | Support | | | Reference |
|---|---|---|---|---|---|---|---|---|---|---|---|---|---|---|---|
| 237 | Singapore CE II, Singapore | Stiff clay | 175 | 11.5 | 1 | 0.09 | 4 | >3 | 100,000 | Contiguous | NA | Multi-prop | 7 | 10 | Wong et al. (1997), Long (2001) |
| 238 | Singapore CE II, Singapore | Stiff clay | 175 | 12 | 1 | 0.08 | 4 | >3 | 100,000 | Contiguous | NA | Multi-prop | 20 | NA | Wong et al. (1997), Long (2001) |
| 239 | Singapore CE II, Singapore | Stiff clay | 175 | 12 | 1 | 0.08 | 4 | >3 | 100,000 | Contiguous | NA | Multi-prop | 20 | NA | Wong et al. (1997), Long (2001) |
| 240 | Singapore CE II, Singapore | Stiff clay | 175 | 12.5 | 1 | 0.08 | 4 | >3 | 100,000 | Contiguous | NA | Multi-prop | 15 | 18 | Wong et al. (1997), Long (2001) |
| 241 | Singapore CE II, Singapore | Stiff clay | 175 | 13.5 | 1 | 0.07 | 4 | >3 | 100,000 | Contiguous | NA | Multi-prop | 7 | 15 | Wong et al. (1997), Long (2001) |
| 242 | Singapore CE II, Singapore | Stiff clay | 175 | 13.5 | 1 | 0.07 | 4 | >3 | 100,000 | Contiguous | NA | Multi-prop | 8 | 20 | Wong et al. (1997), Long (2001) |
| 243 | Singapore CE II, Singapore | Stiff clay | 175 | 13.5 | 1 | 0.07 | 4 | >3 | 100,000 | Contiguous | NA | Multi-prop | 17 | NA | Wong et al. (1997), Long (2001) |
| 244 | Singapore CE II, Singapore | Stiff clay | 175 | 20 | 1 | 0.05 | 4 | >3 | 100,000 | Contiguous | NA | Multi-prop | 15 | NA | Wong et al. (1997), Long (2001) |
| 245 | Singapore CE II, Singapore | Stiff clay | 175 | 21.5 | 1 | 0.05 | 4 | >3 | 100,000 | Contiguous | NA | Multi-prop | 25 | NA | Wong et al. (1997), Long (2001) |
| 246 | Singapore CE II, Singapore | Stiff clay | 175 | 13.5 | 1 | 0.07 | 4 | >3 | 1,280,000 | Diaphragm | NA | Multi-prop | 15 | 15 | Wong et al. (1997), Long (2001) |
| 247 | Singapore CE II, Singapore | Stiff clay | 175 | 14.5 | 1 | 0.07 | 4 | >3 | 1,280,000 | Diaphragm | NA | Multi-prop | 27 | 23 | Wong et al. (1997), Long (2001) |
| 248 | Waterloo, UK | Med gravel | 75 | 5.78 | 2 | 0.35 | 5.8 | >3 | 540,000 | Diaphragm | NA | Berm and support | 20 | NA | Li et al. (1992), Long (2001) |
| 249 | Eastbourne, UK | Gravel/sand | 163 | 11 | 0 | 0.0 | 7 | >3 | 2,500,000 | Diaphragm | NA | Multi-prop | 60 | NA | Fernie et al. (1996), Long (2001) |
| 250 | Buffalo, Canada | Dense sand/gravel | 150 | 11 | NA | 0 | 3.67 | >3 | 63,585 | Sheet | NA | Multi-prop | 63 | NA | Peck (1969a), Long (2001) |
| 251 | Ontario, Canada | VD sand | 250 | 15.2 | NA | 0 | 3.04 | >3 | 42,000 | Sheet | NA | Multi-prop | 230 | NA | Long (2001) |
| 252 | Lyon-P Kleb, France | Sandy gravel | 85 | 6.75 | 3.5 | 0.52 | 6.75 | >3 | 57,700 | Sheet | NA | Single-prop | 7 | NA | Kastner (1982), Long (2001) |

continued

| No. | Location | Soil at dredge level | $S_u$/ kPa | $H$/m | $h$/m | $h/H$ | $s$ | FoS | $EI$/ $(N \cdot m^2)$ | Wall type | Construction method | Support type | $\delta_h$/ mm | $\delta_v$/ mm | Reference |
|---|---|---|---|---|---|---|---|---|---|---|---|---|---|---|---|
| 253 | Lyon-R Ney, France | Sandy gravel | 125 | 9.95 | 6.5 | 0.65 | 3.1 | >3 | 107,000 | Diaphragm | NA | Multi-prop | 14 | NA | Kastner (1982), Long (2001) |
| 254 | Lyon-S. Garn., France | Sandy gravel | 135 | 10.7 | 3 | 0.28 | 5.2 | >3 | 185,000 | Diaphragm | NA | Single-prop | 60 | NA | Kastner (1982), Long (2001) |
| 255 | Karlsruhe, Germany | Sands | 75 | 5 | 0 | 0 | 3.75 | >3 | 2,033 | Sheet | NA | Single-prop | 5 | NA | Josseaume et al. (1997), Long (2001) |
| 256 | Maas, Rotterdam, Netherlands | Silts/sands | 260 | 21 | 0 | 1 | 17 | >3 | 1,717,900 | Sheet | NA | Single-prop | 32 | NA | Bakker and Brinkgrieve (1991), Long (2001) |
| 257 | Lisbon-Carlos, Portugal | Clay/sands | 60 | 13.8 | 5.5 | 0.4 | 2.76 | >3 | 540,000 | Diaphragm | NA | Multi-prop | 43 | NA | Mattos-Fernandes (1985), Long (2001) |
| 258 | Sao Paulo, ES1, Brazil | Residual soil | 40 | 9 | 0 | 1 | 3.6 | >3 | 71,700 | Soldier | NA | Multi-prop | 15 | NA | Massad (1985), Long (2001) |
| 259 | Sao Paulo, ES2, Brazil | Residual soil | 40 | 19 | 0 | 1 | 3.8 | >3 | 71,700 | Soldier | NA | Multi-prop | 18 | NA | Massad (1985), Long (2001) |
| 260 | Argyle Station, Honk Kong, China | Residual soil | 250 | 18.7 | 7 | 1 | 6.2 | >3 | 2,500,000 | Diaphragm | NA | Multi-prop | 29 | 58 | Morton et al. (1980), Long (2001) |
| 261 | Han River, Seoul, Japan | Wth. rock | 175 | 25 | 13.5 | 0.54 | 2.78 | >3 | 60,640 | Secant | NA | Multi-prop | 10 | NA | Choi and Lee (1998), Long (2001) |
| 262 | YMCA, London, UK | London clay | 140 | 16 | 1 | 0.06 | 10 | >3 | 540,000 | Diaphragm | NA | Multi-anchor | 20 | NA | St. John (1975), Long (2001) |
| 263 | Neasden, UK | London clay | 150 | 8 | 1 | 0.13 | 2 | >3 | 540,000 | Diaphragm | NA | Multi-anchor | 52 | 54 | Sills et al. (1977), Long (2001) |
| 264 | Oresund-Sydh, Denmark | Boulder clay | 300 | 10 | 5 | 0.5 | 7 | >3 | 48,125 | Soldier | NA | Single-anchor | 55 | NA | Hess et al. (1997), Long (2001) |

| | | | | | | | | | | | | | |
|---|---|---|---|---|---|---|---|---|---|---|---|---|---|
| 265 | Copenhagen, Denmark | Stiff clays | 175 | 11 | 1 | 0.09 | 5.5 | >3 | 1,000,000 | Contiguous | NA | Single-anchor | 2 | NA | Long (2001) |
| 266 | Lisbon-DD Ave., Portugal | Stiff clay | 175 | 17 | 2.6 | 0.15 | 3.25 | >3 | 19,250 | Soldier | NA | Multi-anchor | 24 | NA | Correia and da Costa Guerra (1997), Long (2001) |
| 267 | Lisbon-Colom, Portugal | Stiff clay | 200 | 14 | 4 | 0.29 | 3.25 | >3 | 14,435 | Soldier | NA | Multi-anchor | 90 | NA | Correia and da Costa Guerra (1997), Long (2001) |
| 268 | Lisbon-Ivens, Portugal | Stiff clay | 140 | 9 | 0 | 0 | 3.25 | >3 | 19,250 | Soldier | NA | Multi-anchor | 4 | NA | Correia and da Costa Guerra (1997), Long (2001) |
| 269 | Colomb., Seattle, USA | Stiff clay | 250 | 37 | 0 | 0 | 1.75 | >3 | 763,400 | Secant | NA | Multi-anchor | 15 | 10 | Grant (1985), O'Rourke and Jones (1990), Long (2001) |
| 270 | University St. Sta., Seattle, USA | Stiff clay | 140 | 18.3 | 0 | 0 | 2.6 | >3 | 1,126,000 | Contiguous | NA | Multi-anchor | 13 | 2 | Borst et al. (1990), Long (2001) |
| 271 | Seattle, USA | Stiff clay | 140 | 23 | 0 | 0 | 1.5 | >3 | 718,690 | Soldier | NA | Multi-anchor | 23 | 5 | Winter (1990), Long (2001) |
| 272 | Houston-Herm, USA | Stiff clay | 75 | 9 | 0 | 0 | 3 | >3 | 203,900 | Contiguous | NA | Multi-anchor | 20 | 5 | Ulrich (1989b), Long (2001) |
| 273 | Houston-Bank, USA | Stiff clay | 130 | 16.8 | 3 | 0.18 | 3.35 | >3 | 203,900 | Contiguous | NA | Multi-anchor | 29 | NA | Ulrich (1989b), Long (2001) |
| 274 | Houston-FCB, USA | Stiff clay | 75 | 9.1 | 0 | 0 | 8 | >3 | 209,650 | Contiguous | NA | Multi-anchor | 20 | NA | Ulrich (1989b), Long (2001) |
| 275 | Houston-Smith, USA | Stiff clay | 150 | 15.5 | 0 | 0 | 3 | >3 | 140,135 | Contiguous | NA | Multi-anchor | 20 | NA | Ulrich (1989b), Long (2001) |
| 276 | Houston-Texas, USA | Stiff clay | 100 | 16 | 3 | 0.19 | 3.2 | >3 | 13,860 | Contiguous | NA | Multi-anchor | 30 | NA | Ulrich (1989b), Long (2001) |
| 277 | Houston-Cullen, USA | Stiff clay | 75 | 8.2 | 0 | 0 | 7 | >3 | 210,470 | Contiguous | NA | Multi-anchor | 20 | NA | Ulrich (1989b), Long (2001) |
| 278 | Houston-321, USA | Stiff clay | 75 | 9.1 | 0 | 0 | 3 | >3 | 315,050 | Contiguous | NA | Multi-anchor | 20 | NA | Ulrich (1989b), Long (2001) |
| 279 | Washington, USA | Stiff clay | 140 | 15.2 | 3 | 0.2 | 3 | >3 | 19,000 | Soldier | NA | Multi-anchor | 51 | NA | Ware et al. (1973), Long (2001) |
| 280 | State Trans., Boston, USA | Stiff clay | 140 | 9.1 | 0 | 0 | 3.36 | >3 | 567,450 | Diaphragm | NA | Multi-anchor | 20 | 31 | Becker and Haley (1990), Long (2001) |

continued

| No. | Location | Soil at dredge level | $S_u$/kPa | $H$/m | $h$/m | $h/H$ | $s$ | FoS | $EI/$ $(N \cdot m^2)$ | Wall type | Construction method | Support type | $\delta_h$/mm | $\delta_v$/mm | Reference |
|---|---|---|---|---|---|---|---|---|---|---|---|---|---|---|---|
| 281 | 60 State St., Boston, USA | Stiff clay | 140 | 9.1 | 0 | 0 | 3.36 | >3 | 567,450 | Diaphragm | NA | Multi-anchor | 19 | 31 | Becker and Haley (1990), Long (2001) |
| 282 | Davis Square, Boston, USA | Stiff clay | 140 | 17.1 | 0 | 0 | 3.36 | >3 | 567,450 | Diaphragm | NA | Multi-anchor | 28 | 43 | Becker and Haley (1990), Long (2001) |
| 283 | 1 Memorial, Boston, USA | Stiff clay | 140 | 8.2 | 0 | 0 | 3.36 | >3 | 567,450 | Diaphragm | NA | Multi-anchor | 25 | 31 | Becker and Haley (1990), Long (2001) |
| 284 | Harvard Square, Boston, USA | Till dense | 300 | 15.7 | 0 | 0 | 5.2 | >3 | 1,908,000 | Diaphragm | NA | Multi-anchor | 10 | NA | Hansmire et al. (1989), Long (2001) |
| 285 | Harvard Square, Boston, USA | Till | 300 | 15.7 | 0 | 0 | 6.2 | >3 | 1,908,001 | Diaphragm | NA | Multi-anchor | 11 | NA | Hansmire et al. (1989), Long (2001) |
| 286 | Boston, USA | Stiff clay | 140 | 18.9 | 2 | 0.11 | 2.7 | >3 | 347,430 | Soldier | NA | Multi-anchor | 89 | NA | Houghton and Dietz (1990), Long (2001) |
| 287 | Salt Lake City, USA | Stiff clay | 140 | 12.5 | 3 | 0.24 | 3.2 | >3 | 127,436 | Soldier | NA | Multi-anchor | 25 | NA | Caliendo et al. (1990), Long (2001) |
| 288 | Singapore CE II, Singapore | Stiff clay | 175 | 12.2 | 1 | 0.08 | 4 | >3 | 26,250 | Soldier | NA | Multi-anchor | 10 | 8 | Wong et al. (1997), Long (2001) |
| 289 | Singapore CE II, Singapore | Stiff clay | 175 | 12.5 | 1 | 0.08 | 4 | >3 | 26,250 | Soldier | NA | Multi-anchor | 18 | 6 | Wong et al. (1997), Long (2001) |
| 290 | Singapore CE II, Singapore | Stiff clay | 175 | 12.5 | 1 | 0.08 | 4 | >3 | 26,250 | Soldier | NA | Multi-anchor | 14 | 12 | Wong et al. (1997), Long (2001) |
| 291 | Singapore CE II, Singapore | Stiff clay | 175 | 16 | 1 | 0.06 | 4 | >3 | 73,500 | Soldier | NA | Multi-anchor | 15 | NA | Wong et al. (1997), Long (2001) |
| 292 | Singapore CE II, Singapore | Stiff clay | 175 | 15 | 1 | 0.07 | 4 | >3 | 100,000 | Soldier | NA | Multi-anchor | 8 | NA | Wong et al. (1997), Long (2001) |

| | | | | | | | | | | | | | | | |
|---|---|---|---|---|---|---|---|---|---|---|---|---|---|---|---|
| 293 | Singapore CE Ⅱ, Singapore | Stiff clay | 175 | 17 | 1 | 0.06 | 4 | >3 | 100,000 | Soldier | NA | Multi-anchor | 35 | NA | Wong et al.（1997）, Long（2001） |
| 294 | Singapore CE Ⅱ, Singapore | Stiff clay | 175 | 18.5 | 1 | 0.05 | 4 | >3 | 100,000 | Soldier | NA | Multi-anchor | 27 | NA | Wong et al.（1997）, Long（2001） |
| 295 | Singapore CE Ⅱ, Singapore | Stiff clay | 175 | 19 | 1 | 0.05 | 4 | >3 | 100,000 | Soldier | NA | Multi-anchor | 23 | NA | Wong et al.（1997）, Long（2001） |
| 296 | A1(M), UK | Sand | 120 | 9.3 | NA | 0 | 10.6 | >3 | 104,785 | Sheet | NA | Single-anchor | 20 | NA | Long（2001） |
| 297 | Hatfield, UK | Gravels | 100 | 9.3 | 3 | 0.32 | 6.8 | >3 | 96,440 | Sheet | NA | Single-anchor | 25 | 27 | Symons et al.（1988）, Long（2001） |
| 298 | Paris-13e, France | Sands | 220 | 17.4 | 0 | 0 | 4.35 | >3 | 540,000 | Diaphragm | NA | Multi-anchor | 20 | NA | Josseaume and Stenne（1979）, Long（2001） |
| 299 | Calais, France | Sands | 360 | 24 | 4.5 | 0.19 | 8 | >3 | 5,881,600 | Diaphragm | NA | Multi-anchor | 58 | NA | Delattre et al.（1995）, Long（2001） |
| 300 | Le Havre, France | Sand/ gravel | 500 | 16.5 | 9 | 0.55 | 10.5 | >3 | 4,320,000 | Diaphragm | NA | Single-anchor | 12 | NA | Maquet（1981）, Long（2001） |
| 301 | Geneva, Le Mail, Switzerland | Sand/ gravel | 190 | 14.8 | 4 | 0.27 | 2.5 | >3 | 2,500,000 | Diaphragm | NA | Multi-anchor | 13 | NA | Monnet et al.（1994）, Long（2001） |
| 302 | Berlin P Platz DB, Germany | Sands | 225 | 18 | 3 | 0.17 | 15 | >3 | 5,184,000 | Diaphragm | NA | Single-anchor | 42 | 10 | Triantafyllidis et al.（1997）, Long（2001） |
| 303 | Berlin-Hofgarten, Germany | Sands | 214 | 17 | 3 | 0.18 | 2.83 | >3 | 1,280,000 | Diaphragm | NA | Multi-anchor | 36 | NA | Nussbaumer（1998）, Long（2001） |
| 304 | Berlin, Germany | Sands | 230 | 18.5 | 3 | 0.16 | 0.75 | >3 | 4,320,000 | Soldier | NA | Multi-anchor | 15 | 5 | Triantafyllidis（1998）, Long（2001） |
| 305 | Berlin, Germany | Sands | 155 | 12.3 | 3 | 0.24 | 3.6 | >3 | 631,000 | Secant | NA | Multi-anchor | 27 | NA | Weibenbach and Gollub（1995）, Long（2001） |
| 306 | SONY, Berlin, Germany | Sand/ gravel | 180 | 14.3 | 5 | 0.35 | 3.58 | >3 | 1,280,000 | Diaphragm | NA | Multi-anchor | 50 | NA | Kudella and Mayer（1998）, Long（2001） |
| 307 | Grauholz, Switzerland | Sands/ silts | 150 | 17 | 0 | 0 | 2.4 | >3 | 483,095 | Contiguous | NA | Multi-anchor | 20 | NA | Steiner and Werder（1991）, Long（2001） |
| 308 | Johannsburg, South Africa | Firm silt | 70 | 18.3 | 0 | 0 | 3.05 | >3 | 2,581 | Soldier | NA | Multi-anchor | 18 | 22 | Day（1990）, Long（2001） |
| 309 | Milwaukee, USA | Sands | 65 | 5 | 0 | 0 | 1.67 | >3 | 60,000 | Deep soil mix | NA | Multi-anchor | 15 | NA | Anderson（1998）, Long（2001） |

·319·

·Appendix □·

continued

| No. | Location | Soil at dredge level | $S_t$/kPa | $H$/m | $h$/m | $h/H$ | $s$ | $FoS$ | $EI$/($N \cdot m^2$) | Wall type | Construction method | Support type | $\delta_h$/mm | $\delta_v$/mm | Reference |
|---|---|---|---|---|---|---|---|---|---|---|---|---|---|---|---|
| 310 | Norwich, UK | Chalk | 75 | 18 | 1.2 | 0.07 | 2.57 | >3 | 919,700 | Contiguous | NA | Multi-anchor | 9 | NA | Grose and Toone (1992), Long (2001) |
| 311 | Dartford, UK | Chalk | 150 | 9 | 0 | 0 | 2 | >3 | 1,280,000 | Diaphragm | NA | Multi-anchor | 15 | NA | Wood et al. (1988), Long (2001) |
| 312 | Bell Common, UK | London clay | 130 | 9 | 4 | 0.44 | 8.9 | >3 | 2,330,250 | Secant | Top-down | Undifferentiated single-support | 25 | 25 | Tedd et al. (1984), Long (2001) |
| 313 | New Palace Yard, UK | London clay | 150 | 18.5 | 2 | 0.11 | 3.08 | >3 | 2,500,000 | Diaphragm | Top-down | Undifferentiated multi-support | 30 | 20 | Burland and Hancock (1977), Long (2001) |
| 314 | British Library, UK | London clay | 200 | 24.4 | 3 | 0.12 | 5 | >3 | 2,571,750 | Secant | Top-down | Undifferentiated multi-support | 30 | 30 | Simpson (1992), Long (2001) |
| 315 | Nat Gal Ext, UK | London clay | 140 | 10 | 4.2 | 0.42 | 7 | >3 | 618,000 | Secant | Top-down | Undifferentiated single-support | 10 | 2 | Long (2001) |
| 316 | Aldersgate, UK | London clay | 250 | 23 | 8 | 0.35 | 3.3 | >3 | 2,500,000 | Diaphragm | Top-down | Undifferentiated multi-support | 33 | 18 | Fernie et al. (1991), Long (2001) |
| 317 | Limehouse, UK | W&R beds | 300 | 16 | 4 | 0.25 | 4 | >3 | 4,320,000 | Diaphragm | Top-down | Undifferentiated multi-support | 5 | NA | Stevenson and De Moor (1994), De Moor and Stevenson (1996), Long (2001) |
| 318 | PO Square Boston, UK | Till | 140 | 23.4 | 4 | 0.17 | 3.3 | >3 | 1,822,500 | Diaphragm | Top-down | Undifferentiated multi-support | 52 | 45 | Whitle et al. (1993), Long (2001) |
| 319 | HK&S Bank, Honk Kong, china | Decom. granite | 200 | 16 | 5 | 0.31 | 4 | >3 | 2,500,000 | Diaphragm | Top-down | Undifferentiated multi-support | 48 | 25 | Humpheson et al. (1986), Long (2001) |
| 320 | Rowes Whr, Boston, USA | Stiff clay | 140 | 16.8 | 5 | 0.3 | 3.36 | >3 | 1,106,125 | Diaphragm | Top-down | Undifferentiated multi-support | 19 | NA | Becker and Haley (1990), Long (2001) |
| 321 | 75 State St., Boston, USA | Stiff clay | 140 | 19.8 | 3 | 0.15 | 3.36 | >3 | 1,106,125 | Diaphragm | Top-down | Undifferentiated multi-support | 51 | 102 | Becker and Haley (1990), Long (2001) |

| | | | | | | | | | | | | | | | |
|---|---|---|---|---|---|---|---|---|---|---|---|---|---|---|---|
| 322 | 125 Sum. St., Boston, USA | Stiff clay | 140 | 18.3 | 0 | 0 | 3.36 | >3 | 1,106,125 | Diaphragm | Top-down | Undifferentiated multi-support | 15 | 10 | Becker and Haley (1990), Long (2001) |
| 323 | Ashford International Station, UK | Atherfield clay | 250 | 20 | 2.8 | 0.14 | 10 | >3 | 800,000 | Contiguous | Top-down | Multi-prop | 25 | 25 | Richards et al. (2007) |
| 324 | Chicago, USA | Glacial Clays | 100 | 12.2 | 4.9 | 0.4 | 4 | >3 | 118,600 | Secant | Top-down | Multi-anchor | 30 | NA | Finno and Calvello (2005) |
| 325 | Chicago, USA | Glacial clays | 100 | 12.2 | 4.9 | 0.4 | 4 | >3 | 118,600 | Secant | Top-down | Multi-anchor | 25 | NA | Finno and Calvello (2005) |
| 326 | Seoul Metro Line 7, Japan | Weathered soil | 250 | 24.8 | 13.1 | 0.53 | 1.7 | >3 | 340,000 | Soldier | Bottom-up | Multi-anchor | 16 | 20 | Lee (2009) |
| 327 | Athens Metro, Greece | Hard sandy clay | 400 | 26 | 6 | 0.23 | 3 | >3 | 285,000 | Contiguous | Top-down | Multi-anchor | 31 | 13 | Kulesza et al. (2008) |
| 328 | Cincon Station, Istanbul, Turkey | Hard-dense silty clay | 213 | 32.5 | 0 | 0 | 5.4 | >3 | 4,320,000 | Diaphragm | Top-down | Multi-prop | 11 | 18 | Sevencan et al. (2013) |
| 329 | Harrods, London, UK | London clay | 130 | 23 | 0 | 0 | 3 | >3 | 680,000 | Diaphragm | Top-down | Undifferentiated multi-support | 27 | 30 | Fong et al. (2014) |
| 330 | Heuston, Dublin, Ireland | Dublin boulder clay | 100 | 14 | 0 | 0 | 7 | >3 | 340,000 | Secant | NA | Single-anchor | 80 | NA | O'Leary et al. (2016) |
| 331 | TOOBA, Tehran, Iran | | 250 | 16.5 | 8 | 0.48 | 4.1 | >3 | 340,000 | Contiguous | Top-down | Multi-prop | 20 | NA | Haeri et al. (2013) |
| 332 | Los Angeles Metro, USA | Sand with silt | 324 | 18 | 0 | 0 | 4 | >3 | 340,000 | Soldier | Top-down | Multi-prop | 51 | NA | Roth et al. (2008) |
| 333 | London, UK | London clay | 130 | 20 | 1 | 0.05 | 5 | >3 | 340,000 | Secant | Top-down | Multi-prop | 10 | NA | Chapman and Green (2004) |
| 334 | Swainswick Bypass, UK | Lias clay | 150 | 7.9 | 4 | 0.51 | 4 | >3 | 680,000 | Diaphragm | Bottom-up | Single-prop | 16 | NA | Gourvenec et al. (2002) |
| 335 | St. Johns Road, Dublin, Ireland | Medium dense gravel | 250 | 8 | 1 | 0.13 | 7.5 | >3 | 245,450 | Contiguous | NA | Single-prop | 23 | NA | Brangan (2007), Long et al. (2012a) |

continued

| No. | Location | Soil at dredge level | $S_u$/kPa | $H$/m | $h$/m | $h/H$ | $s$ | FoS | $EI$/($N \cdot m^2$) | Wall type | Construction method | Support type | $\delta_H$/mm | $\delta_v$/mm | Reference |
|---|---|---|---|---|---|---|---|---|---|---|---|---|---|---|---|
| 336 | South Lotts Road, Dublin, Ireland | Dense gravel | 250 | 6.5 | 3.5 | 0.54 | 6 | >3 | 381,700 | Secant | NA | Single-prop | 19 | NA | Brangan (2007), Long et al. (2012a) |
| 337 | DPT Southern C&C, Dublin, Ireland | Dense sand and gravel | 263 | 22 | 4.5 | 0.2 | 9 | >3 | 4,320,000 | Diaphragm | NA | Multi-prop | 4 | NA | Curtis and Doran (2003), Long et al. (2012a) |
| 338 | DPT Southern C&C, Dublin, Ireland | Loose sand and gravel | 253 | 18 | 8.2 | 0.46 | 7.5 | >3 | 4,320,000 | Diaphragm | NA | Multi-prop | 9 | NA | Curtis and Doran (2003), Long et al. (2012a) |
| 339 | Spencer Dock, Dublin, Ireland | Gravel | 120 | 9.6 | 7.5 | 0.78 | 9.6 | >3 | 644,126 | Secant | NA | Single-anchor | 3 | NA | Looby and Long (2007), Long et al. (2012a) |
| 340 | Smithfield, Dublin, Ireland | Dense gravel | 260 | 11 | 3 | 0.27 | 7 | >3 | 1,130,300 | Diaphragm | NA | Single-anchor | 4 | NA | Long and Dhouwt (2003), Long et al. (2012b) |
| 341 | Railway St., Dublin, Ireland | Dens gravel | 220 | 7.2 | 4 | 0.56 | 8 | >3 | 381,700 | Secant | NA | Single-prop | 11 | NA | Brangan (2007), Long et al. (2012b) |
| 342 | Clancy Barracks, Ireland | Dense gravel | 210 | 7.2 | 4 | 0.56 | 7.2 | >3 | 644,126 | Secant | NA | Single-anchor | 12 | NA | Looby and Long (2010), Long et al. (2012b) |
| 343 | Leinster House, Dublin, Ireland | Dublin boulder clay | 300 | 6 | 0 | 0 | 6 | >3 | 381,700 | Secant | NA | Single-prop | 3 | NA | Brangan (2007), Long et al. (2012c) |
| 344 | Hilton House, Dublin, Ireland | Gravel/Dublin boulder clay | 300 | 6.3 | 4 | 0.63 | 6.3 | >3 | 1,254,800 | Secant | NA | Single-prop | 1 | NA | Long (2002), Long et al. (2012c) |

| | Location | Soil | | | | | | | | | | | | Reference |
|---|---|---|---|---|---|---|---|---|---|---|---|---|---|---|
| 345 | Grand Canal, Dublin, Ireland | Dublin boulder clay | 300 | 4 | 2 | 0.5 | 3.5 | >3 | 347,000 | Secant | NA | Single-anchor | 5 | NA | Long et al. (2012c) |
| 346 | TCD Library-Lecky, Ireland | Dublin boulder clay | 300 | 7.2 | 0 | 0 | 2.5 | >3 | 347,000 | secant | NA | Multi-anchor | 3 | NA | Brangan (2007), Long et al. (2012c) |
| 347 | Pearce St., Dublin, Ireland | Upper Brown boulder clay | 170 | 12 | 3 | 0.25 | 8 | >3 | 372,750 | Secant | NA | Multi-prop | 9 | NA | Long et al. (2012c) |
| 348 | Cherrywood, Dublin, Ireland | Dublin boulder clay | 300 | 8.5 | 1 | 0.12 | 8.5 | >3 | 878,355 | Contiguous | NA | Single-anchor | 8 | NA | Long et al. (2012c) |
| 349 | Mater Hospial, Dublin, Ireland | Dublin boulder clay | 300 | 8 | 0.6 | 0.08 | 7.75 | >3 | 690,135 | Secant | NA | Single-anchor | 6 | NA | Long et al. (2012c) |
| 350 | Rathmines, Dublin, Ireland | Dublin boulder clay | 300 | 7 | 1 | 0.14 | 6 | >3 | 265,000 | Contiguous | NA | Single-anchor | 3 | NA | Long et al. (2012c) |
| 351 | Burlington Rd., Dublin, Ireland | Dublin boulder clay | 300 | 10 | 2 | 0.2 | 2 | >3 | 690,135 | Secant | NA | Single-anchor | 2 | NA | Long et al. (2012c) |
| 352 | Terrenure, Dublin, Ireland | Dublin boulder clay | 300 | 3.2 | 0.3 | 0.09 | 3.2 | >3 | 25,700 | Sheet | NA | Single-prop | 2 | NA | Kearon (2009), Long et al. (2012c) |
| 353 | TCD Library-Nassau, Ireland | Dublin boulder clay | 300 | 8.6 | 0 | 0 | 3.5 | >3 | 347,000 | Secant | NA | Multi-anchor | 7 | NA | Brangan (2007), Long et al. (2012c) |
| 354 | Ely Place, Dublin, Ireland | Dublin boulder clay | 300 | 3.5 | 0 | 0 | 3.5 | >3 | 201,300 | Secant | NA | Single-prop | 3 | NA | Brangan (2007), Long et al. (2012c) |
| 355 | Harcourt St., Dublin, Ireland | Dublin boulder clay | 300 | 4.5 | 0 | 0 | 3 | >3 | 381,700 | Secant | NA | Single-prop | 5 | NA | Brangan (2007), Long et al. (2012c) |

continued

| No. | Location | Soil at dredge level | $S_u$/kPa | $H$/m | $h$/m | $h/H$ | $s$ | FoS | $EI$/(N·m²) | Wall type | Construction method | Support type | $\delta_h$/mm | $\delta_v$/mm | Reference |
|---|---|---|---|---|---|---|---|---|---|---|---|---|---|---|---|
| 356 | Balbriggan, Dublin, Ireland | Gravel and clay | 140 | 5.7 | 2.5 | 0.44 | 5.7 | >3 | 644,126 | Secant | NA | Single-anchor | 9 | NA | Long et al. (2012c) |
| 357 | Westgate, Dublin, Ireland | Dublin boulder clay | 300 | 14 | 2 | 0.14 | 14 | >3 | 690,135 | Secant | NA | Single-anchor | 7 | NA | Looby and Long (2007), Long et al. (2012c) |
| 358 | Tallaght Centre, Dublin, Ireland | Dublin boulder clay | 300 | 11 | 1.5 | 0.14 | 6 | >3 | 293,620 | Contiguous | NA | Single-anchor | 18 | NA | Long et al. (2012c) |
| 359 | TCD-Biosciences, Ireland | Dublin boulder clay | 300 | 10.5 | 4 | 0.38 | 6.7 | >3 | 644,125 | Secant | NA | Single-anchor | 11 | NA | Long et al. (2012c) |
| 360 | TCD-Biosciences, Ireland | Dublin boulder clay | 300 | 10.5 | 4 | 0.38 | 3 | >3 | 644,125 | Secant | NA | Multi-anchor | 9 | NA | Long et al. (2012c) |
| 361 | TCD-Biosciences, Ireland | Dublin boulder clay | 300 | 10.5 | 4 | 0.38 | 3.6 | >3 | 644,125 | Secant | NA | Multi-anchor | 7 | 7 | Long et al. (2012c) |
| 362 | Monte Vetro, Ireland | Dublin boulder clay | 300 | 13.7 | 2 | 0.15 | 12 | >3 | 568,400 | Secant | NA | Single-prop | 3 | NA | Long et al. (2012c) |
| 363 | Monte Vetro, Ireland | Dublin boulder clay | 300 | 17.4 | 8 | 0.46 | 2.9 | >3 | 568,400 | Secant | NA | Multi-anchor | 4 | NA | Long et al. (2012c) |

| No. | Project | Soil | | | | | | | | Wall type | Construction | Support type | | | Reference |
|---|---|---|---|---|---|---|---|---|---|---|---|---|---|---|---|
| 364 | Barrow St., Dublin, Ireland | Dublin boulder clay | 300 | 11.4 | 4 | 0.35 | 2.9 | >3 | 568,400 | Secant | NA | Multi-anchor | 1 | NA | Long et al. (2012c) |
| 365 | TCD Sports Centre, Ireland | Dublin boulder clay | 300 | 6.9 | 1 | 0.14 | 5.6 | >3 | 644,126 | Secant | NA | Single-anchor | 11 | NA | Long et al. (2012c) |
| 366 | TCD-Cramn Building, Ireland | Dublin boulder clay | 300 | 6.9 | 1 | 0.14 | 2.3 | >3 | 644,126 | Secant | NA | Multi-anchor | 7 | NA | Long et al. (2012c) |
| 367 | DPT Northern C&C, Ireland | Dublin boulder clay | 300 | 12 | 1 | 0.08 | 10.2 | >3 | 4,320,000 | Diaphragm | NA | Single-prop | 5 | NA | Curtis and Doran (2003), Long et al. (2012c) |
| 368 | DPT Northern C&C, Ireland | Dublin boulder clay | 300 | 17 | 1 | 0.06 | 5.5 | >3 | 4,320,000 | Diaphragm | NA | Mp | 6 | NA | Curtis and Doran (2003), Long et al. (2012c) |
| 369 | DPT-Shaft WA2, Ireland | Dublin boulder clay | 300 | 25 | 1 | 0.04 | 12 | >3 | 8,437,500 | Diaphragm | NA | Single-prop | 12 | NA | Cabarkapa et al. (2003), Long et al. (2012c) |
| 370 | Mespil Road, Dublin, Ireland | Dublin boulder clay | 300 | 11 | 2 | 0.18 | 8 | >3 | 1,288,250 | Secant | NA | Single-prop | 14 | 6 | Long et al. (2012c) |
| 371 | CTRL-Chart Road, UK | Atherfield clay | 80 | 8 | 2.5 | 0.31 | 6 | >3 | 340,000 | Contiguous | Bottom-up | Undifferentiated single-support | 43 | NA | Roscoe and Twine (2010) |
| 372 | CTRL-Advance, UK | Atherfield clay | 95 | 9.8 | 0 | 0 | 6 | >3 | 340,000 | Contiguous | Top-down | Undifferentiated single-support | 18 | NA | Roscoe and Twine (2010) |
| 373 | CTRL-Chart Road, UK | Atherfield clay | 135 | 14 | 1.4 | 0.1 | 6 | >3 | 340,000 | Contiguous | Bottom-up | Multi-prop | 40 | NA | Roscoe and Twine (2010) |
| 374 | CTRL-Maidstone Railway, UK | Atherfield clay | 135 | 14 | 0 | 0 | 6 | >3 | 340,000 | Contiguous | Top-down | Single-prop | 31 | NA | Roscoe and Twine (2010) |
| 375 | CTRL-Greensands Way, UK | Atherfield clay | 100 | 10 | 0 | 0 | 6 | >3 | 340,000 | Contiguous | Bottom-up | Undifferentiated single-support | 20 | NA | Roscoe and Twine (2010) |
| 376 | CTRL-Gasworks Lane, UK | Atherfield clay | 100 | 10.5 | 5.8 | 0.55 | 4.5 | >3 | 340,000 | Contiguous | Top-down | Multi-prop | 17 | NA | Roscoe and Twine (2010) |

continued

| No. | Location | Soil at dredge level | $S_u/$ kPa | $H/m$ | $h/m$ | $h/H$ | $s$ | FoS | $EI/$ $(N \cdot m^2)$ | Wall type | Construction method | Support type | $\delta_h/$ mm | $\delta_v/$ mm | Reference |
|---|---|---|---|---|---|---|---|---|---|---|---|---|---|---|---|
| 377 | CTRL-Cattlemarket, UK | Atherfield clay | 95 | 9.6 | 4 | 0.42 | 4.5 | >3 | 340,000 | Contiguous | Bottom-up | Multi-prop | 19 | NA | Roscoe and Twine (2010) |
| 378 | Shanghai, China | Silty clay | 25 | 4.5 | 5 | 1 | 3.5 | >3 | 129,000 | Secant | NA | Single-prop | 10 | NA | Xiao et al. (2003) |
| 379 | Shanghai, China | Silty clay | 35 | 7 | 5 | 1 | 3 | >3 | 129,000 | Secant | NA | Multi-prop | 16 | NA | Xiao et al. (2003) |
| 380 | Savoy , Limerick , Ireland | Boulder clay | 300 | 6.4 | 2 | 0.31 | 6.4 | >3 | 191,000 | Secant | NA | Single-anchor | 7 | NA | Long et al. (2013) |
| 381 | Savoy , Limerick , Ireland | Boulder clay | 300 | 5.8 | 2 | 0.34 | 6.4 | >3 | 191,000 | Secant | NA | Single-anchor | 2 | NA | Long et al. (2013) |
| 382 | Main St. , Cavan, Ireland | Boulder clay | 300 | 9 | 2 | 0.22 | 4.5 | >3 | 8,000 | Contiguous | NA | Multi-anchor | 45 | NA | Long et al. (2013) |
| 383 | Dundalk Cellar , Ireland | Boulder clay | 300 | 8.7 | 2 | 0.23 | 8.7 | >3 | 176,000 | Secant | NA | Single-anchor | 6 | NA | Long et al. (2013) |
| 384 | Dundalk Cellar , Ireland | Boulder clay | 300 | 8.1 | 2 | 0.25 | 8.1 | >3 | 176,000 | Secant | NA | Single-anchor | 7 | NA | Long et al. (2013) |
| 385 | Midleton, Ireland | Boulder clay | 300 | 6 | 2 | 0.33 | 6 | >3 | 206,000 | Secant | NA | Single-anchor | 6 | NA | Long et al. (2013) |
| 386 | Kilkenny , Ireland | Boulder clay | 300 | 7.7 | 2 | 0.26 | 7.7 | >3 | 254,000 | Contiguous | NA | Single-prop | 5 | NA | Long et al. (2013) |
| 387 | Kilkenny , Ireland | Boulder clay | 300 | 6.9 | 2 | 0.29 | 6.9 | >3 | 354,000 | Contiguous | NA | Single-prop | 2 | NA | Long et al. (2013) |
| 388 | Portlaoise SC, Ireland | Boulder clay | 300 | 12 | 2 | 0.17 | 6 | >3 | 247,000 | Secant | NA | Multi-anchor | 17 | NA | Long et al. (2013) |
| 389 | Portlaoise SC, Ireland | Boulder clay | 300 | 7 | 2 | 0.29 | 7 | >3 | 247,000 | Secant | NA | Single-anchor | 4 | NA | Long et al. (2013) |

# Reference

Aas, G. (1984). "Stability problems in a deep excavation in clay." *Proceedings of the 1st International Conference on Case Histories in Geotechnical Engineering*, Missouri University of Science and Technology, 315-323.

Addenbrooke, T. I., Potts, D. M., and Dabee, B. (2000). "Dis-placement flexibility number for multipropped retaining wall design." *Journal of Geotechnical and Geoenvironmental Engineering*, 126(8), 718-w726.

Alpan, I. (1970). "The geotechnical properties of soils." *Earth-Science Reviews*, 6, 5-49.

Anderson, T. C. (2014). "Anchored deep soil mixed cutoff/retaining walls for Lake Parkway Project in Milwaukee, Wisconsin." *Design and construction of earth retaining systems*, 1-13.

Atkinson, J. H., and Sallfors, G. (1991). "Experimental determination of soil properties." *10th ECSMFE*, 915-956.

Auvinet, G., and Organista, M. P. R. (1998). "Deep excavations in Mexico City soft clays." *Big digs around the world*, 211-229.

Baggett, J. K., and Buttling, S. (1977). "Design and in-situ performance of a sheet pile wall." *Proceedings of the 9th International Conference on Soil Mechanics and Foundation Engineering*, Tokyo, Japan, 3-8.

Bakker, K. J., and Brinkgrieve, R. B. J. (1991). "Deformation analysis of a sheet pile wall using a 2D model." *Proceedings of the 10th European Conference on Soil Mechanics and Foundation Engineering*, Rotterdam, the Netherlands, 655-658.

Balasubramaniam, A. S., Bergado, D. T., Chai, J. C., and Sutabur, T. (1994). "Deformation analysis of deep excavations in Bangkok." *Proceedings of the 13th International Conference on Soil Mechanics and Foundation Engineering*, Rotterdam, Netherlands, 909-915.

Becker, J. M., and Haley, M. X. (1990). "Up/down construction—Decision making and performance." *Design and performance of earth retaining structures*, 170-189.

Becker, P., Gebreselassie, B., and Kempfert, H. (2008). "Back analysis of a deep excavation in soft lacustrine clays." *Proceedings of the 6th International Conference on Case Histories in Geotechnical Engineering*, Missouri University of Science and Technology.

Benz, T. (2006). "Small-strain stiffness of Soils and its numerical consequences." PhD thesis, Universitat Stuttgart.

Bilotta, E., Ramondini, M., and Viggiani, C. (2004). "Monitoring an excavation in an urban area." *Proceedings of the 5th International Conference on Case Histories in Geotechnical Engineering*,

Missouri University of Science and Technology.

Bjerrum, L., and Eide, O. (1956). "Stability of strutted excavations in clay." *Géotechnique*, 6 (1), 32-47.

Blackburn, J. T., and Finno, R. J. (2007). "Three-dimensional responses observed in an internally braced excavation in soft clay." *Journal of Geotechnical and Geoenvironmental Engineering* 133(11), 1364-1373.

Bolton, M. D. (1986). "The strength and dilatancy of sands." *Geotechnique*, 36(1), 65-78.

Boone, S. J., and Westland, J. (2004). "Failure of an Excavation Support System." *International Conference on Case Histories in Geotechnical Engineering*, Missouri University of Science and Technology.

Borst, A. J., Conley, T. L., Russell, D. P., and Boirum, R. N. (1990). "Subway design and construction for the downtown Seattle transit project." *Design and performance of earth retaining structures*, 510-524.

Bowden, A., and Lees, A. (2010). "Design and performance of a basement in Norwich, UK." *Proceedings of the Institution of Civil Engineers—Geotechnical Engineering*, 55-64.

Bowles, J. E. (1988). *Foundation Analysis and Design*, 4th Ed., New York, USA, McGraw-Hill Book Company.

Brangan, C. (2007). "Retaining walls in Dublin boulder clay." PhD thesis, University College Dublin., Dublin, Ireland.

Brinkgreve, R. B. J. (1994). "Geomaterial Models and Numerical Analysis of Softening." PhD thesis, Delft University of Technology.

Brinkgreve, R. B. J., and Bakker, H. L. (1991). "Non-linear finite element analysis of safety factors." *Proc 7th int conf comp methods and advances in geomech*, Cairns, Australia, 117-122.

Brinkgreve, R. B. J., Engin, E., and Engin, H. (2010). "Validation of empirical formulas to derive model parameters for sands." *7th European Conference Numerical Methods in Geotechnical Engineering*, Trondheim, Norway, 137-174.

Broms, B. B., and Stille, H. (1976). "Failure of anchored sheet pile walls." *Journal of the Geotechnical Engineering Division*, ASCE, 102(3), 235-251.

Broms, B. B., Wong, I. H., and Wong, K. S. (1986). "Experience with finite-element analysis of braced excavations in Singapore." *Proceedings of the 2nd International Symposium on Numerical Methods in Geomechanics*, Rotterdam, the Netherlands, 309-324.

Brooks, N. J., and Spenc, e. J. (1993). "Design and recorded performance of a secant retaining wall in Croydon." Thomas Telford, London, 205-215.

Brown, P. T., and Booker, J. R. (1985). "Finite element analysis of excavation." *Computers and Geotechnics*, 1, 207-220.

Bryson, L. S., and Zapata-Medina, D. G. (2012). "Method for Estimating System Stiffness for Excavation Support Walls." *Journal of Geotechnical and Geoenvironmental Engineering*, 138

(9), 1104-1115.

Burland, J. B. (1965). "The yielding and dilation of clay." *Geotechnique*, 15, 211-214.

Burland, J. B. (1967). "Deformation of Soft Clay." PhD thesis, Cambridge University.

Burland, J. B. (2001). "Assessment methods used in design in building response to tunneling: case studies from construction of the jubilee Line Extension." SP200, CIRIA and Thomas Telfors, London.

Burland, J. B., and Hancock, R. J. R. (1977). "Underground car park at the House of Commons London: Geotechnical aspects." *Structural Engineer*, 55(2), 87-105.

Burland, J. B., and Wroth, C. P. (1974). "Settlement of buildings and associated damage." *Proceedings of Conference on Settlement of Structures*, London, England, 611-654.

Bushridge, W. M. (1974). "Foundations in clay for massive suspension bridge." *Ground Engineering*, 7(3), 24-26.

Busetti, F. (2003). *Simulated Annealing Overview*, JP Morgan, Italy.

C.W.W., N., and Lings, M. L. (1995). "Effects of modeling soil nonlinearity and wall installation on back-analysis of deep excavation in stiff clay." *Journal of Geotechnical Engineering*, 121(10), 687-695.

Cabarkapa, Z., Milligan, G. W. E., Menkiti, C. O., Murphy, J., and Potts, D. M. (2003). "Design and performance of a large diameter shaft in Dublin boulder clay." *British Geotechnical Association International Conference on Foundations: Innovations, Observations, Design and Practice*, London: Thomas Telford, 175-185.

Cacoilo, D., Tamaro, G., and Edinger, P. (1998). "Design and performance of a tied back sheet pile wall in soft clay." *Design and construction of earth retaining systems*, 14-25.

Caliendo, J. A., Anderson, L. R., and Gordon, W. J. (1981). "A filed study of a tie back excavation with FEA." *Design and construction of earth retaining systems*, 747-763.

Calvello, M., and Finno, R. J. (2004). "Selecting parameters to optimize in model calibration by inverse analysis." *Computers and Geotechnics*, 31(5), 411-425.

Carder, D. R. (1995). "Ground movements caused by different embedded retaining wall construction techniques." *Wokingham, UK: Transport Research Laboratory*.

Caspe, M. S. (1966). "Surface settlement adjacent to braced open cuts." *Journal of the Soil Mechanics and Foundations Division*, 92, 51-59.

CGS (2006). *Canadian foundafation engineering manual, fourth edition*, Canadian Geotechnical Society, Bitech Publishers Ltd, Canada.

Chang, J. D., and Wong, K. S. (1996). "Apparent pressure diagram for braced excavations in soft clay with diaphragm wall." *In Proceedings of International Symposium on Geotechnical Aspects of Underground Construction in Soft Ground*, Rotterdam, the Netherlands, 87-92.

Chapman, T., and Green, G. (2004). "Observational Method looks set to cut city building costs." *Proceedings of the Institution of Civil Engineers—Civil Engineering*, 157(3), 125-133.

Chen, S. L., Ho, C. T., and Gui, M. W. (2014). "Diaphragm wall displacement due to creep of soft clay." *Proceedings of the Institution of Civil Engineers—Geotechnical Engineering*, 167 (3), 297-310.

Chen, W., Wu, Y., and Peng, Z. (2006). "Monitoring of Rescue of a Foundation Pit and Technical Analysis of Its Collapse in GuangZhou." *Chinese Journal of Underground Space and Engineering*, 2(6), 1034-1039.

Chen, W. F., and Saleeb, A. F. (1982). *Constitutive Equations for Engineering Materials*, Vol. I *Elasticity and Modeling*, John Wiley & Sons, New York.

Chew, S. H., Yong, K. Y., and Lim, A. Y. K. (1997). "Three-dimensional finite element analysis of a strutted excavation." *Computer Methods and Advances in Geomechanics* 3, 1915-1920.

Choi, C. S., and Lee, I. K. (1998). "Geotechnical aspects of the Seoul subway." *Big digs around the world*, 44-62.

Clough, G. W., and O'Rourke, T. D. (1990). "Construction induced movements of in situ wall." *Geotechnical Special Publication*, (25), 439-470.

Clough, G. W., and Reed, M. W. (1984). "Measured behaviour of braced wall in very soft clay." *Journal of Geotechnical Engineering*, 110(1), 1-19.

Clough, G. W., Smith, E. M., and Sweeney, B. P. (1989). "Movement control of excavation support systems by iterative design." *In Proceedings of Foundation Engineering Congress on Current Principles and Practices*, New York, 869-884.

Cole, K. W., and Burland, J. B. (1972). "Observation of retaining wall movements associated with a large excavation." *Proceedings of the 5th European Conference on Soil Mechanics and Foundation Engineering*, Rotterdam, the Netherlands, 445-458.

Committee of Inquiry (COI). (2005). "Report of the Committee of Inquiry into the Incident at the MRT Circle Line Worksite that led to the Collapse of the Nicoll Highway on 20 April 2004." Ministry of Manpower, Singapore.

Cooley, R. L., and Naff, R. L. (1990). "Regression modeling of groundwater flow." *US Geological Survey techniques in water resources investigations*, USGS, Book 3, Chapter B4, 71-72.

Corbett, B. O., Davies, R. V., and Langford, A. D. (1972). "A load bearing diaphragm wall at Kensington and Chelsea town hall, London." *Diaphragm walls and anchorages. London: Institution of Civil Engineering*, 57-62.

Cording, E. (2010). "Assessment of excavation-induced building damage" *Earth retention conference 3*, Washington DC, USA, 101-120.

Correia, A. G., and da Costa Guerra, N. M. (1997). "Performance of three Berlin type retaining walls." *Proceedings of the 14th International Conference on Soil Mechanics and Foundation Engineering*, 1297-1300.

Coulomb, C. A. (1776). *Essaisurune application des regles de maximis et minimis a quelques problemes de statique*, relatifs a l'architecture, Mem. Roy. des Sciences, Paris, 3, 38.

Cowland, J. W., and Thorley, C. B. B. (1985). "Ground and building settlement associated with adjacent slurry trench excavation." *Ground Movements and Structures*, London, UK, 723-738.

Curtis, P., and Doran, J. (2003). "Retaining wall behaviour at the Dublin Port Tunnel." University College Dublin, Dublin, Ireland.

D., S. J. H. (1975). *Field and theoretical studies of behavior of ground around deep excavations in London Clay.*, University of Cambridge, Cambridge.

D.G., L., and S.M., W. (2007). "Three dimensional analysis of deep excavation in Taipei 101 construction project." *Journal of Geoengineering*, 2(1), 29-42.

Davies, R. V., Fok, P., Norrish, A., and Poh, S. T. (2006). "The Nicoll Highway collapse: field measurements and observations." *Proceeding of the International Conference on Deep Excavations*, Singapore, 15.

Davies, R. V., and Walsh, N. M. (1983). "Excavations in Singapore marine clays " *International Seminar on Construction Problems in Soft Soils*, Singapore: Nanyang Technological Institute.

Dawson, M. P., Douglas, A. R., Linney, L. F., Friedman, M., Abraham, R., and Bermondsey, J. (1996). "Ground movements caused by different embedded retaining wall construction techniques." *Proceedings of the International Symposium Geotechnical Aspects of Underground Construction in Soft Ground*, Rotterdam, the Netherlands, 99-104.

Day, P. (1990). "Design and construction of a deep basement in soft residual soils." *Design and performance of earth retaining structures*, 734-746.

De Moor, E. K., and Stevenson, M. C. (1996). "Evaluation of the performance of a multi propped diaphragm wall during construction." *Proceedings of the International Symposium Geotechnical Aspects of Underground Construction in Soft Ground*, Rotterdam, the Netherlands, 111-116.

Delattre, L., Mespoulhe, L., and Faroux, J. P. (1995). "Monitoring of a cast in place concrete quay wall at the port of Calais." *Proceedings of the 4th International Symposium on Field Measurements in Geomechanics*, Bergamo, Italy, 73-80.

Dennis, J. E., and Schnabel, R. B. (1983). *Numerical method for unconstrained optimization and nonlinear equations*, Prentice-Hall Inc, Pennsylvania, PA.

Desai, C. S., and Nagaraj, B. K. (1988). "Modeling of cyclic normal and shear behavior of interfaces." *Journal of the Engineering Mechanics Division*, 114(7), 1198-1216.

Diakoumi, M., Powrie, W., and Haigh, S. K. (2013). "Mobilisable strength design for flexible embedded retaining walls." *Geotechnique*, 63, 1080-1082.

Ditlevsen, O. (1981). *Uncertainity modeling: With applications to multidimensional civil engineering systems*, McGraw-Hill, New York.

Dong, Y., Harve, H., Houlsby, G., and Xu, Z. (2013). "3D FEM modelling of a deep excavation case considering small-strain stiffness of soil and thermal shrinkage of concrete." *Proceedings of the 7th International Conference on Case Histories in Geotechnical Engineering.*

Duncan, J. M., and Chang, C. Y. (1970). "Nonlinear analysis of stress and strain in soil." *Journal of the Soil Mechnics and Foundation Division*, 96, 1629-1653.

Eisenstein, Z., and Medeiroz, L. V. (1983). "A deep retaining structure in till and sand, Part II: Performance and analysis." *Canadian Geotechnical Journal*, 20(1), 131-140.

EL Shafle, M. (2008). "Effect of building stiffness on excavation-induced displacements." PhD thesis, University of Cambridge, UK.

Endicott, J. (2013). "Lessons learned from the collapse of the Nicoll highway in Singapore April 2004." *IABSE Symposium Report*, 1-6.

Fernie, R., Kingston, P., St. John, H. D., Higgins, K. G., and Potts, D. M. (1996). "Case history of a deep stepped box excavation in soft ground at sea front." *Proceedings of the International Symposium on Geotechnical Aspects of Underground Construction in Soft Ground*, Rotterdam, the Netherlands, 123-129.

Fernie, R., St. John, H. D., and Potts, D. M. (1991). "Design and performance of a 24 m deep basement in London Clay resisting the effects of long term rise in ground water." *Proceedings of the 10th European Conference on Soil Mechanics and Foundation Engineering*, Rotterdam, the Netherlands, 699-702.

Fernie, R., and Suckling, T. (1996). "Simplified approach for estimating lateral movement of embedded walls in UK ground." *Proceedings of the International Symposium on Geotechnical Aspects of Underground Construction in Soft Ground*, Rotterdam, the Netherlands, 131-136.

Finno, R. J., Atmatzidis, D. K., and Perkins, S. B. (1989). "Observed performance of a deep excavation in clay." *Journal of Geotechnical Engineering*, 115, 1045-1064.

Finno, R. J., Blackburn, J. T., and Roboski, J. F. (2007). "Three-dimensional effects for supported excavations in clay." *Journal of Geotechnical and Geoenvironmental Engineering*, 133(1), 30-36.

Finno, R. J., and Bryson, L. S. (2002). "Response of building adjacent to stiff excavation support system in soft clay." *Journal of Performance of Constructed Facilities* 16(1), 10-20.

Finno, R. J., Bryson, S., and Calvello, M. (2002). "Performance of a stiff support system in soft clay." *Journal of Geotechnical and Geoenvironmental Engineering*, 128(8), 660-671.

Finno, R. J., and Roboski, J. F. (2005). "Three-dimensional responses of a tied-back excavation through clay." *Journal of Geotechnical and Geoenvironmental Engineering*, 131(3), 273-282.

Finstad, J. A. (1991). "Royal Christianina Hotel—Basement with permanent sheet pile wall, up and down method." *Proceedings of the 4th International Conference on Piling and Deep Foundations*, 387-392.

Fitzpatrick, B. G. (1991). "Bayesian-analysis in inverse problems." *Inverse Problems*, 5(1), 675-702.

Flaate, K. S. (1996). "Stresses and movement in connection with braced cuts in sand and clay." PhD thesis, University of Illinois, Urbana, USA.

Fletcher, R. (1980). *Practical method of optimization: Volume I. Unconstrained optimization*, John Wiley and Sons.

Fong, F. H., Standing, J. R., and Bourne-Webb, P. J. (2014). "Building response to adjacent deep basement construction." *Proceedings of the Institution of Civil Engineers—Geotechnical Engineering*, 167(2), 130-143.

Forchheimer, P. (1930). *Hydraulik*, Teubner, Berlin, Germany.

Franzius, J. N., Potts, D. M., Addenbrooke, T. I., and Burland, J. B. (2004). "The influence of building weight on tunnelling-induced ground and building deformation." *Soils and foundations*, 45(4), 25-38.

Franzius, J. N., Potts, D. M., and Burland, J. B. (2006). "The response of surface structures to tunnel construction." *Proceedings of the ICE-geotechnical engineering*, 159(1), 9-17.

Fraser, R. A. (1992). "Mobilisation of stresses in deep excavations: The use of earth pressure cells at Sheung Wan Crossover." *Predictive Soil Mechanics, Proceedings of the Wroth Memorial Symposium*, London: Thomas Telford, 279-292.

Gaba, A., Hardy, S., and Doughty, L. (2017). *Guidance on embedded retaining wall design*, CIRIA, Griffin Court, 15 Long lane, London, EC1A 9PN.

Ghaboussi, J., Pecknold, D. A., Zhang, M., and Haj-Ali, R. (1998). "Autoprogressive training of neural network constitutive models." *International Journal for Numerical Methods in Engineering*, 42(1), 105-126.

Ghaboussi, J., and Sidarta, D. E. (1998). "New nested adaptive neural networks (NANN) for constitutive modeling." *Computers and Geotechnics*, 22(1), 29-52.

Gill, S. A., and Lucas, R. G. (1900). "Ground movement adjacent to braced cuts." *Design and performance of earth retaining structures*, 471-488.

Goh, A. T. C., Zhang, F., Zhang, W., and Chew, O. Y. (2017). "Assessment of strut forces for braced excavation in clays from numerical analysis and field measurements." *Computers and Geotechnics*, 86, 141-149.

Goh, A. T. C., Zhang, F., and Zhang, W. G. (2017). "A simple estimation model for 3D braced excavation wall deflection." *Computers and Geotechnics*, 83, 106-113.

Gong, X. N., and Zhang, X. C. (2012). "Excavation collapse of Hangzhou subway station in soft clay and numerical investigation based on orthogonal experiment method." *Journal of Zhejiang Univ Sci A (Appl Phys Eng)*, 13(10), 760-767.

Goodman, R. E., Taylor, R. L., and Brekke, T. L. (1968). "A model for mechanics of jointed rock." *Journal of the Soil Mechanics and Foundation Division*, 94(3), 637-658.

Gourvenec, S., Powrie, W., and De Moor, E. K. (2002). "Three-dimensional effects in the construction of a long retaining wall." *Proceedings of the Institution of Civil Engineers—Geotechnical Engineering*, 155(3), 163-173.

Graf, E. D. (1992). *Compaction grouting, Grouting/Soil Improvement and Geosynthetics*, ASCE Special Publication.

Grant, W. P. (1985). "Performance of Columbia Centre shoring wall." *Proceedings of the 11th International Conference on Soil Mechanics and Foundation Engineering*, Rotterdam, the Netherlands, 2079-2082.

Gronin, H., Depauw, J., and N'Guyen, M. (1991). "Singapore's expressway: Temporary retaining structures." *Proceedings of the 10th European Conference on Soil Mechanics and Foundation Engineering*, 805-808.

Grose, W. J., and Toone, B. H. (1992). "The selection, design and performance of a multi propped contiguous pile retaining wall in Norwich." *Proceedings of the International Conference on Retaining Structures*, London: Thomas Telford, 24-36.

Haeri, M., Sasar, M., and Afshari, K. (2013). "Deep excavation on 3 sides of a 21 story building: Accounts of a successful deep excavation project." *Proceedings of the 7th International Conference on Case Histories in Geotechnical Engineering*, Missouri University of Science and Technology.

Hansmire, W. H., Russell, H. A., Rawnsley, R. P., and Abbott, E. L. (1989). "Field performance of a structural slurry wall." *Journal of Geotechnical Engineering*, 115(2), 141-156.

Hardin, B. O., and Black, W. L. (1969). "Closure to vibration modulus of normally consolidated clays." *Journal of the Soil Mechanics and Foundations Division*, 95(SM6), 1531-1537.

Hardin, B. O., and Drnevich, V. P. (1972). "Shear modulus and damping in soils: Design equations and curves." *Journal of the Soil Mechanics and Foundations Division*, 98(SM7), 667-692.

Harry, H., Emilio, G.-T., and Raul, F. (2019). "Meta-analysis of ground movements associated with deep excavations using a data mining approach." *Journal of Rock Mechanics and Geotechnical Engineering*, 11(2), 409-416.

Hashash, Y. M. A., Osouli, A., and Marulanda, C. (2008). "Central artery/tunnel project excavation induced ground deformations." *Journal of Geotechnical and Geoenvironmental Engineering*, 134(9), 1399-1406.

Hasofer, A. M., and Lind, N. (1974). "An exact and invariant first-order reliability format." *Journal of Engineering Mechnics*, 100(1), 111-121.

Hastings, W. K. (1970). "Monte Carlo sampling methods using Markov chains and their applications." *Biometricka*, 57(1), 97-109.

Hausman, M. R. (1990). *Engineering Principles of Ground Modification*, McGraw-Hill Publishing Company, New York, USA.

Hess, U. H., Pedersen, S. K., and Ladefoged, L. M. (1997). "Fixed link across the Oresund: Large excavation for tunnel ramp in freight railway yard." *Proceedings of the 14th International Conference on Soil Mechanics and Foundation Engineering*, Rotterdam, the Netherlands, 1369-1372.

Hill, M. C. (1998). "Methods and guidelines for effective model calibration." U.S. Geological Survey Water-Resources investigations, Rep. 98-4005, USGS, 90.

Himmelblau, D. M. (1968). *Applied nonlinear programming*, McGraw-Hill Inc., New York, USA.

Holland, J. (1992). *Adaptation in Natural and Artificial Systems*, MIT Press, Cambridge, MA.

Hong, Y., Ng, C., Liu, G., and Liu, T. (2015). "Three-dimensional deformation behaviour of a multi-propped excavation at a 'greenfield' site at Shanghai soft clay." *Tunnelling and Underground Space Technology*, 45, 249-259.

Honjo, Y., Wentsung, L., and Guha, S. (2010). "Inverse analysis of an embankment on soft clay by extended Bayesian method." *International Journal for Numerical & Analytical Methods in Geomechanics*, 18(10), 709-734.

Houghton, R. C., and Dietz, D. L. (1990). "Design and performance of a deep excavation support system in Boston." *Design and performance of earth retaining structures*, 795-816.

Hsieh, P. G., and Ou, C. Y. (1998). "Shape of ground surface settlement profiles caused by excavation." *Canadian Geotechnical Journal*, 35(6), 1004-1017.

Huang, W. (1992). "Case histories of underpinning during the construction of the Singapore mass transit system." *Sino-Geotechnics*, 40, 77-90.

Hulme, T. W., Potter, J., and Shirlaw, N. (1989). "Singapore mass rapid transit system: Construction." *Proceedings of the Institution of Civil Engineers—Geotechnical Engineering*, 86 (4), 709-770.

Humpheson, C., Fitzpatrick, A. J., and Anderson, J. M. D. (1986). "The basement and substructure for the new headquarters of the Hong Kong and Shanghai Banking Corporation, Hong Kong." *Proceedings of the Institution of Civil Engineers*, 80(4), 831-858.

Hwang, R. N., and Moh, Z. C. (2007). "Performance of floor slabs in excavations using top-down method of construction and correction of inclinometer readings." *Journal of Geoengineering*, 2(3), 111-121.

Janbu, N. (1963). "Soil compressibility as determined by oedometer and triaxial tests." *Proc. ECSMFE*, Wiesbaden, 19-25.

Janbu, N. (1985). "Soil models in offshore engineering (25th rankine lecture)." *Geotechnique*, 35, 241-280.

Jia, J., Wang, J. H., Liu, C. P., Zhang, L. L., and Xie, X. L. (2008). "Behaviour of an excavation adjacent to a historical building and metro tunnels in Shanghai soft clays." *Proceedings of the 6th International Conference on Case Histories in Geotechnical Engineering*, Missouri University of Science and Technology.

Josseaume, H., Dellattre, L., and Mespoulhe, L. (1997). "Interprétation par le calcul aux coefficients de réaction du comportement du rideau de palplanches expérimental de Hochstetten." *Revue Française de Géotechnique*, 79, 59-72.

Josseaume, H., and Stenne, R. (1979). "Étude expérimentale d'une paroi moulée ancrée par quatre nappes de tirants." *Revue Française de Géotechnique*, 8, 51-64.

JSA (1988). "Guidelines of Design and Construction of Deep Excavations." Japanese Society of Architecture.

Juang, C. H., Luo, Z., Atamturktur, S., and Huang, H. (2013). "Bayesian updating of soil parameters for braced excavations using field observations." *Journal of Geotechnical and Geoenvironmental Engineering*, 139(3), 395-406.

Karaboga, D. (2005). "An idea based on honey bee swarm for numerical optimization." *Technical report-tr 06*, Erciyes university, engineering faculty, computer engineering department.

Karlsrud, K. (1981). "Performance and design of slurry walls in soft clay." *Proceedings of ASCE Spring Convention*, New York, 81-147.

Karlsrud, K. (1983). "Performance and design of slurry walls in soft clay." *Norwegian Geotechnical Institute Publication*, 149, 1-9.

Karlsrud, K. (1986). "Performance monitoring in deep supported excavations in soft clay." *Proceedings of the 4th International Geotechnical Seminar on Field Instrumentation and in Situ Measurements*, Singapore: Nanyang Technological Institute, 187-202.

Kastner, R. (1982). "Excavations profondes en site urbain." PhD thesis, Laboratoire de Geotechnique-Insa Lyon, Lyon, France.

Kastner, R., and Ferrand, J. (1992). "Performance of a cast in situ retaining wall in a sandy silt." *Proceedings of the International Conference on Retaining Structures*, London: Thomas Telford, 237-247.

Kearon, B. (2009). "The use of LiDAR to monitor retaining wall movements and the engineering geology of the Dublin docklands." Civil Engineering Department, University College Dublin, Dublin, Ireland.

Kennedy, J. (2011). "Particle swarm optimization." *Encyclopedia of Machine Learning*, USA.

Koiter, W. T. (1960). "General theorems for elastic-plastic solids." *Progress in Solid Mechanics*, North-holland, Amsterdam, 165-221.

Kondner, R. L. (1963). "A hyperbolic stress strain formulation for sands." *2nd Pan. Am. ICOSFE Brazil*, 289-324.

Konstantakos, D. C., Whittle, A. J., Regalado, C., and Scharner, B. (2004). "Control of ground movements for a multi-level-anchored, diaphragm wall during excavation." *Proceedings of the 5th International Conference on Case Histories in Geotechnical Engineering* New York.

Kozeny, J. (1953). *Hydraulik*, Springer, Verlag.

Kudella, P., and Mayer, P. M. (1998). "Calculation of deformations using hypoplasticity—Demonstrated by the SONY-Centre excavation in Berlin." *Proceedings of International Conference on Soil/Structure Interaction*, Darmstadt, Germany, 151-164.

Kulesza, R., Boussoulas, N., and Marr, W. A. (2008). "Deep excavations in hard sandy clays for stations and shafts of the Athens Metro Stavros Extension." *Proceedings of the 6th*

*International Conference on Case Histories in Geotechnical Engineering*, Missouri University of Science and Technology.

Kung, G. T., Juang, C. H., Hsiao, E. C., and Hashash, Y. M. (2007). "Simplified model for wall deflection and ground-surface settlement caused by braced excavation in Clays." *Journal of Geotechnical and Geoenvironmental Engineering*, 133(6), 731-747.

Kung, G. T. C., Hsiao, E. C. L., Schuster, M., and Juang, C. H. (2007). "A neural network approach to estimating excavation induced wall deflection in soft clays." *Computers and Geotechnics*, 35(5), 385-396.

Ledesma, A., Gens, A., and Alonso, E. E. (1996). "Parameter and variance estimation in geotechnical back analysis using prior information." *Numerical & Analytical Methods in Geomechanics*, 20, 119-141.

Lee, F. H., Yong, K. Y., Quan, K. C. N., and Chee, K. T. (1998). "Effect of corners in strutted excavations: Field monitoring and case histories." *Journal of Geotechnical and Geoenvironmental Engineering*, 124(4), 339-349.

Lee, J. S. (2009). "An application of three-dimensional analysis around a tunnel portal under construction." *Tunnelling and Underground Space Technology*, 24(6), 731-738.

Lee, S. L., Karunaratne, G. P., Lo, K. W., Yong, K. Y., and Choa, V. (1985). "Developments in soft ground engineering in Singapore" *Proceedings of the 11th International Conference on Soil Mechanics and Foundation Engineering*, Rotterdam, the Netherlands, 1661-1666.

Leonard, M., Wong, H., and DeLabrusse, P. (1987). "The design and performance of the temporary works for Somerset and Tiong Bahru stations." *Mass Rapid Transit System: Proceedings of the Singapore Mass Rapid Transit Conference*, Singapore: Mass Rapid Transit Corp, 141-145.

Leonards, G. A. (1975). "Discussion of differential settlement of buildings." *Journal Geotechnical Engineering Division*, 101(7), 700-702.

Levasseur, S., Malécot, Y., Boulon, M., and Flavigny, E. (2008). "Soil parameter identification using a genetic algorithm." *International Journal for Numerical and Analytical Methods in Geomechanics*, 32(2), 189-213.

Li, E. S. F., Nyirenda, Z. M., and Pickles, A. R. (1992). "Design and measured performance of diaphragm walls at Waterloo International Terminal." *Proceedings of the International Conference on Retaining Structures*, London: Thomas Telford, 237-247.

Li, X. S., and Dafalias, Y. F. (2000). "Dilatancy for cohesionless soils." *Geotechnique*, 50(4), 449-460.

Liao, H. J., and Hsieh, P. G. (2002). "Tied-back excavations in alluvial soil of Taipei." *Journal of Geotechnical and Geoenvironmental Engineering*, 128(5), 435-441.

Lim, K. W., Wong, K. S., Orihara, K., and Ng, P. B. (2003). "Comparison of results of excavation analysis using WALLUP, SAGE CRISP, and EXCAV97." *Proceedings of*

*Underground Singapore. Singapore*, Nanyang Technological University, 83-94.

Liu, C. C., Hsieh, H. S., and Huang, C. S. (1997). "A study of the stability analysis for deep excavations in clay." *The 7th Geotechnical Conference*, Taipei, 629-638.

Liu, G., Ng, C., and Wang, Z. (2005). "Observed performance of a deep multistrutted excavation in Shanghai soft clays." *Journal of Geotechnical and Geoenvironmental Engineering* 131(8), 1004-1013.

Long, M. (1997). "Design and construction of deep basements in Dublin, Ireland." *Proceedings of the 14th International Conference on Soil Mechanics and Foundation Engineering*, 1377-1380.

Long, M. (2001). "Database for retaining wall and ground movements due to deep excavations." *Journal of Geotechnical and Geoenvironmental Engineering*, 127(3), 203-224.

Long, M. (2002). "Observations of ground and structure movements during site re-development in Dublin." *Proceedings of the Institution of Civil Engineers—Geotechnical Engineering*, 155 (4), 229-242.

Long, M., Brangan, C., Menkiti, C., Looby, M., and Casey, P. (2012). "Retaining walls in Dublin boulder clay, Ireland. " *Proceedings of the Institution of Civil Engineers—Geotechnical Engineering*, 165(4), 247-266.

Long, M., Daynes, P., Donohue, S., and Looby, M. (2012). "Retaining wall behaviour in Dublin's fluvio-glacial gravel, Ireland." *Proceedings of the Institution of Civil Engineers— Geotechnical Engineering*, 165(50), 289-307.

Long, M., and Dhouwt, B. (2003). "Behaviour of a diaphragm wall in dense gravels." *Proceedings of the 6th International Symposium on Field Measurements in Geomechanics*, Lisse, the Netherlands, 209-214.

Long, M., Menkiti, C., Skipper, J., Brangan, C., and Looby, M. (2012). "Retaining wall behaviour in Dublin's estuarine deposits, Ireland." *Proceedings of the Institution of Civil— Engineers Geotechnical Engineering*, 165(6), 351-365.

Long, M., O'Leary, F., Ryan, M., and Looby, M. (2013). "Deep excavation in Irish glacial deposits." *Proceedings of the 18th International Conference on Soil Mechanics and Geotechnical Engineering. International Society for Soil Mechanics and Geotechnical Engineering* 2039-2042.

Looby, M., and Long, M. (2007). "Deep excavations in Dublin—Recent developments." *Transactions of Engineers Ireland*, 31, 1-15.

Looby, M., and Long, M. (2010). "Behaviour of cantilever retaining walls in Dublin soils." *Proceedings of the DFI/EFFC 11th International Conference: Geotechnical Challenges in Urban Regeneration*, London.

Low, B. K., and Tang, W. H. (1997). "Efficient reliability evaluation using spreadsheet." *Jouranl Engineering Mechanics*, 123(7), 749-752.

Low, B. K., and Tang, W. H. (2007). "Efficient spreadsheet algorithm for first-order reliability method." *Journal of Engineering Mechanics*, 133(12), 1378-1387.

Mair, R. J., and Padfield, C. J. (1984). "Design of retaining walls embedded in stiff clay." *Publication of Construction Industry Research & Information Assoc*, 104, 1-16.

Maldonado, R. (1998). "Big digs in the lacustrine soil of Bogotá, Colombia." *Big digs around the world*, 252-272.

Mana, A. I., and Clough, G. W. (1981). "Prediction of movements for braced cuts in clay." *Journal of the Geotechnical Engineering Division*, 107(6), 759-777.

Maquet, J. F. (1981). "Quai en paroi moulee du port autonome du Havre." *Bulletin de Liaison des Laboratoires des Ponts et Chaussées*, 113, 109-134.

Marquardt, D. W. (1963). "An algorithm for least-squares estimation of nonlinear parameters." *Journal of the Society for Industrial & Applied Mathematics*, 11(2), 431-441.

Massad, M. (1985). "Braced excavations in lateritic and weathered sedimentary soils." *Proceedings of the 11th International Conference on Soil Mechanics and Foundation Engineering*, Rotterdam, the Netherlands, 113-116.

Mattos-Fernandes, M. (1985). "Performance and analysis of a deep excavation in Lisbon." *Proceedings of the 11th International Conference on Soil Mechanics and Foundation Engineering*, Rotterdam, the Netherlands, 127-132.

Metropolis, N., Rosenbluth, A. W., Rosenbluth, M. N., Teller, A. H., and Teller, E. (1953). "Equations of state calculations by fast computing machines." *Journal of chemical physics*, 21 (6), 1087-1092.

Monnet, J., Khlif, J., and Biard, C. (1994). "The diaphragm wall—Le Mail—Experimental and numerical study." *Proceedings of the 13th International Conference on Soil Mechanics and Foundation Engineering*, Rotterdam, the Netherlands: A.A. Balkema, 39-44.

Moormann, C. (2004). "Analysis of wall and ground movements due to deep excavations in soft soil based on a new worldwide database." *Soils and Foundations*, 44, 87-98.

Morton, K., Leonard, M. S. M., and Cater, R. W. (1980). "Building settlements and ground movements associated with the construction of two stations of the modified initial system of the mass transit railway, Hong Kong." *Proceedings of the 2nd International Conference on Ground Movements and Structures*, Wales, UK: University of Cardiff, 788-800.

Mothersille, D., Duzceer, R., Gokalp, A., and Okumusoglu, B. (2015). "Support of 25 m deep excavation using ground anchors in Russia." *Proceedings of the Institution of Civil Engineers—Geotechnical Engineering* 168(4), 281-295.

Mu, L., and Huang, M. (2016). "Small strain based method for predicting three-dimensional soil displacements induced by braced excavation." *Tunnelling and Underground Space Technology*, 2, 12-22.

Muir Wood, D. (1990). *Soil Behaviour and Critical State Soil Mechanics*, Cambridge University Press, London, UK.

Murphy, D. J., Woolworth, R. S., and Clough, G. W. (1975). "Temporary excavation in varved clay." *Journal of the Geotechnical Engineering Division*, 101(3), 279-295.

Nalcakan, M. S., Tekin, M., Tönuk, G., and Ergun, U. (2004). "Behaviour of a watertight anchored retaining wall in soft soil conditions." *Proceedings of the 5th International Conference on Case Histories in Geotechnical Engineering*, Missouri University of Science and Technology.

NAVFAC (1982). "Foundations and Earth Structures, Design Manual 7.2." Department of the Navy, USA.

Ng, C. W. W. (1992). "An Evaluation of Soil-Structure Interaction Associated with a Multi-propped Excavation." *Ph D Thesis University of Bristol*, 13(1), 31-35.

Ng, C. W. W., Hong, Y., Liu, G. B., and Liu, T. (2012). "Ground deformations and soil-structure interaction of a multi-propped excavation in Shanghai soft clays." *Géotechnique*, 62 (10), 907-921.

Ng, C. W. W., and Yan, R. W. M. (1998). "Stress transfer and deformation mechanisms around a diaphragm wall panel." *Journal of Geotechnical and Geoenvironmental Engineering*, 124 (7), 638-648.

Nicholson, D. P. (1987). "The design and performance of the retaining walls at Newton Station." *Mass Rapid Transit System: Proceedings of the Singapore Mass Rapid Transit Conference*, Singapore, Mass Rapid Transit Corp, 47-54.

Nonveiller, E. (1989). *Grouting Theory and Practice*, Elsevier, Amersterdam.

Norwegian Geotechnical Institute (NGI). (1962). "Measurements at a strutted excavation, Oslo Subway, Enerhaugen South, km 1982. " *Technical Report No. 3.*, Oslo, Norway.

Norwegian Geotechnical Institute (NGI). (1962b). "Measurements at a strutted excavation, the new headquarters building for the Norwegian Telecommunications Administration, Oslo. " Technical *Report No. 4.*, Oslo, Norway.

Norwegian Geotechnical Institute (NGI). (1962c). "Measurements at a strutted excavation, Oslo Subway, Grønland 2, km 1692." *Technical Report No. 5.*, Oslo, Norway.

Norwegian Geotechnical Institute (NGI). (1962d). "Measurements at a strutted excavation, Oslo Subway, Vaterland 1, km 1373." *Technical Report No. 6.*, Oslo, Norway.

Norwegian Geotechnical Institute (NGI). (1962e). "Measurements at a strutted excavation, Oslo Subway, Vaterland 2, km 1408." *Technical Report No. 7.*, Oslo, Norway.

Norwegian Geotechnical Institute (NGI). (1962f). "Measurements at a strutted excavation, Oslo Subway, Vaterland 3, km 1450." *Technical Report No. 8.*, Oslo, Norway.

Nussbaumer, M. F. (1998). "Massive big digging in the center of Berlin." *Big digs around the world*, 333-357.

O'Leary, F., Long, M., and Ryan, M. (2016). "The long-term behaviour of retaining walls in Dublin." *Proceedings of the Institution of Civil Engineers—Geotechnical Engineering*, 169(2), 99-109.

O'Rourke, T. D. (1992). "Base stability and ground movement prediction for excavations in soft clay." *Proceedings of the International Conference on Retaining Structures*, London: Thomas TelforD, 657-686.

O'Rourke, T. D., and C.J.P.F., J. (1990). "Overview of earth retention systems: 1970-1990." *Design and performance of earth retaining structures*, 22-51.

O'Rourke, T. D., and McGinn, A. J. (2006). "Lessons learned for ground movements and soil stabilization from the Boston Central Artery." *Journal of Geotechnical and Geoenvironmental Engineering*, 132(8), 966-989.

Onishi, K., and Sugawara, T. (1999). "Behaviour of an earth retaining wall during deep excavation in Shanghi soft ground." *Soils and Foundations*, 39(3), 89-97.

Ou, C. Y. (2006). *Deep excavation: theory and practice*, Taylor and Francis.

Ou, C. Y., Chiou, D. C., and Wu, T. S. (1996). "Three-dimensional finite element analysis of deep excavations." *Journal of Geotechnical Engineering-Asce*, 122(5), 337-345.

Ou, C. Y., Hsieh, P. G., and Chiou, D. C. (1993). "Characteristics of ground surface settlement during excavation." *Canadian Geotechnical Journal*, 30(5), 758-767.

Ou, C. Y., Liao, J. T., and Lin, H. D. (1998). "Performance of diaphragm wall constructed using top-down method." *Journal of Geotechnical and Geoenvironmental Engineering*, 124(9), 798-808.

Ou, C. Y., Lin, Y. L., and Hsieh, P. G. (2006). "Case record of an excavation with cross walls and buttress walls." *Journal of Geoengineering*, 1(2), 79-86.

Ou, C. Y., and Lin, Y. W. (1999). "Application of Cross Wall to Deep Excavations, Geotechnical Research Report No. GT99006." National Taiwan University of Science and Technology, Taipei.

Ou, C. Y., Shiau, B. Y., and Wang, I. W. (2000). "Three dimensional deformantion behavior of the TEC excavation case history." *Canadian Geotechnical Journal*, 37(2), 438-448.

Ou, C. Y., and Wang, I. W. (1997). "Measures for Reducing Wall Deformation in Deep Excavations, Geotechnical Research Report No. GT97006." Taipei, National Taiwan University of Science and Technology.

Ou, C. Y., and Wu, C. H. (1990). "Deformation behavior of excavations in sandy soils due to grouting." *Journal of the Civil and Hydraulic Engineering*, 2(2), 169-182.

Padfield, C. J., and Mair, R. J. (1984). "Design of Retaining Walls Embedded in Stiff Clay, CIRIA Report No. 104." England, 83-84.

Pande, G. N., and Sharma, K. G. (1979). "On the joint/interface elements and associated problems of numerical Ⅲ-conditioning." *International Journal for Numerical and Analytical Methods in Geomechanics*, 3, 301-312.

Peck, R. B. (1969a). "Deep excavation and tunneling in soft ground." *Proceedings of the 7th International Conference on soil Mechanics and Foundation Engineering*, Mexico City, 225-290.

Peck, R. B. (1969b). "Advantages and limitations of the observational method in applied soil mechanics." *Geotechnique*, 19(2), 171-187.

Pei, L. J., Qu, B. N., and Qian, S. G. (2010). "Uniformity of slope instability criteria of strength reduction with FEM." *Rock and Soil Mechanics*, 31(10), 3337-3341.

Poeter, E. P., and Hill, M. C. (1998). *Documentation of UCODE, a computer code for universal inverse modeling*, U.S. Geological Survey Water-Resources Investigations Rep.

Potts, D. M., and Addenbrooke, T. I. (1997). "A structure's influence on tunnelling-induced groundmovements." *Proceedings of the ICE-geotechnical engineering*, 125(2), 117-125.

Powell, M. (1962). "An iterative method for finding stationary value of a function." *The Computer Journal*, 5, 147-151.

Powrie, W., and Batten, M. (1997). *Prop loads in large braced excavations*, Construction Industry Research & Information Association (CIRIA), London.

Powrie, W., and Kantartzi, C. (1996). "Ground response during diaphragm wall installation in clay: centrifuge model tests." *Geotechnique*, 46(4), 725-739.

Preene, M., Roberts, T. O. L., and Powrie, W. (2016). *Groundwater control: design and practice, second edition*, C750, CIRIA, London.

Price, K., and Storn, R. (1997). "Differential evolution: a simple evolution strategy for fast optimization." *Dr Dobb's journal*, 22(4), 18-24.

Puller, M. (2003). *Deep excavation-a practical manual, second edition*, ICE Publishing, London.

Puzrin, A. M., Alonso, E. E., and Pinyol, N. M. (2010). *Geomechanics of Failures*, Springer, Netherlands.

Rahimi, M. M., Karwaj, C., and Deb, P. K. (1993). "Failure of Sewerage Mains Constructed in Sof Estuarine Deposit." *International Conference on Case Histories in Geotechnical Engineering*.

Rampello, S., Tamagini, C., and Calabresi, G. (1992). "Observed and predicted response of a braced excavation in soft to medium clay." *Proceedings of the Wroth Memorial Symposium*, London, Thomas Telford, 544-561.

Rankine, W. M. J. (1857). *On stability on loose earth*, Philosophic Transactions of Royal Society, UK.

Reddy, A. S., and Srinivasan, R. J. (1967). "Bearing capacity of footing on layered clay." *Journal of the Soil Mechanics and Foundations Division*, 93(2), 83-99.

Richards, D. J., Homes, G., and Beadman, D. R. (1999). "Measurement of temporary prop loads at Mayfair park." *Proceedings of the ICE-geotechnical engineering*, 137(3), 165-174.

Richards, D. J., Powrie, W., Roscoe, H., and Clark, J. (2007). "Pore water pressure and horizontal stress changes measured during construction of a contiguous bored pile multi-propped retaining wall in Lower Cretaceous clays." *Géotechnique*, 57(2), 197-205.

Roboski, J. F., and Finno, R. J. (2006). "Distributions of Ground Movements Parallel to Deep Excavations." *Canadian Geotechnical Journal*, 43(1), 43-58.

Rodriguez, J. M., and Flamand, C. L. (1969). "Strut loads recorded in a deep excavation in clay." *Proceedings of the 7th International Conference on Soil Mechanics and Foundation Engineering*, Mexico City, 459-468.

Roscoe, H., and Twine, D. (2010). "Design and performance of retaining walls." *Proceedings of the Institution of Civil Engineers—Geotechnical Engineering*, 163(5), 279-290.

Roth, W., Su, B., Vanbaarsel, J., and Lindquist, E. (2008). "Effect of high in-situ stress on braced excavations." *Proceedings of the 6th International Conference on Case Histories in Geotechnical Engineering*, Missouri University of Science and Technology.

Roti, J. A., and Friis, J. (1985). "Diaphragm wall performance in soft clay excavation." *Proceedings of the 11th International Conference on Soil Mechanics and Foundation Engineering*, Rotterdam, the Netherlands, 2073-2078.

Rowe, P. W. (1952). "Anchored sheet pile wall." *Proceedings Civil Engineering*, 1(1), 27-70.

Rowe, P. W. (1986). "The potentially latent dominance of groundwater in ground engineering, Journal of engineering geology and hydrogeology." *Engineering geology special publication*, 3, 27-42.

Sachdeva, T. D., and Ramakrishnan, C. V. (1981). "A finite element solution for the two-dimensional elastic contact problems with friction." *International Journal for Numerical and Analytical Methods in Engineering*, 1257-1271.

Saleem, M. (2015). "Application of numerical simulation for the analysis and interpretation of pile-anchor system failure." *Geomechanics and Engineering*, 9(6), 689-707.

Santos, J., A., and Correia, A. G. (2001). "Reference threshold shear strain of soil: Its application to obtain a unique strain-dependent shear modulus curve for soil." *Proceedings 15th International Conference on Soil Mechanics and Geotechnical Engineering*, Istanbul, Turkey, 267-270.

Sanzeni A, C. F., Mino, M., and Merline, A. (2013). "Behavior prediction and monitoring of a deep excavation in the historic center of Brescia." *Proceedings of the 7th International Conference on Case Histories in Geotechnical Engineering*, Missouri University of Science and Technology.

Schanz, T. (1998). *Zur Modellierung des Mechanischen Verhaltens von Reibungsmaterialen.*, Habilitation, Stuttgart Universitat.

Sen, K. K., Alostaz, Y., Pellegrino, G., and Hagh, A. (2004). "Support of deep excavation in soft clay: A case history study." *Proceedings of the 5th International Conference on Case Histories in Geotechnical Engineering*, Missouri University of Science and Technology.

Sevencan, O., Ozaydin, K., and Kilic, H. (2013). "Numerical analysis of soil deformations around deep excavations." *Proceedings of the 7th International Conference on Case Histories in Geotechnical Engineering*, Missouri University of Science and Technology.

Shao, Y., Macari, E. J., and Cai, W. (2005). "Compound deep soil mixing columns for retaining structures in excavations." *Journal of Geotechnical and Geoenvironmental Engineering*, 131(11), 1370-1377.

Sharma, K. G., and Desai, C. S. (1992). "Analysis and implementation of thin-layer element for interfaces and joints." *Journal of Engineering Mechanics*, 118(2), 2444-2461.

Shi, J., Liu, G., Huang, P., and Ng, C. W. W. (2015). "Interaction between a large-scale triangular excavation and adjacent structures in Shanghai soft clay." *Tunnelling and Underground Space Technology*, 50, 282-295.

Sichart, W. (1928). *Das Fassungsvermogen von Rohrbrunnen*, Julius Springer, Berlin.

Sills, G. C., Burland, J. B., and Czechowski, M. K. (1997). "Behaviour of an anchored diaphragm wall in stiff clay." *Proceedings of the 9th International Conference on Soil Mechanics and Foundation Engineering*, Tokyo, Japan, 147-155.

Simic, M., and French, D. J. (1998). "Three dimensional analysis of deep underground stations." *Seminar on the value of geotechnics in construction*, 93-100.

Simpson, B. (1992). "Retaining structures: Displacement and design." *Géotechnique*, 42(4), 541-576.

Skempton, A. W. (1951). "The bearing capacity of clays." *Proceeding of Building Research Congress*, 1, 180-189.

Smith, I. M., and Griffiths, D. V. (1982). *Programming the Finite Element Method, second edition*, John Wiley Sons, Chisester, U. K.

St. John, H. D., Potts, D. M., Jardine, R. J., and Higgins, K. G. (1992). "Prediction and performance of ground response due to the construction of a deep basement at 60 Victoria Embankment." *Predictive Soil Mechanics, Proceedings of the Wroth Memorial Symposium*, London: Thomas Telford, 449-465.

Steiner, W., and Werded, F. (1991). "Performance of a 17 m deep tie back wall under large surcharge loads." *Proceedings of the 10th European Conference on Soil Mechanics and Foundation Engineering*, Rotterdam, the Netherlands.

Stevens, A., Corbett, B. O., and Steele, A. J. (1977). "Barbican arts centre: The design and construction of the substructure." *Structural Engineer*, 55(11), 473-485.

Stevenson, M. C., and De Moor, E. K. (1994). "Limehouse Link: Cut and cover tunnel, design and performance." *Proceedings of the 13th International Conference on Soil Mechanics and Foundation Engineering*, Rotterdam, the Netherlands, 887-890.

Storn, R., and Price, K. (1997). "Differential evolution—a simple and efficient heuristic for global optimization over continuous spaces." *Journal of Global Optimization*, 11(4), 341-359.

Swanson, P. G., and Larson, T. W. (1990). "Shoring failure in soft clay." *Design and performance of earth retaining structures*, 551-561.

Symons, I. F., and Carder, D. R. (1991). "The behaviour in service of a propped retaining wall in stiff clay." *Proceedings of the 10th European Conference on Soil Mechanics and Foundation Engineering*, Rotterdam, the Netherlands, 761-766.

Symons, I. F., and Carder, D. R. (1992). "Field measurement on embedded retaining walls." *Geotechnique*, 42(1), 117-126.

Symons, I. F., Little, J. A., and Carder, D. R. (1988). "Ground movements and deflections of an anchored sheet pile wall in granular soil." *Engineering Geology Special Publications*, 5 (1), 117-127.

Tait, R. G., and Taylor, H. T. (1975). "Rigid and flexible bracing systems on adjacent sites." *Journal of the Construction Division*, 101(2), 365-376.

Tamano, T., Fukui, S., Mizutani, S., Tsuboi, H., and Hisatake, M. (1996). "Earth and water pressures acting on a braced excavation in soft ground." *Proceedings of the International Symposium Geotechnical Aspects of Underground Construction in Soft Ground*, Rotterdam, the Netherlands, 207-212.

Tan, S. B., Tan, S. L., and Chin, Y. K. (1985). "A braced sheet pile excavation in Singapore marine clay." *Proceedings of the 11th International Conference on Soil Mechanics and Foundation Engineering*, Rotterdam, the Netherlands, 1671-1674.

Tan, Y., and Wei, B. (2012). "Observed behaviours of a long and deep excavation constructed by cut-and-cover technique in Shanghai soft clay." *Journal of Geotechnical and Geoenvironmental Engineering*, 138(1), 69-88.

Tanaka, H. (1996). "Undrained shear strength for passive earth pressure in an excavation in soft clay." *Proceedings of the International Symposium Geotechnical Aspects of Underground Construction in Soft Ground*, Rotterdam, the Netherlands, 213-218.

Tang, Y. G., and Kung, T. C. (2009). "Application of nonlinear optimization technique to back analyses of deep excavation." *Computers and Geotechnics*, 36(1-2), 276-290.

Tarantola, A. (1987). *Inverse Problem Theory: Methods for Data Fitting and Model Parameter Estimation*, Elsevier Amsterdam.

Tedd, P., Chard, B. M., Charles, J. A., and Symons, I. F. (1984). "Behaviour of a propped embedded retaining wall in stiff clay at Bell Common Tunnel." *Géotechnique*, 34 (4), 513-532.

Terzaghi, K. (1943). *Theoretical Soil Mechanics*, John Wiley & Sons, New York, USA.

Terzaghi, K., and Peck, R. B. (1967). *Soil Mechanics in Engineering Practice*, John Wiley & Sons, New York.

TGS (2001). "Design Specifications for the Foundation of the Building." Taiwanese Geotechnical Society.

Thasnanipan, N., Aye, Z. Z., and Submaneewong, C. (2004). "Construction of diaphragm wall support underground car park in historical area of Bangkok." *Proceedings of the 5th International Conference on Case Histories in Geotechnical Engineering*, Missouri University of Science and Technology.

Theis, C. V. (1935). "The relation between the lowering of the piezometric surface and the rate and discharge of a well using ground water storage." *Transactions of the American Geophysical Union 16th Annual Meeting*.

Thiem, G. (1906). *Hydrologische Methoden*, JM Gephardt, Leipzig.

Triantafyllidis, T. (1998). "Neue erkenntnisse an tiefen baugruben am Potsdamer Platz in

Berlin." *Bautechnik*, 75(3), 133-154.

Triantafyllidis, T., Brem, G., and Vogel, U. (1997). "Construction monitoring and performance of the deep basement excavation at Potsdamer Platz, Berlin." *Proceedings of the 14th International Conference on Soil Mechanics and Foundation Engineering*, 1347-1350.

Tse, C. M., and Nicholson, D. P. (1992). "Design, construction and monitoring of the basement diaphragm wall at Minster Court, London." *Proceedings of the International Conference on Retaining Structures*, London: Thomas Telford, 323-332.

Twine, D., and Roscoe, H. (1999). "Temporary Propping of Deep Excavations-Guidance on Design." CIRIA C517, CIRIA London.

Ulrich, E. J. (1989). "Internally braced cuts in overconsolidated soils." *Journal of Geotechnical Engineering*, 115(4), 504-520.

Ulrich, E. J. (1989). "Tieback supported cuts in overconsolidated soils." *Journal of Geotechnical Engineering*, 115(44), 521-545.

Van Langen, H., and Vermeer, P. A. (1990). "Automatic step size correction for non-associated plasticity problems." *International Journal for Numerical Methods in Engineering*, 29(3), 579-598.

Vanderplaats, G. (1984). *Numerical optimization techniques for engineering design with applications*, McGraw-Hill Inc, USA.

Veneziano, D. (1974). *Contributions to second moment reliability*, Dept. of Civil Engineering, MIT, Cambridge, Mass.

Vezinhet, M., Brucy, M., and Balay, J. (1989). "Behaviour of a quay in compressible silts." *Proceedings of the 12th International Conference on Soil Mechanics and Foundation Engineering*, Rotterdam, the Netherlands, 1527-1528.

Von Rosenvinge, T. I. (2013). "Tied-Back Wall Failure, Boston, MA." *International Conference on Case Histories in Geotechnical Engineering*.

Von Soos, P. (1990). "Properties of soil and rock." *Grundbautaschenbuch Part 4*, Berlin.

Vucetic, M., and Dobry, R. (1991). "Effect of soil plasticity on cyclic response." *Journal of Geotechnical Engineering*, 117(1), 89-107.

Vuillemin, R. J., and Wong, H. (1991). "Deep excavation in urban environment: 3 examples." *Proceedings of the 10th European Conference on Soil Mechanics and Foundation Engineering*, Rotterdam, the Netherlands, 843-847.

Wallace, J. C., Ho, C. E., and Long, M., M. (1992). "Retaining wall behaviour for a deep basement in Singapore marine clay." *Proceedings of the International Conference on Retaining Structures*, London: Thomas Telford, 195-204.

Ward, K. (1992). "The design and performance during construction of the propped secant pile wall at Holborn Bars London." *Proceedings of the International Conference on Retaining Structures*, London: Thomas Telford, 216-226.

Ware, K. R., Mirsky, M., and Leuniz, W. E. (1973). "Tieback wall construction—Results and controls." *Journal of the Soil Mechanics and Foundations Division*, 99(12), 1136-1152.

Watson, G. V. R., and Carder, D. R. (1994). "Comparison of the measured and computed performance of a propped bored pile retaining wall at Walthamstow." *Proceedings of the Institution of Civil Engineers—Geotechnical Engineering*, 107(3), 127-133.

WeiBenbach, A., and Gollub, P. (1995). "Neue Erkenntnisse über mehrfach verankerte Ortbetonwände." *Bautechnik*, 72(12), 780-799.

Whittle, A. J., and Davies, R. V. (2006). "Nicoll Highway Collapse: Evaluation of Geotechnical Factors Affecting Design of Excavation Support System." *International Conference on Deep Excavations*, 28-30.

Whittle, A. J., Hashash, Y. M. A., and Whitman, R. V. (1993). "Analysis of deep excavation in Boston." *Journal of Geotechnical Engineering*, 119(1), 69-90.

Wikipedia, <http://en.wikipedia.org/wiki/File:Nicoll_Highway_Mrt_Locator.png>.

Winkler, E. (1867). *Die Lehre Von Elasticitaet Und Festigkeit*, Pray (H. Dominicus).

Winter, D. G. (1990). "Pacific first center performance of the tie back shoring wall." *Design and performance of earth retaining structures*, 764-777.

Wong, I. H., and Chua, T. S. (1999). "Ground movements due to pile driving in an excavation in soft soil." *Canadian Geotechnical Journal*, 36(1), 152-160.

Wong, I. H., Poh, T. Y., and Chuah, H. L. (1997). "Performance of excavations for depressed expressway in Singapore." *Journal of Geotechnical and Geoenvironmental Engineering*, 123 (7), 617-625.

Wong, K. S., and Broms, B. B. (1989). "Lateral wall deflections of braced excavations in clay." *Journal Geotechnical Engineering*, 115(6), 853-870.

Wong, L. W., Shau, M. C., and Chen, H. T. (1996). "Compaction grouting for correcting building settlement." *Grouting and Deep Mixing*, Rotterdam, T he Netherlands.

Woo, S. M. (1992). "Method, design and construction for building protection during deep excavation." *Sino-Geotechnics*, 40, 51-61.

Wood, L. A., Maynard, A., and Forbes-King, C. J. (1988). "The instrumentation and performance of an anchored retaining wall." *Instrumentation in Geotechnical Engineering*, London: Thomas Telford, 137-154.

Wood, L. A., and Perrin, A. J. (1984). "Observations of a strutted diaphragm wall in London Clay: A preliminary assessment." *Géotechnique*, 34(4), 563-579.

Xanthakos, P. (1994). *Slurry Walls as Structural System, 2nd Edition*, McGraw-Hill, Inc, New York.

Xiao, H. B., Tang, J., Li, Q. S., and Luo, Q. Z. (2003). "Analysis of multi-braced earth retaining structures." *Proceedings of the Institution of Civil Engineers—Structures and Buildings*, 156(3), 307-318.

Xu, J., and Zheng, Y. R. (2001). "Random back analysis of field geotechnical parameter by response surface method." *Rock and Soil Mechanics*, 22(2), 167-170.

Xu, Z. H., Zhang, J., and Chen, C. (2013). "A case history of a deep foundation pit constructed by zoned excavation method in Shanghai soft deposit." *Proceedings of the 7th International Conference on Case Histories in Geotechnical Engineering*, Missouri University of Science and Technology.

Young, D. K., and Ho, E. W. L. (1994). "The observational approach to design of a sheet piled retaining wall." *Géotechnique*, 44(4), 637-654.

Zekkos, D. P., Athanasopoulos, A. G., and Athanasopoulos, G. A. (2004). "Deep supported excavation in difficult ground conditions in the city of Patras, Greece—Measured vs. predicted behavior." *Proceedings of the 5th International Conference on Case Histories in Geotechnical Engineering*, Missouri University of Science and Technology.

Zhang, W., Goh, A. T. C., and Xuan, F. (2015). "A simple prediction model for wall deflection caused by braced excavation in clays." *Computers and Geotechnics*, 63, 67-72.

Zhang, W., Goh, A. T. C., and Zhang, Y. (2015). "Updating soil parameters using spreadsheet method for predicting wall deflections in braced excavations." *Geotechnical and Geological Engineering*, 33(6), 1489-1498.

Zhang, W., Z., H., Goh, A. T. C., and R., Z. (2019). "Estimation of strut forces for braced excavation in granular soils from numerical analysis and case histories." *Computers and Geotechnics* 106, 286-295.

Zhang, W. G., and Goh, A. T. C. (2013). "Multivariate adaptive regression splines for analysis of geotechnical engineering systems." *Computers and Geotechnics*, 8, 82-95.

Zhang, W. G., Wang, W., Zhou., D., Z., R. H., Goh, A. T. C., and Hou, Z. J. (2018). "Influence of groundwater drawdown on excavation responses—A case history in Bukit Timah granitic residual soils." *Journal of Rock Mechanics and Geotechnical Engineering*, 10(5), 856-864.

Zhang, W. G., Zhang, R. H., Fu, Y. R., Goh, A. T. C., and Zhang, F. (2018). "2D and 3D numerical analysis on strut responses due to one-strut failure." *Geomechanics and Engineering*, 15, 965-972.

Zhao, X. H., Gong, J., Chen, Z. M., and Bao, Y. (1999). "Design and practice on special deep and large excavation engineering in Shanghai." *Proceedings of the 12th European Conference on Soil Mechanics and Geotechnical Engineering*, Rotterdam, the Netherlands, 239-244.